Molecular Cell Biology

Molecular Cell Biology

Edited by **Gloria Doran**

SYRAWOOD
PUBLISHING HOUSE

New York

Published by Syrawood Publishing House,
750 Third Avenue, 9ᵗʰ Floor,
New York, NY 10017, USA
www.syrawoodpublishinghouse.com

Molecular Cell Biology
Edited by Gloria Doran

International Standard Book Number: 978-1-68286-098-4 (Hardback)

Printed in the United States of America.

Contents

Preface   IX

Chapter 1   **Regenerative action of *Cochlospermum tinctorium* aqueous root extract on experimentally induced hepatic damage in rats**   1
Etuk E. U, Francis U. U and Garba I

Chapter 2   **Effect of certain insect growth regulators on the lipid content of some tissues of the desert locust *Schistocerca gregaria***   5
Hamadah Kh.Sh., Ghoneim K.S. and Tanani M.A.

Chapter 3   **Ginger extract (*Zingiber officinale*) triggers apoptosis and G_0/G_1 cells arrest in HCT 116 and HT 29 colon cancer cell lines**   13
Shailah Abdullah, Siti Amalina Zainal Abidin, Noor Azian Murad, Suzana Makpol, Wan Zurinah Wan Ngah and Yasmin Anum Mohd Yusof

Chapter 4   **Changes in liver and serum transaminases and alkaline phosphatase enzyme activities in *Plasmodium berghei* infected mice treated with aqueous extract of *Aframomum sceptrum***   22
George, B. O., Osioma, E., Okpoghono, J. and Aina, O. O.

Chapter 5   **Nerve growth factor (NGF) combined with oxygen glucose deprivation OGD induces neural ischemia tolerance in PC12 cells**   27
Chunli Mei, Jinting He, Jing Mang, GuihuaXu, Zhongshu Li, Wenzhao Liang and Zhongxin Xu

Chapter 6   **Effect of 50 Hz electromagnetic fields on acid phosphatase activity**   33
K. S. Prashanth, T. R. S. Chouhan and Snehalatha Nadiger

Chapter 7   **Thermogenic response of guinea pig adipocytes to noradrenaline and β3 AR agonists**   39
Ghorbani Masoud and Mehdi Shafiee Ardestani

Chapter 8   **Deficiency of vitamin B12 among Jordanian people with psychological and biological activity**   45
Mohammad A. Al-Fararjeh, Nader Jaradat and Abdulrahim Aljamal

Chapter 9   **Measurement of cholesterol sub-fractions, high density lipoprotein 2 and high density lipoprotein 3, in HIV infected patients treated with highly active antiretroviral therapy in Burkina Faso**   50
Jean Sakandé, Josiane B Kaboré, Elie Kabré, Boblwendé Sakandé and Mamadou Sawadogo

Chapter 10   **Adsorption, metabolism and degradation of erythromycin in giant freshwater prawn and tilapia aquaculture in Mekong River Delta**
N. P. Minh, T. B. Lam and T. T. D. Trang   **56**

Chapter 11   **Selected heavy metals and electrolyte levels in blood of workers and residents of industrial communities**
Babalola O. O and Babajide S. O   **65**

Chapter 12   **Metabolic engineering of an ethanol-tolerant *Escherichia coli* MG1655 for enhanced ethanol production from xylose and glucose**
Ruiqiang Ma, Ying Zhang, Haozhou Hong, Wei Lu, Wei Zhang, Min Lin and Ming Chen   **69**

Chapter 13   **Changes in the tissue antioxidant enzyme activities of palm weevil (*Rynchophorous phoenicis*) larva by the action of 2, 2-dichlorovinyl dimethyl phosphate**
Olufemi Bamidele, Joshua Ajele, Ayodele Kolawole and Akinkuolere Oluwafemi   **75**

Chapter 14   **Effects of ethanolic and aqueous leaf extracts of *Landolphia owariensis* on the serum lipid profile of rats**
Nwangwu Spencer C., Ike Francisca, Olley Misan, Oke James M., Uhunmwangho Esosa, Amegor, O. F. Ubaoji Kingsley and Nwangwu Udoka   **85**

Chapter 15   **Oxidative stress biomarkers in young male rats fed with stevioside**
Hala A. Awney   **89**

Chapter 16   **Synergistic effects of glucan and resveratrol**
Vaclav Vetvicka and Zuzana Vancikova   **97**

Chapter 17   **Effect of *Hibiscus sabdariffa* anthocyanins on 2, 4- dinitrophenylhydrazine-induced hematotoxicity in rabbits**
A. Ologundudu, A. O. Ologundudu, I. A. Ololade and F. O. Obi   **103**

Chapter 18   **Cytoplasmic, peroxisomal and mitochondrial membrane phospholipid alteration in 3^1-diamino-azobenzene (3^1-DAB)-induced hepatocellular carcinoma**
Omotuyi I. O., Ologundudu A., Okugbo T. O., Oluyemi K. A. and Omitade, O. E.   **108**

Chapter 19   **The effects of mannan-oligosaccharides on cecal microbial populations, blood parameters, immune response and performance of broiler chicks under controlled condition**
Saeed Khalaji, Mojtaba Zaghari and Somaye Nezafati   **115**

Chapter 20   **Dyslipidemic and atherogenic effects of academic stress**
Adekunle Adeniran S.   **120**

Chapter 21   **Radio-protective chelating agents against DNA oxidative damage**
M. H. Awwad, Samy A. Abd El-Azim, F. A. M. Marzouk, E. A. El-Ghany, and M. A. Barakat   **124**

Chapter 22   **Assessment of current iodine status of pregnant women in a suburban area of Imo State Nigeria, twelve years after universal salt iodization**
Cosmas O. Ujowundu, Agwu I. Ukoha, Comfort N. Agha, Ngwu Nwachukwu and Kalu O. Igwe   **131**

Chapter 23   Trace metals and oxidative metabolic changes in malignant prostate cancer patients   138
Akiibinu M. O., Ogundahunsi A. O., Kareem O. I., Adesiyan A. A., Idonije B. O. and Adeniyi F. A. A.

Chapter 24   Unutilized energy reserves and mineral contents of fibroid tissues suggesting perturbed membrane transport processes   142
Ibegbulem C. O., Agha N. C. and Emeka-Nwabunnia I.

Chapter 25   Multipotent mesenchymal stem cells (MSCs) from human umbilical cord: Potential differentiation of germ cells   147
Jinlian Hua, Pubin Qiu, Haijing Zhu, Hui Cao, Fang Wang and Wei Li

Chapter 26   Biochemical markers in semen and their correlation with fertility hormones and semen quality among Sudanese infertile patients   158
Abdelmula M. Abdella, Al-Fadhil E. Omer and Badruldeen H. Al-Aabed

Chapter 27   Peripheral blood and C-reactive protein levels (CRP) in chronic periodontitis   164
Balwant Rai, Jasdeep Kaur, Simmi Kharb, Rajnish Jain, S. C. Anand and Jaipaul Singh

Chapter 28   Polymerization of human sickle cell haemoglobin (HbS) in the presence of three antimalarial drugs   168
Chikezie P. C., Chikezie C. M. and Amaragbulem P. I.

Chapter 29   Influence of the bioinsecticides, NeemAzal, on main body metabolites of the 3rd larval instar of the house fly *Musca domestica* (Diptera: Muscidae)   172
Mohammad A. Kassem, Tarek A. Mohammad and Ahmad S. Bream

Chapter 30   Effect of essential oil of the leaves of *Eucalyptus globulus* on heamatological parameters of wistar rats   177
Oyesomi, Tajudeen Oyesina, Ajao, Moyosore Salihu, Olayaki, luquman Aribidesi and Adekomi, Damilare Adedayo

Chapter 31   Screening of antioxidant activity, total phenolics and gas chromatograph and mass spectrometer (GC-MS) study of delonix regia   181
P. Maria jancy Rani, P. S. M. Kannan and S. Kumaravel

Chapter 32   Changes in haemorrheologic and fibrinolytic activities upon hypertension and diabetic chemotherapy in Calabar diabetic residents, Nigeria   188
M. S. Edem, A. O. Emeribe and J. O. Akpotuzor

Chapter 33   Impact of *Anthocleista vogelii* root bark ethanolic extract on weight reduction in high carbohydrate diet induced obesity in male wistar rats   193
Anyanwu, G. O., Onyeneke, E. C., Usunobun, U. and Adegbegi, A. J.

Permissions

List of Contributors

Preface

Over the recent decade, advancements and applications have progressed exponentially. This has led to the increased interest in this field and projects are being conducted to enhance knowledge. The main objective of this book is to present some of the critical challenges and provide insights into possible solutions. This book will answer the varied questions that arise in the field and also provide an increased scope for furthering studies.

This book aims to elucidate the concepts and recent advances in the fields of molecular and cell biology. Molecular biology is concerned with the study of molecular structures and processes that take place within cells, while cell biology involves the study of physiological properties, structures and functions of cells. It is a compilation of relevant topics such as types of enzyme, protein structures and their functions, metabolic engineering, effect of various substances and factors on cellular activities, etc. which will provide a comprehensive understanding of the subject. Various up-to-date researches and case studies have been included in this book by experts from across the globe that explores the latest developments in these fields. Students, researchers, experts and all associated with molecular cell biology will benefit, alike, from this book.

I hope that this book, with its visionary approach, will be a valuable addition and will promote interest among readers. Each of the authors has provided their extraordinary competence in their specific fields by providing different perspectives as they come from diverse nations and regions. I thank them for their contributions.

Editor

Regenerative action of *Cochlospermum tinctorium* aqueous root extract on experimentally induced hepatic damage in rats

Etuk E. U[1*], Francis U. U[2] and Garba I[3]

[1&3]Department of Pharmacology, College of Health Sciences, Usmanu Danfodiyo University, Sokoto, Nigeria.
[2]Department of Pharmacognosy and natural medicine, University of Uyo, Nigeria.

The hepatoprotective effect of aqueous root extract of *Cochlospermum tinctorium* on carbon tetrachloride (CCl_4) on induced hepatic damage in rats was reported. The present study examined the curative action of the plant extract on experimentally induced hepatic damage in rats. Wistar rats were divided into normal control, induction control, extract and prednisolone treated groups. Hepatotoxicity was induced in rats by intraperitoneal administration of CCl_4 (30% in olive oil) for 5 days. Treatment group received 200 mg/kg of extract post hepatotoxicity induction orally for 7 days. The animals were sacrificed on the 8th day, blood and hepatic tissue collected for liver function test and histopathological analysis respectively. Administration of carbon tetrachloride induced hepatic damage in the rats was evidenced by a significant increase ($P < 0.05$) in the blood clotting time, serum levels of alanine aminotransferase (ALT), aspartate aminotransferase (AST), alkaline phosphatase (AP) and bilirubin as compared to the control. There was also a significant reduction in the serum total protein, serum albumin and reduced glutathione levels. Treatment with the extract reversed the values of all the biochemical parameters to near normal values in control. The histopathological reports collaborate with the biochemical analysis results. Oral administration of aqueous root extract of *Cochlospermum tinctorium* for 7 days has significantly reversed hepatic damage produced by CCl_4 in wistar rats.

Key words: *Cochlospermum tinctorium*, carbon tetrachloride, hepatotoxicity, wistar rats.

INTRODUCTION

Cochlospermacac is a well known plant family in herbal medicine. The species *Cochlospermum tinctorium* A. Rich (CTR), the plant of interest in this study is widely distributed in the savannah area of west and central Africa. An inventory on African hepatoprotective remedies puts *C. tinctorium* in third position based on the number of countries in which its use is cited (Abondo et al., 1990). A recent study has shown that the plant has heaptoprotective effect against carbon tetrachloride induced toxicity in rats. But apart from the preventive actions, drugs are also needed for the treatment of existing pathological conditions. The aetiological factors of hepatic diseases are multiple and developing a single agent capable of preventing hepatic diseases at all times appears elusive. Alternatively, finding a potent drug that can regenerate hepatic functions irrespective of the initial cause of the damage appears more feasible.

Approximately two million people die annually from hepatic related disorders in the world (Roger et al., 2001). There are no reliable curative drugs for the treatment of hepatic diseases in modern medicine. But a number of medicinal plants have been recommended for the treatment of liver disorders (Sanmugapriya and Venkataraman, 2006). The efficacies of most of these medicinal plants are not yet validated. Thus investigation of medicinal plants with potential hepatic regenerative activity becomes very important. The present study examined the ability of *C. tinctorium* root extract to reverse hepatic damage produced by administration of carbon tetrachloride and restore normal hepatic functions in rats.

* Corresponding author. E-mail: etuk2005@yahoo.co.uk.

Table 1. Effect of post-treatment with CTR extract on blood clotting time and serum enzymes in CCl_4 induced hepatotoxicity in rats.

Group	Treatment	Clotting time (s)	AST (uL^{-1})	ALT (uL^{-1})	ALP (uL^{-1})
A	CONTROL	275.0 ± 6.9	39.4 ± 1.1	54.6 ± 0.8	126.2 ± 1.0
B	CCl_4	620.6 ± 2.1	109.0 ± 1.5	121.8 ± 1.1	300.8 ± 1.1
C	CTR (200 mg/kg) + CCl_4	290.3 ± 5.4	50.0 ± 0.7	80.1 ± 0.5	130.8 ± 1.4
D	Pred (2 mg/kg) + CCl_4	382.5 ± 6.0	68.8 ± 0.9	62.5 ± 0.9	206.2 ± 1.7
One-way ANOVA	F	865.97	789.31	1236.4	3789.0
	df	23	23	23	23
	P	0.001	0.001	0.001	0.001

Values are mean ± SEM; n = 6 rats in each group. Comparison was between control vs treatment groups; and then CCl_4-treated Vs CTR + CCl_4-treated groups. $P < 0.05$; all values are significant.

MATERIALS AND METHODS

This study was conducted in the Department of Pharmacology, College of Health Sciences, Usmanu Danfodiyo University, Sokoto (UDUS), Nigeria between the months of March and September, 2007.

Experimental animals

Wistar rats weighing 280 – 300 g of either sex were obtained and kept in the animal facility of Department of Pharmacology, UDUS, for about two weeks before the commencement of the study. They were kept in a well ventilated room with free access to rat feeds and tap water. Animals were randomly assigned to the treatment groups (n = 5). Permission from the departmental ethical committee for laboratory use of animals was duly obtained before the animals were put into use.

Preparation of plant extract

The dried powdered root (200 g) of *C. tinctorium* was extracted with distilled water using a Soxhlet apparatus. The filtrate was concentrated to dryness in an oven at $45°C$ and the yield calculated. The extract was stored in a close container and preserved at - $17°C$ until required for use in the study. On the day of the experiment, fresh extract was reconstituted in distilled water at required concentration and put into use.

Treatment of hepatotoxic rats with extract

Carbon tetrachloride (CCl_4) hepatotoxicity was induced in rats according to the method of Rao et al. (2006) with little modifications. Animals were divided into six groups (n = 5). Group A (normal control) were treated with daily dose of olive oil (1 ml/kg body weight p.o.) for 5 days. Group B (induction control) were dosed intraperitoneally daily with CCl_4 (30% in olive oil) for 5 days.

Group C received prednisolone (2 mg/kg b.w., p.o.) a standard anti-inflammatory drug for 7 days after hepatotoxicity induction with CCl_4 as in group B.

The animals in group D were treated orally with 200 mg/kg (body weight) of the extract for 7 days after hepatotoxicity induction with CCl_4 as in group B. The dose of the test extract was selected based on the basis of an earlier work. On the 8th day, clotting time for each rat was determined by the method of Lee et al. (1996). Thereafter, the animals were anaesthetized with chloroform and sacrificed. Blood sample was withdrawn by cardiac puncture, centrifuged and serum separated and preserved for biochemical analysis. The liver samples were collected and preserved in 10% formalin for histopathological analysis.

In the biochemical analysis, the method of Reitman and Frankel (1957) was used in determining aspartate amino transferases (AST) and alanine amino transferase (ALT) activity in the serum. Alkaline phosphatase (AP) level was estimated by Randox kit Colorimetric method. The serum total bilirubin was obtained by using Jendrassik and Graf method (1997) while Doumas method (1997) was used to estimate the serum total protein and albumin. The tissue sample obtained from the liver of each rat was divided into two portions. The first portion was perfused with cold 0.86% KCL, homogenized and centrifuged to obtain post mitochondrial supernatant for estimation of reduced liver glutathione (Slack and Lindsay, 1996). The remaining portion was fixed with 10% formalin; and stained with haematoxylin and eosin before the slides were examined under a microscope.

Statistical analysis

The results of biochemical analysis were expressed as mean plus standard error of mean (MEAN ± S.E.M). The control and treatment groups were compared by using one way analysis of variance (ANOVA). Further differences were detected by Turkey-Kramer multiple comparison test. The level of significance was taken at probability less than 5%.

RESULTS

The results shown in Table 1 revealed that, the transaminases (ALT and AST), alkaline phosphatase (AP) enzymes and serum total bilirubin were significantly increased ($P < 0.05$) in CCl_4 – treated rats (group B) when compared to the control (group A). Administration of the extract returned the enzymes and bilirubin levels in the intoxicated rats near to the normal values in control group. Also a significant decrease in serum total protein, serum albumin and reduced liver glutathione levels were observed in the CCl_4– treated rats when compared to control group (Table 2). The treatment with 200 mg/kg body weight of the extract raised the values of the affected parameters in the hepatotoxic rats to near the normal values in the control rats. The effect of the extract in restoring the biochemical values in the treated animals

Table 2. Effect of post-treatment with CTR extract on serum total protein, albumin, and reduced liver glutathione in CCl_4 induced hepatotoxicity in rats.

Group	Treatment	STP (100g/ml)	SA (mg/dl)	STB (g/dl)	GSH (mM/g tissue)
A	CONTROL	6.9 ± 0.9	2.9 ± 0.8	1.7 ± 0.1	6.1 ± 0.1
B	CCl_4	6.3 ± 0.3	2.6 ± 0.8	4.8 ± 0.0	1.5 ± 0.0
C	CTR(20 mg/kg) + CCl_4	8.1 ± 0.7	3.1 ± 0.6	2.2 ± 0.1	4.1 ± 0.1
D	Pred(2 mg/kg) + CCl_4	9.0 ± 0.6	4.2 ± 0.6	1.8 ± 0.3	3.9 ± 0.4
One-way ANOVA	F	3.343	0.9733	78.152	78.851
	df	23	23	23	23
	P	0.4249	0.4249	0.001*	0.001*

Values are mean ± SEM; n = 6 rats in each group. Comparison was between control vs treatment groups, and then CCl_4- treated versus CTR + CCl4-treated groups. With $P < 0.05$; values marked* are significant; g = gram; ml = millilitre; dl = decillitre ; mM = millimole.

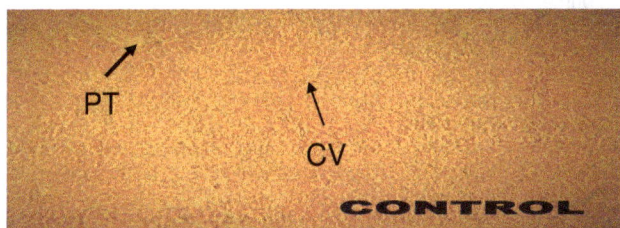

Plate 1. Cross section of rat liver showing normal architecture (× 50). CV: central vein, PT: portal tract.

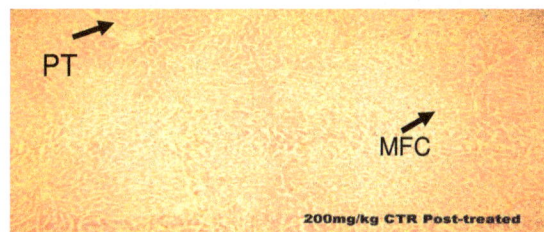

Plate 3. 200 mg/kg CTR post treated liver section showing mild fatty change (MFC).

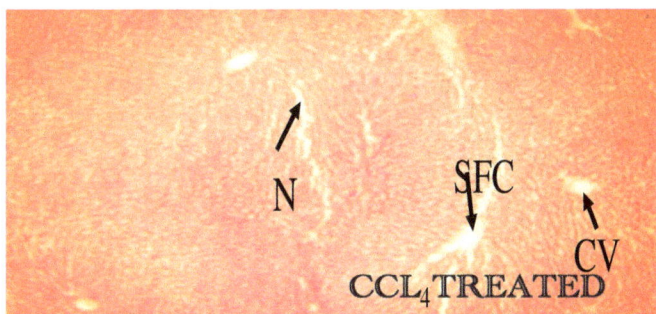

Plate 2. CCl_4 treated liver section showing severe fatty change (SFC) and necrosis (N) (× 50).

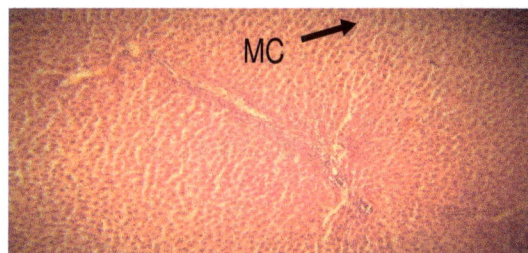

Plate 4. 2 mg/kg PRD post treated liver section showing moderate congestion (MC).

to near normal levels was more significant compared to the prednisolone treatment. Treatment with the extract also reduced the blood clotting time in the rats from 620.6± 2.1 s in the CCl_4 treated rats to 290.3 ± 5.4 s.

The histopathological findings collaborated with the biochemical results. Plate 1 displayed normal hepatic architecture in the rats. Administration of CCl_4 produced severe distortion of the hepatic architecture in the rats (Plate 2). A severe vacuolar fatty change and moderate number of lymphocytes infiltrations were observed. Treatment with the extract reversed the fatty change, restored the normal architecture and reduced lymphocytes infiltration (Plate 3). There was moderate congestion retained in the prednisolone treated group (Plate 4).

DISCUSSION

In this study, the experimental induction of liver damage was achieved by CCl_4 administration in rats. This was evidenced by the elevation in the liver transaminases activity, bilirubin and the histopathological lesions observed. Bilirubin, albumin, transaminases, phosphorlipids and cholesterol assay are sensitive tests to substantiate the functional integrity of the liver and severity of necrosis (Edmundstone et al., 1985). Abdel-Hamid (2006) used a similar method to induce fatty liver in male rats. Chemicals and drugs such as CCl_4 catabolized radicals induced lipid peroxidation, damaged the membranes of liver cells and organelles, caused the swelling and necrosis of hepatocytes and resulted to the release of cytosolic enzymes such as AST, ALT, AP and

Gamma – Glutamate transpeptidase into the circulating blood (Venukumar et al., 2002). The present result showed that CCl_4 administration significantly increased serum transaminase activity, alkaline phosphatase and bilirubin levels while the protein synthesis in the liver was concomitantly inhibited. These observations are in accordance with what was previously reported (Venukumar et al., 2002).

Treatment with *C. tinctorium* aqueous root extract significantly reversed the changes and recovered the five parameters near to normal control values. The effect of this extract might be as a result of the presence of some phytochemical compounds with inherent antioxidant properties capable of inhibiting the free radicals scavenging activity or its ability to regenerate the depressed endogenous antioxidant substances such as reduced liver glutathione as seen in this result. It was reported that, antioxidant activity or inhibition of the generation of free radicals is important in the protection against CCl_4 induced liver lesion (Behattacharyya et al., 2003). It is well recognized that, free radicals are critically involved in various pathological conditions such as arthritis, inflammation and liver diseases (Quambo et al., 1998).

Prednisolone was used as a standard drug in this study not because it is a known hepatic curative agent but because it is one of the drugs reported to have modulatory actions on hepatic disorders irrespective of the cause. Steroids have marked anti-inflammatory and immunosuppressive effects which makes them useful for disease conditions such as lupus erythematosus, arthritis, ulcerative colitis, hepatitis and nehphritic syndrome (Graig and Stitzel, 1986). The ability of this extract to reduce clotting time in the rats to less than half the period in the CCl_4 treated rats is very significant. Uncontrolled hemorrhage is a major complication of hepatic diseases. The reduction in clotting time by the extract may be associated with the increase in serum protein levels as recorded in this study.

Overall, the seven days oral treatment with aqueous root extract of *C. tinctorium* has restored normal hepatic functions in CCl_4 intoxicated rats and this makes the plant extract a potential curative agent for liver disorders.

REFERENCES

Abdel – Hamid NM (2006). Diphenyl dimethyl bicarboxylate as an effective treatment for chemical-induced fatty liver in rats. Afr. J. Biomed. Res. 9: 77-81.

Abondo A, Mbenkum F, Thomas D (1990). Ethnobotany and the medicinal plants of the Korup rainforest project area, Cameroon. Mshigeni KE (ed) Proceedings of International Conference on Traditional Medicinal Plants. Arusha, Tanzania, Feb 18 – 23. Dares Salam Uni. Press. ISBN – 9976 60 2294. pp 112-124.

Bhattacharyya D, Mukherjee R, Pandit S, Das N, Sur TK (2003). Prevention of carbon tetrachloride induced hepatotoxicity in rats by Himoliv, a polyherbal formulation. Indian J. Pharmacol. 35:133-135.

Doumas BT, Watson W, Homer G (1997). Albumin standards and the measurement of serum albumin with bromcresol green. Clinica Chimica Acta. 258(1): 21-30.

Edmonson HA. Peters RL (1985). Anderson's pathology. 8th edition Kissane K M.(ed.), C.V. Mosby, St. Louis Press, USA. pp 1096 - 1212.

Graig J, Stittzel D (1986). Delta Viral hepatitis Histopathology and Course. Pathol. Ann. J. 27(1): 12-17.

Jendanssik L, Goffan P (1997). Bilirubin, Colorimetric method. Biochem. Z. 297: 81-93.

Lee J, Son KH, Chang HW, Do JC, Yang KY, Kang SS, Kim HP (1996). Haematological profile following chemical induced hepatotoxicity in rats. Arch. Pharm Res. 16: 25 -31.

Rao GMR, Chandana P, Palpu Annie S (2006). Hepatoprotective effects of rubiadin, a major constituent of Rubia cordifolia (Linn) J. Ethnopharmacol. 103(3): 484-490.

Reitman S, Frankel S (1957). Hepatic disorders. Am. J. Clin. Pathol. 28(1):53-56.

Sanmugapriya E, Venkataraman S (2006). Studies on the hepatoprotective and antioxidant actions of Strychnos potatorum Linn seeds on CCl_4 - induced hepatic acute injury in experimental rats. J. Ethnopharmacol. 105: 154-160.

Slack J, Lindsay RH . Estimation of blood proteins bound sulphydril groups in tissue with Eliman's reagent. Anal. Biochem. 1968. 25: 192-197.

Roger D. Pamplona R (2001). Liver toxicity Encyclopedia of Medicinal plants.. (1): 392 – 395.

Venukumar MR, Latha MS (2002). Hepatoprotective effect of the methanolic extract of Curculigo orchioides in CCl_4 – treated male rats. Int. J. Pharmacol. 34: 29 – 75.

Effect of certain insect growth regulators on the lipid content of some tissues of the desert locust *Schistocerca gregaria*

Hamadah Kh.Sh.*, Ghoneim K.S. and Tanani M.A.

Department of Zoology, Faculty of Science, Al-Azhar University, Madenit Nasr, Cairo, Egypt.

The present study was carried out to investigate the metabolic effects of pyriproxyfen, tebufenozide or lufenuron on the lipid content in two different tissues: hemolymph and fat body of the early-, mid- and late-aged old nymphs as well as 1- and 4-days old adult females. Hemolymph lipid content of the early-aged nymphs had been subjected to a reducing effect after treatment with high concentration of insect growth regulators (IGRs). With the age of nymphs, all IGRs could significantly or non-significantly reduce the lipid content of hemolymph. Concerning the lipid content in fat bodies of nymphs, a predominant inhibitory effect of all IGRs was detected. With regard to the adults, nymph treatments led to remarkable or slightly decreased lipids in the hemolymph, as an exceptional case; because the lipid content non-significantly increased in 4-day old adults after treatment with low concentration level of lufenuron. The current IGRs unexceptionally exhibited inhibitory effects on the lipids of adult fat bodies.

Key words: *Schistocerca gregaria*, lipids, fat body, hemolymph, nymph, adult, pyriproxyfen, tebufenozide, lufenuron.

INTRODUCTION

Nowadays, alternative methods for pest control are being appreciated. One of the alternatives is the so called insect growth regulators (IGRs). These compounds are highly effective against various insects and other pests that have become resistant to organic insecticides. Meanwhile, all these compounds are less toxic to mammals and non target organisms because of their non-toxic effect and their quick disintegrating abilities (Carter, 1975; Staal, 1975; Zurfleuh, 1976; Oberlander et al., 1978, 1979; Ishaaya et al., 1987; Kostyukovsky et al., 2000; Ghasemi et al., 2010).

Since the target sites of common insecticides on insects and mammals are known to be similar, it is desirable to develop insecticides whose primary target site does not exist in mammals for selective toxicity. IGRs may belong to this type of (selective) insecticides and can be grouped according to their mode of action, as follows

chitin synthesis inhibitors (that is, of cuticle formation) and substances that interfere with the action of insect hormones (Tunaz and Uygun, 2004).

Pyriproxyfen is a pyridine-based juvenile hormone agonist that competes for juvenile hormone binding site receptors in insects, mimicking the action of juvenile hormone and thus, maintaining an immature state (Sullivan and Goh, 2008). This compound has a relatively low toxicity for mammals and was first registered in Japan in 1991 for controlling public health pests (Miyamoto et al., 1993).

The non-steroidal ecdysone agonist tebufenozide (RH-5992) is a novel caterpillar control agent with unusually high target selectivity (Carlson, 2000). It was found to be much more potent against and selective towards larval Lepidoptera than was RH-5849 (Carlson et al., 1994; Dhadialla et al., 1998). Under the trade names Confirm®, Mimic®, and Romdan®, it is now widely used on several crops in the world Spruce budworm larvae (*Choristoneura fumiferana*) upon ingesting tebufenozide (RH-5992) stop feeding and go into a precocious, incomplete molt, leading eventually to death (Retnakaran et al., 2001). Tebufenozide could disrupt the growth and development of *Spodoptera exigua* because the larval period was elongated and the fecundity was decreased and the ratios of survival, pupation, emergence and hatch of the larvae were also decreased (Linrui et al., 2006).

Lufenuron (RS)-1-[2,5-dichloro-4-(1,1,2,3,3,3-hexafluoropropoxy)-phenyl]-3-(2,6-difluorobenzoyl)-urea is known as an insect development inhibitor / insect growth regulator. It is active against larval developmental stages causing cuticular lesions and interfering in the chitin biosynthesis (Dean et al., 1998; Moriello et al., 2004). It is principally used for controlling the cat flea *Ctenocephalides felis* (Dean et al., 1998), being also active against diptera (Wilson and Cryan, 1997). It is also employed to control pests of several vegetal crops, including the citrus rust mite *Phyllocoptruta oleivora* (Bueno and Freitas, 2004) and adult predators of cotton pests, such as earwings, ladybugs, spiders, mirids and green lacewings (Castane et al., 1996; Angeli and Forti, 1997; Javaid et al.,, 1999) Lufenuron is used for controlling fungal infections in several animal species, due to its capacity to tear chitin from fungal cell wall, as well as, to inhibit the synthesis polymerization and deposition of chitin Based on results obtained by Alves et al. (2011), lufenuron was not interfered in the entomopathogenic fungus and *Metarhizium anisopliae* conidia germination when used in a concentration of 1 mg/ml and increased it to 700 µg/ml.

Lipids are important source of energy for insects. Lipid turnover in insects is regulated by neuroendocrine-controlled feed-back loops (Downer, 1985). Quantity of lipids available for the reserves seems to be the result of a balance between the catch of food and the requests for reserves by processes such as maintenance, growth and reproduction, and this balance is disturbed by any toxic product (Canavoso et al., 2001). Although the first site of action of JHAs in particular and IGRs in general is the endocrine system, many biochemical and physiological changes have been reported to occur in different metabolism pathways (Leonardi et al., 2001; Kim et al., 2002; Etebari et al., 2007). The present study is an extension of previous studies which dealt with the effects of certain IGRs on some metabolic parameters in the desert locust *S. gregaria*. It particularly aims to assess the action of pyriproxyfen, tebofenuzide and lufenuron on the lipid content of hemolymph and fat bodies of nymphs and adults.

MATERIALS AND METHODS

The experimental insect

Successive generations of the desert locust *S. gregaria* (Forskål) (Orthoptera: Acrididae) were maintained for several years under the gregarious conditions in Department of Zoology, Faculty of Science, Al-Azhar university, Cairo, Egypt. It was originated from Locust and Grasshopper Res. Division, Plant Protection Research Institute, Giza, Egypt. The culture was raised and handled under crowded

breeding conditions described by Hassanein (1965). The hoppers were reared in wooden cages with wire-gauze sides (40×40×60 cm). Each cage was equipped internally with 60 W electric bulb for lightening (17:7 LD) and warming (32 ± 2°C). The relative humidity varied from 30 to 70%. Nymphs and adults were allowed to feed on fresh leaves of leguminous plant *Medicago sativa*.

Insect growth regulators

Some of the insect growth regulators used is as follows:

- Pyriproxyfen (S-31183) is a product of Sumitomo Chemical Co. Ltd., Pesticides Division, Osaka, Japan, with the chemical formula: 2-{1-methyl-2-(4-phenoxy-phenoxy) ethyl} pyridine.
- A technical grade of Tebufenozide (RH-5992) was used. Its chemical name is 1-N-t-butyl-1 (3, 5-dimethyl benzoyl)-2-(4-ethylbenzoyl) hydrazine (Rohm and Haas Company, Philadelphia, PA).
- Lufenuron (Match, CGA-184699) was used. Its chemical formula is: N-{{{ 2,5-dichloro-4-(1,1,2,3,3-hexafluoro-propoxyl)-phenyl}amino}-2,6-difluorobenzamide (CA)}}}.

All insect growth regulators were provided from Institute of Plant Protec., Ministry of Agric., Giza, Egypt.

Nymphal treatments

Two concentration levels of each IGR were prepared using the distilled water: 1000 and 62.5 ppm. The concentration range was chosen depending on some preliminary trials carried out on the present insect species. Feeding technique was applied using fresh clean clover leaves (*M. sativa*) after dipping for 3 min in the concentration level and then offered to the newly molted last fifth instar nymphs. The control nymphs had been provided with fresh clean clover leaves after dipping them in distilled water. 30 nymphs were used for each treatment or controls. Each individual nymph was kept in a suitable glass vial whose bottom was covered with a thin layer of sterilized sand. All vials were carefully located in a cage provided with a suitable electric bulb for lightening and warming.

Lipid determination

Hemolymph of 1, 4 and 7 days old of instar nymphs was drawn out from the coxal joint into Eppendorff Pipetman containing few milligrams of phenoloxidase inhibitor (phenylthiourea) to prevent tanning or darkening and then diluted 5× with saline solution 0.7%. The hemolymph samples were then centrifuged at 2000 r.p.m. for 5 min, and only the supernatant were used for assay directly or frozen until use. 3 replicates (1 nymph/replicate) were used for sampling the hemolymph.

The same nymphs (treated or control) have been dissected to collect their fat body (Visceral and parietal) and then homogenized in a saline solution (the fat body of one insect / 1 ml saline solution 0.7 %) using a fine electric homogenizer tissue grinder for 2 min. Homogenates were centrifuged at 4000 r.p.m. for 15 min. The supernatant was used directly or frozen until use for the enzymatic determination. Three replicates (one nymph/replicate) were used for sampling the fat body.

Dealing with the adult females of 0-day old (newly emerged) and 4-day old, the same work for hemolymph and fat bodies was carried out.

Quantitative determination of the total lipid content of hemolymph or fat body was conducted according to the technique of Folch et al. (1957) and lipid estimation was taken place by phosphovanilin reagent depending on Knight et al. (1972) and using the

Spectrophotometer at 520 nm.

Analysis of data

Data obtained were analyzed using the Student t-distribution and were refined by Bessel's correction (Moroney, 1969) for testing the significance of difference between means.

RESULTS

Newly molted last (5th) instar nymphs of the desert locust *S. gregaria* were treated (through the fresh food plant) with a high (1000.0 ppm) or a low (62.5 ppm) concentration level of pyriproxyfen (juvenoids), tebufenozide (ecdysone agonist) or lufenuron (chitin synthesis inhibitor). The present study was carried out to investigate the metabolic effects on the lipid content of two different tissues: hemolymph and fat body. Lipid content was determined for the early- (1-day old), mid- (4-day old) and late-aged old (7-day old) nymphs as well as for 1- and 4-day old adult females.

With respect to the control insects, the current results indicated that the lipid content of the fat body was higher than that of the hemolymph. In other words, when the lipid content of the fat body is high, the lipid content of the hemolymph is low and *vice versa*.

Influenced lipid content of nymphs by the IGRs

The disturbance of lipid content of nymphal hemolymph was varied according to the potency of IGR as well as the developmental age (Table 1). Hemolymph lipid content of the early-aged nymphs had been subjected to a reducing effect after treatment with the high concentration level of pyriproxyfen, tebufenozide or lufenuron (20.54 ± 3.25, 20.17 ± 3.48 and 20.54 ± 3.22 mg/ml by pyriproxyfen, tebufenozide and lufenuron, respectively, as compared to 22.25 ± 3.21 mg/ml of controls). Reversely, treatment with the low concentration level of each IGR resulted in increasing lipid content. The strongest promoting effect on hemolymph lipids of these early-aged nymphs was exhibited by tebufenozide (34.87 ± 7.52 mg/ml, at low concentration level, in comparison with 22.25 ± 3.21 mg/ml of control nymphs). With the age of nymphs, all IGRs could significantly or non-significantly reduce the lipid content of hemolymph. The maximal reducing effects on hemolymph lipids was exhibited by pyriproxyfen (Change%: -59.34, at the high concentration level) in the mid-aged nymphs and by lufenuron (Change%: -59.85, at the high concentration level) in the late-aged nymphs while the minimal reducing effects was exhibited by pyriproxyfen (Change%: -7.64 and -2.61 at the high concentration level) in both early- and late-aged nymphs, and lufenuron (Change%: -8.19 at the low concentration level) in mid-aged nymphs.

With regard to the lipid content in fat bodies of nymphs,

data assorted in Table 2 obviously show a predominant inhibitory effect of all IGRs, regardless of the concentration level or the nymphal age. Moreover, the most drastic inhibitory effect was exhibited by pyriproxyfen (Change% s: -28.0, - 63.38 and – 64.71, at the high concentration level, in early-, mid- and late-aged nymphs, respectively). Also, the strongest prohibiting action for both tebufenozide and lufenuron was recorded in the late-aged nymphs (Change% s: -50.59, at the high concentration level of tebufenozide) and in the mid-aged nymphs (Change% s: -31.69, at the high concentration level of lufenuron), respectively (Table 2).

Influenced lipid content of adults by the IGRs

The estimation of lipid content in the adult stage was carried out for 1- and 4-day old females (within the ovarian maturation period). It suffices to see the data of Table 3 for concluding that pyriproxyfen, tebufenozide and lufenuron-nymphal treatments resulted in remarkably or slightly decreased lipids in the hemolymph with an exceptional case because the lipid content non-significantly increased in 4-day old adults after treatment with the low concentration level of lufenuron (29.45 ± 3.50 vs. 28.16 ± 5.11 mg/ml of control adults). However, adults died just after emergence as affected by the extended lethal effect of tebufenozide (at the high concentration level) but 4-day old adults died as affected by the extended lethal effect of pyriproxyfen (at the high concentration level). Therefore, the lipid content could not de determined in these two cases. It can be concluded from data of Table 3, also, that lufenuron (at the high concentration level) exerted the greatest prohibiting action on the hemolymph lipids (Change% s: - 54.46 and – 54.44 in 1- and 4-day old adults, respectively).

As clearly shown in Table 4, the current IGRs unexceptionally exhibited inhibitory effects on the lipids of adult fat bodies whatever the concentration level or the adult age. For some details, the most deteriorated lipid content in fat body was caused by pyriproxyfen at high concentration level (27.41 ± 3.54 mg/gm, as compared to 97.33 ± 5.24 mg/gm of adult congeners) for 1-day old adults, while a similar effect was exhibited by lufenuron at high concentration level only for 4-day old adults (55.62 ± 4.22 mg/gm, as compared to 87.25 ± 2.42 mg/gm of control congeners).

DISCUSSION

The lipids are necessary as a source of energy in live creatures such as insects. Insects obtain lipids from the food sources or synthesize them from within the bodies (Gilbert, 1967). It has been reported that the lipid accumulation is more likely to be related to lack of juvenile hormone (Hill and Ezzat, 1974). Hence, the lipid turnover in insects is regulated by neuroendocrine-

Table 1. Total lipid content (ng/ml ± SD) of the hemolymph of the desert locust *Schistocerca gregaria* nymphs as influenced by some IGRs after treatment of the early last instar nymphs.

IGRs	Conc. (ppm)	Last instar nymphs (Age in days)					
		1-day old	Change %	4-day old	Change %	7-day old	Change %
Pyriproxyfen	1000	20.54 ± 3.25^a	-7.64	7.44 ± 2.81^d	-59.34	19.77 ± 2.42^a	-2.61
	62.5	25.43 ± 2.65^a	14.41	16.34 ± 2.48^a	-10.92	16.34 ± 2.22^a	-19.7
	Controls	22.25 ± 3.21	-	18.35 ± 4.32	-	20.34 ± 3.42	-
Tebufenozide	1000	20.17 ± 3.48^a	-9.43	12.45 ± 1.43^b	-34.2	8.33 ± 1.77^d	-58.96
	62.5	34.87 ± 7.52^d	57.07	15.13 ± 4.24^a	-17.55	19.32 ± 2.26^a	-4.92
	Controls	22.25 ± 3.21	-	18.35 ± 4.32	-	20.34 ± 3.42	-
Lufenuron	1000	20.54 ± 3.22^a	-7.86	15.66 ± 3.22^a	-14.42	8.15 ± 2.41^d	-59.85
	62.5	25.61 ± 4.85^a	15.31	9.82 ± 2.22^c	-8.19	17.88 ± 4.50^a	-12.31
	Controls	22.25 ± 3.21	-	18.35 ± 4.32	-	20.34 ± 3.42	-

Conc.: concentration, mean ± SD followed with the same letter (a): is not significantly different (P>0.05), (b): significantly different (P<0.05), (c): highly significantly different (P<0.01), (d): very highly significantly different (P<0.001).

Table 2. Total lipid content (mg/g ± SD) of the fat body of the desert locust *Schistocerca gregaria* nymphs as influenced by some IGRs after treatment of the early last instar nymphs.

IGRs	Conc. (ppm)	Last instar nymphs (Age in days)					
		1-day old	Change %	4-day old	Change (%)	7-day old	Change (%)
Pyriproxyfen	1000.0	63.51 ± 2.45^d	-28	34.25 ± 2.33^d	-63.38	26.54 ± 2.45^d	-64.71
	62.5	81.53 ± 2.42^b	-7.29	87.34 ± 4.89^a	-6.62	70.52 ± 3.54^a	-6.62
	Controls	88.26 ± 4.22	-	93.48 ± 4.85	-	75.17 ± 5.12	-
Tebufenozide	1000.0	84.35 ± 5.34^a	-4.42	82.54 ± 3.34^c	-11.57	37.15 ± 4.22^d	-50.59
	62.5	82.15 ± 4.62^a	-6.92	88.59 ± 4.28^a	-5.24	59.33 ± 2.57^d	-24.99
	Controls	88.26 ± 4.22	-	93.48 ± 4.85	-	75.17 ± 5.12	-
Lufenuron	1000.0	85.66 ± 6.40^a	-2.94	63.85 ± 2.54^d	-31.69	66.61 ± 5.32^c	-11.31
	62.5	80.35 ± 2.51^c	-8.95	74.88 ± 5.48^d	-19.82	57.65 ± 3.88^d	-23.5
	Controls	88.26 ± 4.22	-	93.48 ± 4.85	-	75.17 ± 5.12	-

Conc., concentration, mean ± SD followed with the same letter (a): is not significantly different (P>0.05), (b): significantly different (P<0.05), (c): highly significantly different (P<0.01), (d): very highly significantly different (P<0.001).

controlled feedback loops (Downer, 1985).

In the present study on the desert locust *Schistocerca gregaria*, the disturbance of lipid content in hemolymph of last instar nymphs was varied according to the potency of the IGR (pyriproxyfen, tebufenozide or lufenuron) as well as the developmental age (1-, 4- or 7-day old nymphs). Hemolymph lipid content of the early-aged nymphs had been subjected to a reducing effect after treatment with the high concentration level (1000.0 ppm) of pyriproxyfen (juvenoids or JHA), tebufenozide (ecdysone agonist) or lufenuron (chitin synthesis inhibitor). Reversely, treatment with the low concentration level (62.5 ppm) of each IGR resulted in an increase of lipids, especially by the action of tebufenozide. However, the lipid content of hemolymph was significantly or non-significantly depleted in mid- and late-aged nymphs. Concerning the lipid content in fat bodies of nymphs, a dominant inhibitory effect of all IGRs was detected, regardless of the concentration level or the nymphal age. The strongest inhibitory effect was exhibited on nymphs of all ages by pyriproxyfen but on mid-aged nymphs by tebufenozide and on late-aged nymphs by lufenuron.

Results of the present study to some extent agree with those reported results on different insects by various IGRs in a side but disagree with others on the other side. Decreased lipid content was recorded for the rice moth *Corcyera cephalonica* by the action of pyriproxyfen (Mandal and Chaudhuri, 1992), for the cotton leaf worm *Spodoptera littoralis* by (Bay Sir-8514) (Ahmad et al., 1989) and for the same insect by pyriproxyfen (Ahmed,

Table 3. Total lipid content (mg/ml ± SD) of the hemolymph of the desert locust *Schistocerca gregaria* adults as influenced by some IGRs after treatment of the early last instar nymphs.

| IGRs | Conc. (ppm) | Adult stage (Age in days) | | | |
		1-day old	Change (%)	4-days old	Change (%)
Pyriproxyfen	1000	20.52 ± 3.24[a]	-8.48	=	-
	62.5	19.54 ± 4.41[a]	-12.94	22.64 ± 2.25[a]	-19.16
	Controls	22.41 ± 5.48	-	28.16 ± 5.11	-
Tebufenozide	1000	=	-	-	-
	62.5	21.15 ± 4.73[a]	-5.8	20.43 ± 5.84[a]	-27.4
	Controls	22.41 ± 5.48	-	28.16 ± 5.11	-
Lufenuron	1000	10.25 ± 3.49[d]	-54.46	12.85 ± 4.25[d]	-54.44
	62.5	20.51 ± 1.82[a]	-8.48	29.45 ± 3.50[a]	4.62
	Controls	22.41 ± 5.48	-	28.16 ± 5.11	-

Conc., concentration, mean ± SD followed with the same letter (a): is not significantly different (P>0.05), (b): significantly different (P<0.05), (c): highly significantly different (P<0.01), (d): very highly significantly different (P<0.001). =: adults died.

Table 4. Total lipid content (mg/g ± SD) of the fat body of the desert locust *Schistocerca gregaria* adults as influenced by some IGRs after treatment of the early last instar nymphs.

| IGRs | Conc. (ppm) | Adult stage (Age in days) | | | |
		1-day old	Change (%)	4-day old	Change (%)
Pyriproxyfen	1000	27.41 ± 3.54[d]	-71.87	=	-
	62.5	88.53 ± 3.22[c]	-9.04	84.85 ± 4.66[a]	-3.85
	Controls	97.33 ± 5.24	-	87.25 ± 2.42	-
Tebufenozide	1000	=	-	-	-
	62.5	88.76 ± 3.27[c]	-8.8	80.15 ± 6.54[b]	-8.14
	Controls	97.33 ± 5.24	-	87.25 ± 2.42	-
Lufenuron	1000	56.81 ± 2.63[d]	-41.62	55.62 ± 4.22[d]	-36.23
	62.5	88.67 ± 6.73[c]	-8.9	85.33 ± 4.85[a]	-2.14
	Controls	97.33 ± 5.24	-	87.25 ± 2.42	-

Conc., concentration, mean ± SD followed with the same letter (a): is not significantly different (P>0.05), (b): significantly different (P<0.05), (c): highly significantly different (P<0.01), (d): very highly significantly different (P<0.001).=: =: adults died.

2001). Also, lipid levels in the hemolymph and fat body of 6th instar larvae of the spruce budworm *Choristoneura fumiferana* were severely depleted as a result of fenoxycarb treatments (Mulye and Gordon, 1993). The decrease lipid content in the pupae of *S. littoralis* was estimated after larval treatment with mevalonic acid (Ghoneim, 1994). However, lufenuron and diofenolan treatments resulted in decreasing lipids at the start and end days of pupae of the red palm weevil *Rhynchophora ferrugineus* but increasing lipids were estimated for mid-aged pupae (Ghoneim et al., 2003).

In addition, diofenolan treatments remarkably reduced the lipid content along the pupal stage of the house fly *Musca domestica* except the last day at which pupae gained more lipids (Amer et al., 2005). Lipid levels in the hemolymph of silkworm *Bombyx mori* larvae were declined throughout the experimental period but were elevated initially in the fat body and then lowered (Etebari et al., 2007). Injection of 0.1 pmol of the synthetic adipokinetic hormone (Peram-AKH II) onto the American cockroach *Periplaneta americana* led to a significant reduction of the levels of neutral lipids and phospholipids in hemolymph (Michitsch and Steele, 2008). Lipid content in the whole body of *Plodia interpunctella* larvae was reduced as a response to the action of 20-hydroxyecdysone and azadirachtin (Rharrabe et al., 2008) or to pyriproxyfen (Ghasemi et al., 2010). Pyriproxyfen treatments resulted in decreasing lipid content in hemolymph and fat body of the sunn pest *Eurygaster integriceps* nymphs (Zibaee et al., 2011).

Instead of the inhibition of lipid content by IGRs, activation of this metabolite was reported by some authors using different IGRs against various insect species (Amer, 1990; Ghoneim, 1994; Soltani-Mazouni et al., 1999; Bouaziz et al., 2011).

The current results, for *S. gregaria*, evidently show dominant inhibitory effects of pyriproxyfen, tebufenozide and lufenuron on the lipid content of fat body of nymphs, regardless of their age or the concentration level. Also, similar inhibitory effects of these IGRs were recorded on the lipid content of hemolymph of nymphs after 2 days of treatment. On the other hand, the hemolymph lipid content of 1-day old nymphs was variably affected depending upon the concentration level of all IGRs because increasing lipids were detected at the high concentration (1000.0 ppm) but decreasing lipids were estimated at the low concentration level (62.5 ppm). Hence, it is reasonable to conclude that the present IGRs generally prohibited the nymphs to attain normal lipid level in hemolymph or fat body in spite of the difference in the nature of each tissue. These findings are controversial to the trend of lipids in control nymphs, where the lipid content had been increased in hemolymph as it is decreasing in fat body and *vice versa*. Therefore, the present IGRs pronouncedly interfered with not only the synthesis of lipids but also their mobilization as promoted to convert into other metabolites or fatty acids. This suggestion may be supported by the increasing cholesterol in the mid gut brush border membrane of silkworm *B. mori* larva after treatment with fenoxycarb (JHA) (Leonardi et al., 2001) or in the hemolymph of 120 h post-treatment of silkworm larvae with pyriproxyfen (JHA) (Etebari et al., 2007).

Although pyriproxyfen and tebufenozide are classified in the group interfering with the action of insect hormones, and lufenuron is classified in the chitin synthesis inhibitors Tunaz and Uygun, 2004), all compounds vigorously acted as lipid inhibitors on the desert locust *S. gregaria*, in the present study, since the intense metabolic modifications are related to the various hormonal systems and under neuroendocrine control (Gade et al., 1997) or are transported from the storage site (fat body) *via* the hemolymph towards the user organs, in particular cuticular synthesis (Dapporto et al., 2008) and vitellogenesis (Zhou and Miesfeld, 2009). In addition, the odd case of lipid increasing a day after treatment with pyriproxyfen, tebufenozide or lufenuron, in the present study, may be attributed to the accumulation of carbohydrates which might lead to an inverse in their conversion rate to lipids as a reverse material (Tanani et al., 2012).

With regard to the adult females in the present study, pyriproxyfen, tebufenozide and lufenuron-treatments of nymphs resulted in considerably or slightly decreased lipids in the hemolymph of both 1- and 4-day old adults. There was an exceptional case because lipid content non-significantly increased in 4-day old adults after nymphal treatments with 62.6 ppm lufenuron whereas it exhibited the strongest prohibiting effect on hemolymph lipids at the concentration level 1000.0 ppm. Furthermore, these IGRs unexceptionally exerted inhibitory actions on the lipid content in fat bodies of adults, irrespective of their age.

The reported works on the lipids in hemolymph or fat body of adults of other insect species, as affected by IGRs, are unfortunately scarce. On the other hand, lipids were detected in other tissues such as ovaries and testes after treatment with some IGRs. For example, lipid content in ovaries of the cut worm *Spodoptera litura*, *T. molitor*, the Mediterranean flour moth *Ephestia kuehniella*, *P. interpunctella* and *S. litura* was depleted as a response to chlorfluazuron, flucycloxuron, tebufenozide, pyriproxyfen and chlorfluazuron, respectively (Perveen and Miyata, 2000; Hami et al., 2004; Kebbeb et al., 2008; Ghasemi et al., 2010; Perveen, 2012). The depletion of lipids in ovaries of treated insects could be understood in the light of reduced ovaries due to the effects of IGRs on oogenesis or/and vitellogenesis (Kanost et al., 1990; Shaaya et al., 1993; Ghasemi et al., 2010).

However, the exact interpretation of lipid depletion in hemolymph and fat body of adult *S. gregaria*, in the present study, remains speculative. On the other hand, this desert locust, like nearly all other insects capable of flying long distances, mobilizes lipids stored in the fat body to meet the energy requirements for sustained flight. These lipids are transported *via* the hemolymph to flight muscles, where they are completely oxidized to CO_2 and H_2O. If it is possible to intervene and reduce this metabolic activity, then this has two consequences for *S. gregaria*. One is that the insect then lacks the fuel it needs for flight activity, and the other is that its water regime is disturbed. In the extremely arid regions inhabited by these insects, the water from oxidation of lipids and other foodstuffs is needed by them to regulate processes at the cellular level (Al-Fifi, 2009). So, the lipid depletion in hemolymph and fat body of adults, as recorded by the current results, will reflect on the insect migration defect and this will result in control of this migratory destructive insect. Frankly, the transportation problems between different organs, such as hemolymph, fat body, flight muscles and gonads, still need further studies to accurately investigate the lipid mobilization.

ACKNOWLEDGMENT

The authors thank Prof. A.M. El-Gammal, Institute of Plant Protec. Ministry of Agric., Giza, Egypt, for providing the insect growth regulators used in the present work.

REFERENCES

Ahmad YM, Ali FA, Mostafa AMA (1989). Effects of two benzoylphenyl urea derivatives on hemolymph constituents of *Spodoptera littoralis* (Boisd.) larvae. Alex. Sci. Exch., 10: 209-220.

Ahmed AM (2001). Biochemical studies on the effect of some insect growth regulators on the cotton leafworm. Unpublished M.Sc. Thesis, Faculty of Agriculture, Cairo University, Egypt.

Al-Fifi ZIA (2009). Effect of different Neem products on the mortality and fitness of adult *Schistocerca gregaria* (Forskål). JKAU: Sci., 21(2): 299-315.

Alves MMTA, Orlandeli RC, Lourenco DAL, Pamphile JA (2011). Toxicity of the insect growth regulator lufenuron on the entomopathogenic fungus *Metarhizium anisopliae* (Metschnikoff) Sorokin assessed by conidia germination speed parameter. Afr. J. Biotech., 10(47): 9661-9667.

Amer MS (1990). Effects of the anti-JH synthesis (Mevalonic acid) on the main metabolites of *Spodoptera littoralis*. Egypt. J. Appl. Sci., 5(1): 82-91.

Amer MS, Ghoneim KS, Al-Dali AG, Bream AS, Hamadah KhSh (2005). Effectiveness of certain IGRs and plant extracts on the lipid metabolism of *Musca domestica* (Diptera: Muscidae). (Entomology), J. Egypt. Acad. Soc. Environ. Dev., 6(1): 53-67.

Angeli G, Forti D (1997). Effetti collaterali di inseticidi regolatori della crescita (IGRs) verso l'emittero *Orius laevigatus* (Fieber) (Heteroptera: Anthocoridae). Difesa-delle-Piante, 20: 1-2.

Bouaziz A, Boudjelida H, Soltani N (2011). Toxicity and perturbation of the metabolite contents by a chitin synthesis inhibitor in the mosquito larvae of Culiseta longiareolata. Ann. Biol. Res., 2(3): 134-143.

Bueno AF, Freitas S (2004). Effect of the insecticides abamectin and lufenuron on eggs and larvae of *Chrysoperla externa* under laboratory conditions. BioControl, 49: 277-283.

Canavoso LE, Jouni Z E, Karnas KJ, Pennington JE, Wells MA (2001). Fat metabolism in insects. Annu. Rev. Nutr., 21: 23-46.

Carlson GR (2000). Tebufenozide: A novel caterpillar control agent with unusually high target selectivity. In "Green Chemical Syntheses and Processes" (ed.s: P. T. Anastas, L. G. Heine, T. C. Williamson). Am. Chem. Soc., 767(2): 8-17.

Carlson GR, Dhadialla TS, Thompson C, Ramsey R, Thirugnanam M, James W, Slawecki R (1994). Insect toxicity, metabolism and receptor binding characteristics of the non-steroidal ecdysone agonist, RH- 5992. Proc. XIth Ecdysone Workshop, p. 43.

Carter SW (1975). Laboratory evaluation of three novel insecticides inhibiting cuticle formation against some susceptible and resistant stored product beetles. J. Stored Prod. Res., 11: 187-193.

Castane C, Arino J, Arno J (1996). Toxicity of some insecticides and acaricides to the predatory bug *Dicyphus tamaninii* (Het.: Miridae). Entomophaga, 41: 211-216.

Dapporto L, Lambardi D, Turillazzi S (2008). Not only cuticular lipids: first evidence of differences between foundresses and their daughters in polar substances in the paper wasp *Polistes dominulus* J. Insect Physiol., 54(1):89-95.

Dean SR, Meola RW, Meola SM, Sittertz-Bhatkar H, Schenker R (1998). Mode of action of lufenuron on larval cat fleas (Siphonaptera: Pulicidae). J. Med. Entomol., 35: 720-724.

Dhadialla TS, Carlson GR, Le DP (1998). New Insecticides with ecdysteroidal and juvenile hormone activity. Annu. Rev. Entomol., 43: 545-69.

Downer RGH (1985). Lipid metabolism. In "Comprehensive Insect Physiology, Biochemistry, and Pharmacology" (eds.: G.A. Kerkut, L.I. Gilbert), vol. 10, Pergamon Press, Oxford, , pp. 75–114.

Etebari K, Bizhannia AR, Sorati R, Matindoost L (2007). Biochemical changes in haemolymph of silkworm larvae due to pyriproxyfen residue. Pesticide Biochem. Physiol., 88: 14–19.

Folch J, Less M, Sloane-Stanley GH (1957). A simple method for the isolation and purification of total lipids from animal tissues. J. Biol. Chem., 26: 497-509.

Gade G, Hoffmann KH, Spring JH (1997). Hormonal regulation in insets: Facts, Gaps, and future directions. Physiol. Rev., 77: 963-1032.

Ghasemi A, Sendi JJ, Ghadamyari M (2010). Physiologicak and biochemical effect of pyriproxyfen on Indian meal moth *Plodia interpunctella* (Hubner)(Lepidoptera: Pyralidae). J. Plant Prot. Res., 50(4): 416- 422.

Ghoneim KS (1994). Synergistic and antagonistic action of Chlorfluazuron and mevalonic acid against the main body metabolism of the Egyptian cotton leafworm *Spodoptera littoralis* Boisd. (Lepidoptera: Noctuidae) (1994). J. Egypt. Ger. Soc. Zool., 14(D): 89-115.

Ghoneim KS, Al-Dali AG, Abdel-Ghaffar AA (2003). Effectiveness of Lufenuron (CGA-184699) and Diofenolan (CGA-59205) on the general body metabolism of the red palm weevil, *Rhynchophorus ferrugineus* (Curculionidae: Coleoptera) (2003). Pakistan J. Biol. Sci., 6(13): 1125-1129.

Gilbert LI (1967). Biochemical correlation in insect metamorphosis. Comp. Biochem., 28: 199-252.

Hami M, Taibi F, Soltani-Mazouni N (2004). Effects of flucycloxuron, chitin synthesis inhibitor on reproductive events and thickness of chorion in mealworms. Commun. Agric. Appl. Biol. Sci., 69(3): 249-255.

Hassanein MS (1965). Laboratory and outdoor cultures and breeding of locusts and grasshoppers. FAO Publ., 5/31901, p. 10.

Hill L, Ezatt EG (1974). The relationships between corpora allata, fat body and hemolymph lipids in the adult female desert locusts. J. Insect Physiol., 20: 2143-2156.

Ishaaya I, Yablonski S, Ascher KRS (1987). Toxicological and biochemical aspects of novel acylureas on resistant and susceptible strains of *Tribolium castaneum*. In: Conference on Stored- Product Protection. Tel Aviv, Israeel, 21–26 September 1986. pp. 613–622.

Javaid I, Uaine RN, Massua J (1999). The use of insect growth regulators for the control of insect pests of cotton. Int. J. Pest Manag., 45: 245-247.

Kanost MR, Kawooya JK, Law JH, Ryan RO, Heusden VMC, Ziegler R (1990). Insect hemolymph proteins. In: "Advances Insect physiology" (eds. P. D. Evans and V.B. Wigglesworth), Academic Press, London. 22: 299-396.

Kebbeb MEH, Gaouaoui R, Bendjeddou F (2008). Tebufenozide effects on the reproductive potentials of the Mediterranean flour moth, *Ephestia kuehniella*. Afr. J. Biotech., 7(8): 1166-1170.

Kim K, Kim Y, Kim Y (2002). A biochemical evidence of the inhibitory effect of difubenzuron on the metamorphosis of the silkworm, *Bombyx mori*, J. Asia-Pacific Entomol., 5: 175-180.

Knight JA, Anderson S, Jams MR (1972). Chemical basis of the sulphovanilin reaction of estimating total lipid. J. Clin. Chem., 18: 199.

Kostyukovsky MB, Chen AS, Shaaya E (2000). Biological activity of two juvenoids and ecdysteroids against three stored product insects. Insect Biochem. Mol., 30 (8–9): 891-897.

Leonardi MG, Marciani P, Montorfono PG, Cappellozza S, Giordana B, Monticalli G (2001). Effects of fenoxycarb on leucine uptake and lipid composition of midgut brush border membrane in the silkworm, *Bombyxo moryi* (Lepidoptera: Bombycidae). Pest. Biochem. Physiol., 70(1): 42-51.

Linrui H, Chengju W, Jiazhen J, Xuefeng L, Hingqi Z, Wenji Z, Lihong Q (2006). Studies on the bioactivity of tebufenozide to beet armyworm, *Spodoptera exigua* (Hübner). Plant Prot., 32(2): 48-52.

Mandal D, Chaudhuri DR (1992). Studies on carbohydrate, protein and lipid levels in normal and stress conditions in fat body and integument as compared to whole body during development of moth *Corcyra cephalonica*. Insect Sci. Appl., 13: 121-128.

Michitsch J, Steele JE (2008). Carbohydrate and lipid metabolism in cockroach (*Periplaneta americana*) fat body are both activated by low and similar concentrations of Peram-AKH II. Pept., 29: 226-234.

Miyamoto J, Hirano M, Takimoto Y, Hatakoshi M (1993). Insect growth regulators for pest control, with emphasis on juvenile hormone analogs: present and future prospects. In: "Pest Control with Enhanced Environmental Safety" (eds. Duke, S.O.; Menn, J.J. ; Plimmer, J.R.). Washington D.C., ACS Symp. Ser., 524: 144–168.

Moriello KA, Deboer DJ, Schenker R, Blum JL, Volk LM (2004). Efficacy of pre-treatment with lufenuron for the prevention of *Microsporum canis* infection in a feline direct topical challenge model. Vet. Dermatol., 15: 357-362.

Moroney MJ (1969). Facts from figures (3rd ed.). Penguin Books Ltd., Harmondsworth. Middle Sex.

Mulye H, Gordon R (1993). Effects of two juvenile hormone analogs on hemolymph and fat-body metabolites of the eastern spruce budworm, *Choristoneura fumiferana* (Clemens) (Lepidoptera: Tortricidae). Can. J. Zool., 71: 1169-1174.

Oberlander H, Nickle D, Silhacek DL, Hagstrum DW (1978). Advances in insect growth regulators research with grain insects. In: "Symposium on the Prevention and Control of Insects in Stored Foods Products". Manhattan, Kansas, 1987: 247-263.

Oberlander H, Silhacek DL, Shaaya E, Ishaaya I (1979). Current status and future perspective of the use of insect growth regulators for the control of stored product insects. J. Stored Prod. Res., 33(1): 1-6.

Perveen F (2012). Biochemical Analyses of Action of Chlorfluazuron as Reproductive Inhibitor in *Spodoptera litura*. Insecticides – Adv. Integ. Pest Manag., pp. 293-326.

Perveen F, Miyata T (2000). Effects of sublethal dose of chlorfluazuron on ovarian development and oogenesis in the common cutworm, *Spodoptera litura* (F.) Lepidoptera: Noctuidae). Ann. Entomol. Soc. Am., 93: 1131-1137.

Retnakaran A, Gelbic I, Sundaram M, Tomkins W, Ladd T, Primavera M, Feng Q, Arif B, Palli R, Krell P (2001). Mode of action of the ecdysone agonist tebufenozide (RH-5992), and an exclusion mechanism to explain resistance to it. Pest Manag. Sci., 57(10):951-7.

Rharrabe K, Amri H, Bouayad B, Sayah F (2008). Effects of azadirachtin on post-embryonic development, energy reserves and α-amylase activity of *Plodia interpunctella* Hübner (Lepidoptera: Pyralidae). J. Stored Prod. Res., 44 (3): 290-294.

Shaaya E, Shirk RD, Zimowska G, Plotkin S, Young NJ, Rees HH, Silhacek DL (1993). Declining ecdysteroid levels are temporally correlated with the initiation of vitellogenesis during pharate adult development in Indian meal moth, Plodia interpunctella. Insect Biochem. Mol. Biol., 23(1): 153-158.

Soltani-Mazouni N, Khebbeb MEH, Soltani N (1999). Production of ovarian ecdysteroids in oocyte maturation in *Tenebrio molitor*. Ann. Soc. Ent. France, 35: 82-86.

Staal GB (1975). Insect growth regulators with juvenile hormone activity. Annu. Rev. Entomol., 20: 417-460.

Sullivan JJ, Goh KS (2008). Environmental fate and properties of pyriproxyfen. J. Pestic. Sci., 33: 339-350.

Tanani MA, Ghoneim KS, Hamadah KhSh (2012). Comparative effectiveness of certain IGRs on the carbohydrates of hemolymph and fat body of Schistocerca gregaria (Orthoptera: Acrididae). In press.

Tunaz H, Uygun N (2004). Insect Growth Regulators for Insect Pest Control. Turk. J. Agric., 28: 377-387.

Wilson TG, Cryan JR (1997). Lufenuron, a chitin-synthesis inhibitor, interrupts development of *Drosophila melanogaster*. J. Exp. Zool., 278: 37-44.

Zhou G, Miesfeld RL (2009). Energy metabolism during diapause in *Culex pipiens* mosquitoes. J. Insect Physiol., 55: 40-46.

Zibaee A, Zibaee I, Sendi JJ (2011). A juvenile hormone analog, pyriproxyfen, affects some biochemical components in the hemolymph and fat bodies of *Eurygaster integriceps* Puton (Hemiptera: Scutelleridae). Pestic. Biochem.Physiol., 100(3): 289-298.

Zurfleuh RC (1976). Phenyl ethers as insect growth regulators: laboratory and field experiments. In: "The Juvenile Hormones" (L.I. Gilbert, ed.). Plenum Press, New York. pp. 61–74.

3

Ginger extract (*Zingiber officinale*) triggers apoptosis and G_0/G_1 cells arrest in HCT 116 and HT 29 colon cancer cell lines

Shailah Abdullah[1], Siti Amalina Zainal Abidin[1], Noor Azian Murad[2], Suzana Makpol[1], Wan Zurinah Wan Ngah[1] and Yasmin Anum Mohd Yusof[1]*

[1]Department of Biochemistry, Faculty of Medicine, Universiti Kebangsaan Malaysia, Jalan Raja Muda Abdul Aziz, 50300 Kuala Lumpur, Malaysia.
[2]Centre of Lipids and Engineering and Applied Research, Universiti Teknologi Malaysia, Jalan Semarak, 50300 Kuala Lumpur, Malaysia.

Although many studies have shown the antitumor properties of ginger extract (*Zingiber officinale*), little is known regarding the mechanism of its effects. This study was conducted to determine the mechanism of antitumor effects of ginger extract by evaluating apoptosis rate and cell cycle progression status in colon cancer cell lines HCT 116 and p53 defective HT 29. HCT 116 and HT 29 cells were cultured in the presence of ginger extract at various concentrations for 24 h. The percentage of cell viability was determined by 3-(4, 5-dimethylthiazol-2-yl)-2, 5-di phenyl tetrazolium bromide (MTT) assay. Our results showed that ginger extract inhibited proliferation of HCT 116 and HT 29 cells with an IC_{50} of 496 ± 34.2 µg/ml and 455 ± 18.6 µg/ml, respectively. We also found that ginger extract at increasing concentrations induced apoptosis dose dependently in both colon cancer cells. Apoptosis rates were 11.15, 35.05 and 57.49% for HCT 116 and 4.39, 19.81 and 28.09% for HT 29 at 200, 500 and 800 µg/ml of ginger extract, respectively. Ginger extract arrested HCT 116 and HT 29 cells at G_0/G_1 and G_2/M phases with corresponding decreased in S-phase. This study suggests that ginger extract may exert its antitumor effects on colon cancer cells by suppressing its growth, arresting the G_0/G_1-phase, reducing DNA synthesis and inducing apoptosis.

Key words: *Zingiber officinale*, HCT 116, HT 29, G_0/G_1 phase, S phase, apoptosis.

INTRODUCTION

Colorectal cancer accounts for 10 to 15% of all cancers and is the second leading cause of cancer-related death in industrialized countries (Virginie et al., 2007). According to National Cancer Registry of Malaysia (Gerard and Halimah, 2003), colon cancer is recognized as the commonest cancer among men and the third most common cancer among women. Colon cancer occurrence is commonly ascribed to the transformation of normal colon epithelium to adematous polyps and ultimately invasive cancer. Since cancer is a disorder of deregulated cell proliferation and cell survival (Evan and Vousden, 2001), inhibiting cell proliferation and increasing apoptosis in tumors are effective strategies for preventing tumor growth (Kim et al., 2005). Recent studies showed that curcumin, an active component of turmeric inhibits the growth of human colon cancer cells independent of COX-2 expression (Hanif et al., 1997). Lee et al. (2009) showed that thiosulfinates from *Allium tuberosum* L. inhibited cell proliferation and activated both the caspase-dependent and caspase-independent apoptotic pathways in colon cancer cells (Lee et al., 2009). Other study demonstrated that luteolin induced G_1 and G_2/M cell-cycle arrest and promoted apoptosisthrough down regulation of several antiapoptotic proteinsin colon cancer cells.

Emerging evidence has demonstrated that many natural products isolated from plant sources possess antitumor properties (Sakamoto et al., 1991). Since nature has provided many effective anticancer agents, plant derived drug research has made significant progress in anticancer therapies. Many components of medicinal or dietary plants have been identified as possessing potential chemopreventive properties capable of inhibiting, retarding or reversing the multi-stage carcinogenesis process (Surh et al., 1998; Surh et al., 2002). Plants of the ginger family (*Zingiber officinale*) are one of the most heavily consumed dietary substances in the world (Surh, 1999). It is one of the most widely used spices and has been used in traditional oriental medicines for centuries. Its extract and major pungent compounds have been shown to exhibit a variety of biological activities (Wei et al., 2005). The oleoresin from rhizome of ginger contains pungent ingredients including gingerol, shogaol and zingerone (Surh et al., 1998). Ginger's active components have been reported to exhibit cancer-preventive activity in several experimental carcinogenesis models including skin carcinogenesis (Murillo et al., 2008; Katiyar et al., 1996). Ginger supplementation suppresses colon carcinogenesis in the presence of procarcinogen (DMH) (Manju and Nalini, 2005). Recent study showed that [6]-gingerol a major phenolic compound derived from ginger, inhibited TRAIL-induced NF-κB activation by impairing the nuclear translocation of NF-κB, suppresses cIAP expression and increased TRAIL-induced caspase-3/7 activation in gastric cancer cells. On the other hand, [6]-shogaol alone reduced viability by damaging microtubules, arrested cell cycle in G_2/M phase and reduced viability in a caspase-3/7-independent manner in gastric cancer cells (Kazuhiro, 2007). Lee et al. (1998) showed that [6]-gingerol exerted inhibitory effects on the cell viability and DNA synthesis while inducing apoptosis of human promyelocytic leukemia HL-60 cells (Lee and Surh, 1998). Ginger was also shown to have anticancer effect by inducing apoptosis in rat liver cancer cells via up-regulation of the expression of pro-apoptotic protein, caspase-8 and downregulation of the expression of anti-apoptotic protein Bcl-2 (Yasmin et al., 2008). Exposure of Jurkat human T cell leukemia cells to various ginger constituents galanals A and B (isolated from the flower buds of Japanese ginger) resulted in apoptosis mediated through the mitochondrial pathway (Miyoshi et al., 2003). β-Elemene, an anticancer drug extracted from the ginger plant, triggered apoptosis in non-small-cell lung cancer cells through a mitochondrial release of the cytochrome c-mediated apoptotic pathway (Wang et al., 2005).

We have selected *Z. offinale* in this study to determine its mechanism as a potential chemopreventive agent against colon cancer cells HCT 116 and HT 29. Our results demonstrated that the inhibition of colon cancer cell growth by ginger involves both interference with cell cycle progression and apoptotic event.

MATERIALS AND METHODS

Ginger extract

Z. officinale (ginger) crude extract was obtained by ethanol extraction as provided by Noor Azian Murad from Center for Lipids Engineering Applied Research (CLEAR), Universiti Teknologi Malaysia. The sequence of pre-treatment for ginger extract include: peeling, slicing, washing of ginger rhizomes followed by bleaching, blanching, drying and grinding prior to solvent extraction. The sample calculation to obtain the moisture loss percent based on the initial and final weight of the sample is given by the equation:

$$\% \text{ moisture loss} = \frac{\text{Initial weight} - \text{current sample weight}}{\text{Initial weight}} \times 100\%$$

The dried ginger was fibrous and tough and therefore grinding of samples was done manually by pounding using 'mortar'. The pounded samples were then shredded manually. The ground ginger was dried prior to extraction using rotary evaporator in ethanol (1 L) for 6 h. The solvent was removed under vacuum at 500 mbar in the first hour, followed by 400 and 300 mbar in the next two hours to yield oleoresin, a brown viscous liquid (9.80 g, 4.9%). Refractive index reading of pure oleoresin is 1.5100.

Cell culture and treatment

HCT 116 and HT 29 cell lines were obtained from the American Type Culture Collection (Rockville, MD USA) and were cultured in RPMI 1640 medium (Flowlab, Australia) supplemented with 10% Fetal Calf Serum (FCS) (PAA Laboratories GmbH), 100 U/ml of penicillin and streptomycin (Thermo Scientific) at 37°C in 5% CO_2. Rates of cells proliferation, apoptosis and cell cycle were performed when cells reached 70% confluence density. Ginger extract was dissolved in 0.01% DMSO and added to cell lines after an overnight incubation.

Cell viability assay: 3-(4,5-dimethylthiazol-2-yl)-2,5-diphenyl tetrazolium bromide (MTT) assay

For cell viability assay, 2×10^4 cells (HCT 116 and HT 29)/well were plated in 100 μl of RPMI 1640 media. HCT 116 and HT 29 cells were incubated overnight at 37°C in humidified atmosphere of 5% CO_2 for cells attachment. Ginger extract was added at various concentrations ranging from 0, 10, 50, 100, 200, 500, 800, 1000 and 1500 μg/ml after 24 h incubation. After 24 h incubation, MTT solution (2 mg/ml) was added to the plate at the final concentration of 0.5 mg/ml. The resulting MTT-products were determined by measuring the absorbance at 550 nm with ELISA reader. Each point represents the mean of triplicate experiments.

$$\% \text{ viability} = \frac{\text{Optical density of sample}}{\text{Optical density of control}} \times 100$$

Cell cycle assay

The HCT 116 and HT 29 cells were seeded into 25 cm^2 flask (1 x 10^5 cells/ml) and were incubated with 200, 500 and 800 μg/ml ginger extract for 24 h. Following treatment, cells were trypsinized and incubated for 30 min at room temperature in staining solution consisting of propidium iodide (PI; 50 μg/ml), sodium citrate (0.1%), Triton X-100 (0.1%) and DNase-free RNase (20 μg/ml). Stained cells were then analyzed for DNA content by flow cytometer within

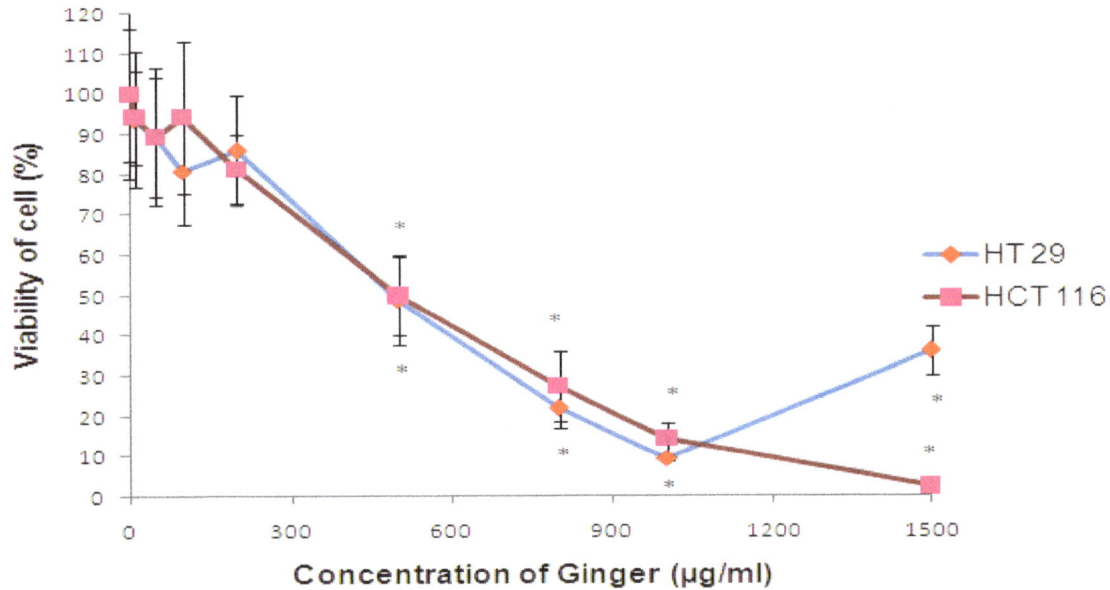

Figure 1. Effect of ginger extract on the proliferation of HCT 116 and HT 29 cells. Cells were treated with different concentration of ginger extract, incubated at 37°C and harvested after 24 hour. Cell viability was determined by the MTT assay. Data represent the mean ± S.D. (n = 3).

3 h of staining. Percentages of cells in each phase were calculated using Cell Modfit software programs (Becton Dickson, Canada).

Apoptosis–annexin V-FITC Staining

Apoptosis was detected with an annexin V-FITC kit (Beckton-Dickinson, Canada). All adhering and floating cells were harvested after incubation for 24 h with 200, 500 and 800 µg/ml ginger extract. Cells were collected, washed with ice-cold PBS and centrifuged. The cell pellet was resuspended in ice-cold binding buffer (100 mM HEPES/ NaOH, pH 7.5 containing 1.4 M NaCl and 25 mM CaCl$_2$) at a density of 1×10^6 cells per ml. Five hundred µl of this cell suspension was transferred to a 5 ml culture tube, to which 5 µl of Annexin V-FITC conjugate and 10 µl of propidium iodide were added. The cells were gently vortexed and incubated for 10 min at room temperature in the dark. The fluorescence of the cells was immediately determined by flow cytometry (Becton Dickson, Canada). The results were analysed using Cell Modfit software programs (Becton Dickson, Canada).

Statistical analysis

The experiments were repeated at least 3 times and the results were expressed as mean ± S.D. Statistical evaluation was done using the ANOVA (SPSS 16.0) where $p < 0.05$ was considered significant.

RESULTS

Figure 1 showed that ginger extract significantly inhibited the proliferation of both colon cancer cells in a dose-dependent manner. Proliferation of HCT 116 cells decreased when treated with ginger extract resulting in a 50% reduction at 496 ± 34.2 µg/ml, while HT 29 cells showed significant decrease in proliferation with 50% reduction at 455 ± 18.6 µg/ml of ginger extract.

To elucidate the mechanisms involved in the reduction in cell number by ginger, we examined whether such reduction was associated with cytostatic effect due to changes in cell cycle progression. Figures 2a and 2b summarized the relative percentages of HCT 116 and HT 29 cells in each phase of the cell cycle, following a 24-h treatment with varying ginger extract concentrations (200, 500 and 800 µg/ ml). The cell cycle alterations were concentration dependent. As shown in Figures 2a and 2b, in the absence of ginger extract, 48.82% HCT 116 cells and 54.22% of HT 29 were in the G$_0$/G$_1$-phase. However, treatment with ginger extract arrested both cells at G$_0$/G$_1$-phase: 63.15% for HCT 116 and 70.98% for HT 29 at 200 µg/ml. Interestingly, we found a unique characteristic whereby higher doses of ginger extract (800 µg/ml) did not arrest cells at G$_0$/G$_1$-phase but instead they were arrested at G$_2$/M-phase. The G$_0$/G$_1$ and G$_2$/M arrests were accompanied by corresponding reduction in the percentages of cells in the S-phase as shown in Figure 2a and 2b. Reduction in DNA synthesis was dependent on the dose of ginger extract: 15.06, 9.52 and 8.66%, when treated with 200, 500 and 800 µg/ml of ginger extract for HCT 116 cells and 8.81, 14.63 and 3.21% for HT 29 cells when treated with the same concentrations of ginger extract, respectively.

Figure 3a showed apoptotic effect of ginger extract on both HCT 116 and HT 29 cells using double staining method FITC-conjugated annexin V and PI. Ginger treatment increased the number of early and late apoptotic cells, respectively. Ginger extract at increasing concentrations induced higher rate of poptosis

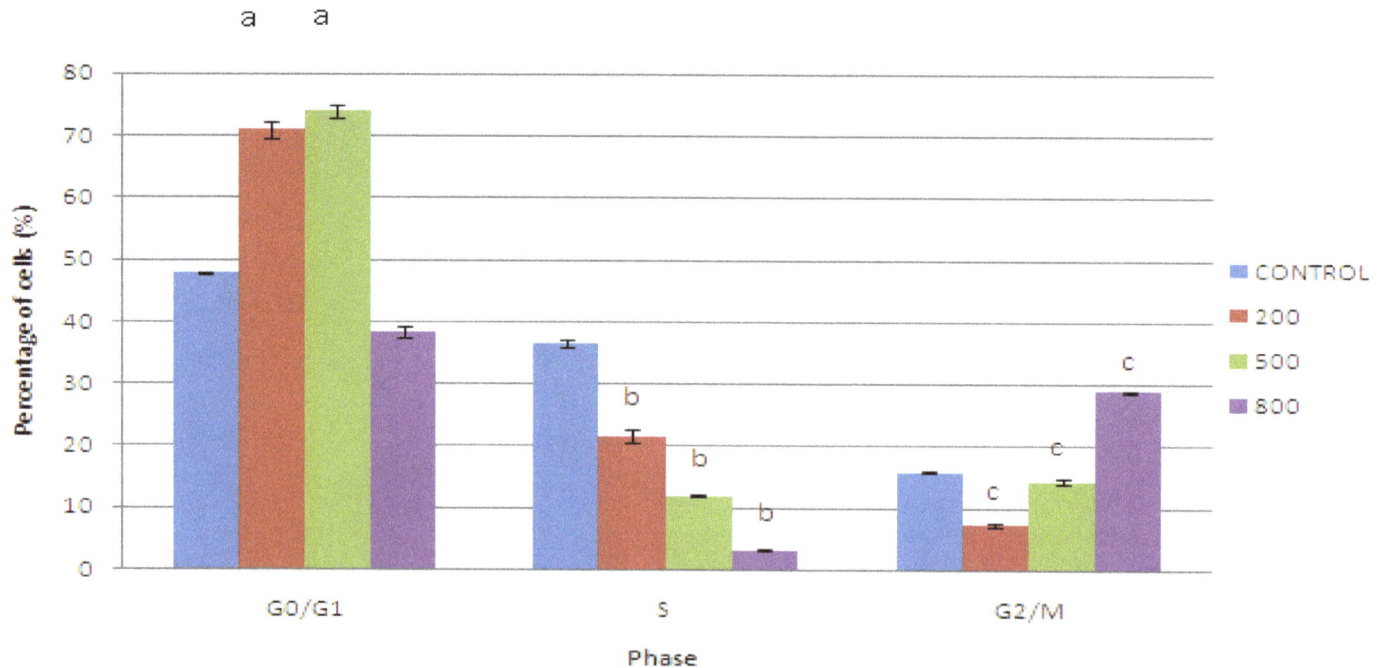

Figure 2a. Cell cycle profile of HCT 116 colon cancer cells in the presence of ginger extract. After a 24 h incubation with varying ginger extract concentrations. DNA content was evaluated with propidium iodide (PI) staining and fluorescence measured and analyzed. The values are the means ± SD of three independent experiments. a; significant compared to G0/G1 phase of control cell, b; significant compared to S phase of control cell, c; siginificant compared to G2/M phase of control cells..

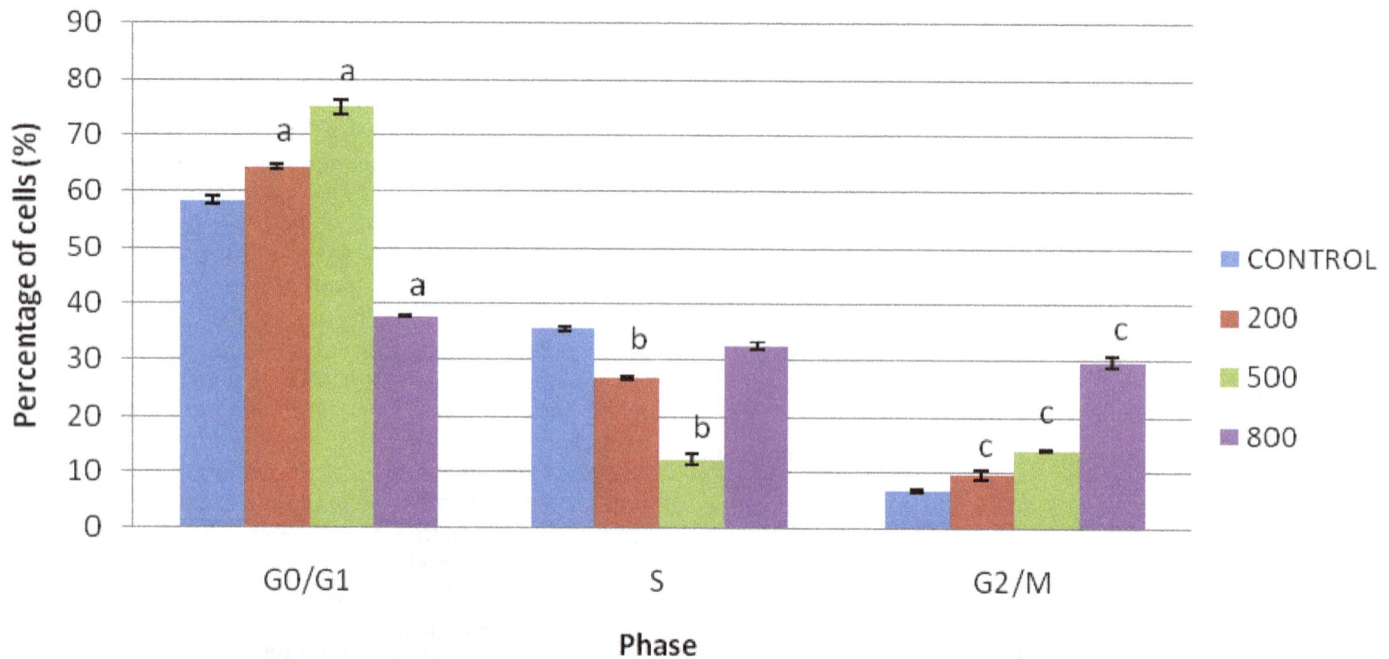

Figure 2b. Distribution of cell cycle phases of HT 29 cells after treatment with ginger extract. After a 24 hr incubation with varying ginger extract concentrations (0, 200, 500 and 800 μg/ml), DNA content were evaluated with propidium iodide staining and fluorescence measured and analyzed. The values are the means ± SD of three independent experiments. a; significant compared to G0/G1 phase of control cell, b; significant compared to S phase of control cell, c; siginificant compared to G2/M phase of control cells.

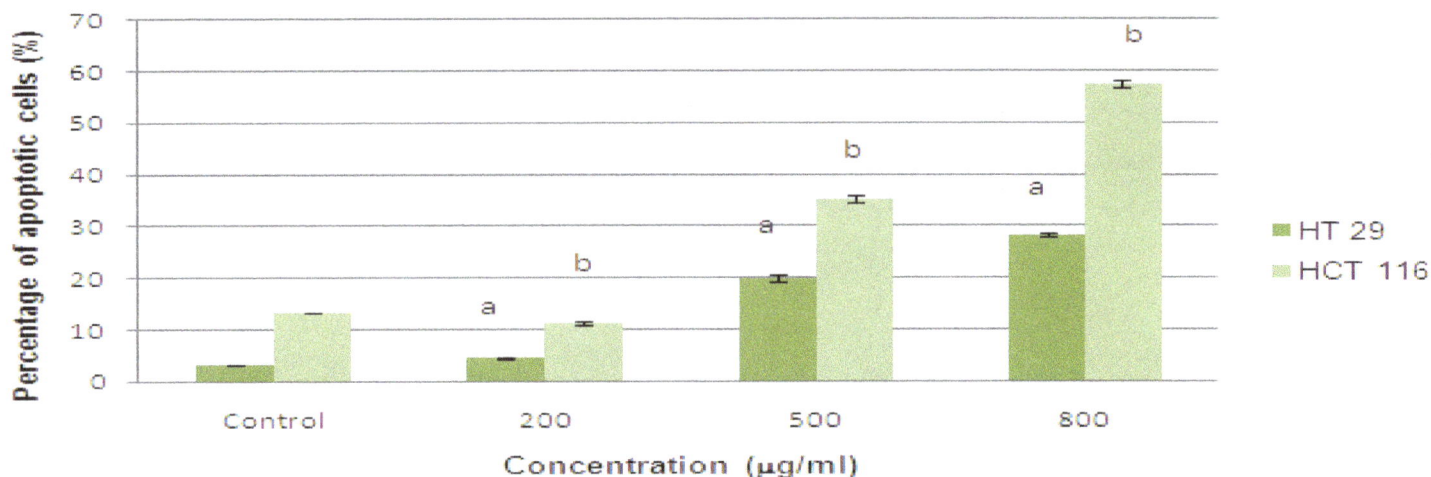

Figure 3a. Induction of apoptosis by ginger extract in human colon cancer cell lines, HT 29 and HCT 116. Percentages of apoptotic cells after a 24 h incubation with varying ginger extract concentrations. a: significant compared to control HT 29, b: significant compared to control HCT 116.

significantly ($p < 0.05$) compared to control for both HCT 116 and HT 29 cells as shown by cytograms of apoptosis for HCT 116 (Figure 3b) and HT 29 cells (Figure 3c).

DISCUSSION

There has been substantial interest in the search for potential chemopreventive agents in the past years for treatment against many cancers. Understanding how dietary components regulate proliferation and cell survival could play a critical role in the development of new agents that can prevent and treat cancer with lo w toxicity (Lee and Surh, 1998). Cancer chemoprevention was defined as the utilization of chemically active compoundsto reverse, suppress and prevent progression of disease from pre-invasive cancer to frank malignancy (Sporn et al., 1976).

Numerous diet-derived agents are included among promising agents and agent combination that are being evaluated clinically as chemopreventive agents for major cancer targets including breast, prostate, colon and lung cancers (Kellof et al., 2000). Ginger (*Z. officinalis* Roscoe, Zingiberaceae) is not only widely used as a dietary condiment but it has also been extensively utilized as a traditional oriental medicine (Taraphdar et al., 2001). Gingerols (that is, 6-gingerol, 8-gingerol and zingerone) and [6]-paradol have been identified as the main active ingredients of ginger and are responsible for the antioxidant activity and its characteristic pungent taste (Gruenwald et al., 2000; Keum et al., 2002). Antioxidative capacity of ginger has been associated with the ability of ginger to inhibit carcinogenesis by reducing oxidative stress and inducing apoptosis (Shukla and Singh, 2007; Katiyar, 1996). Bode and his co-workers, reported that

[6]-paradol exerted its primary inhibitory effect on cell transformation through the induction of apoptosis (Bode et al., 2001). [6]-Paradol and other structurally related derivatives induced apoptosis in oral squamous carcinoma cell line in a dose- dependent manner through a caspase-3-dependent mechanism (Keum at al., 2002). Ginger has been shown to inhibit cell growth in several tumor cells including liver (HepG2) (Yasmin et al., 2008), gastrointestinal (Yoshimi et al., 1992) and breast (Nagasawa et al., 2002). Ginger was also found to possess an anti-tumor promoting potential as determined by inhibition of phorbol ester-induced inhibition of Epstein- Barr virus (EBV) activation in Raji cells (Koshimizu et al., 1988).

The present study provides evidence that ginger may act as a potent growth inhibitory compound in human colon adenocarcinoma cells and supports the possibility of a chemopreventive potential in colon cancer cells. Our study showed that HT 29 cells were more sensitive to ginger treatment with an IC_{50} much lower [455 µg/ml] than HCT 116 cells [496 µg/ml]. The cytotoxic effect could be as a result of the active component, gingerol that has been reported to inhibit the growth of HCT 116 human colon colorectal and liver $HepG_2$ cancer cells (Bode, 2003; Harliansyah et al., 2007). Azoxymethane-induced intestinal carcinogenesis in rats was significantly suppressed by dietary administration of gingerol (Yoshimi et al., 1992).

We also reported in this study that the inhibition mechanism of colon cancer cells growth by ginger involved interference in both cell cycle progression and apoptosis. Ginger was able to induce cell cycle progression in G_0/G_1 phase and apoptotic death in a dose-dependent manner in both cells. Deregulated cell cycle progression driven by activation of growth-stimulating oncogenes, is one of the

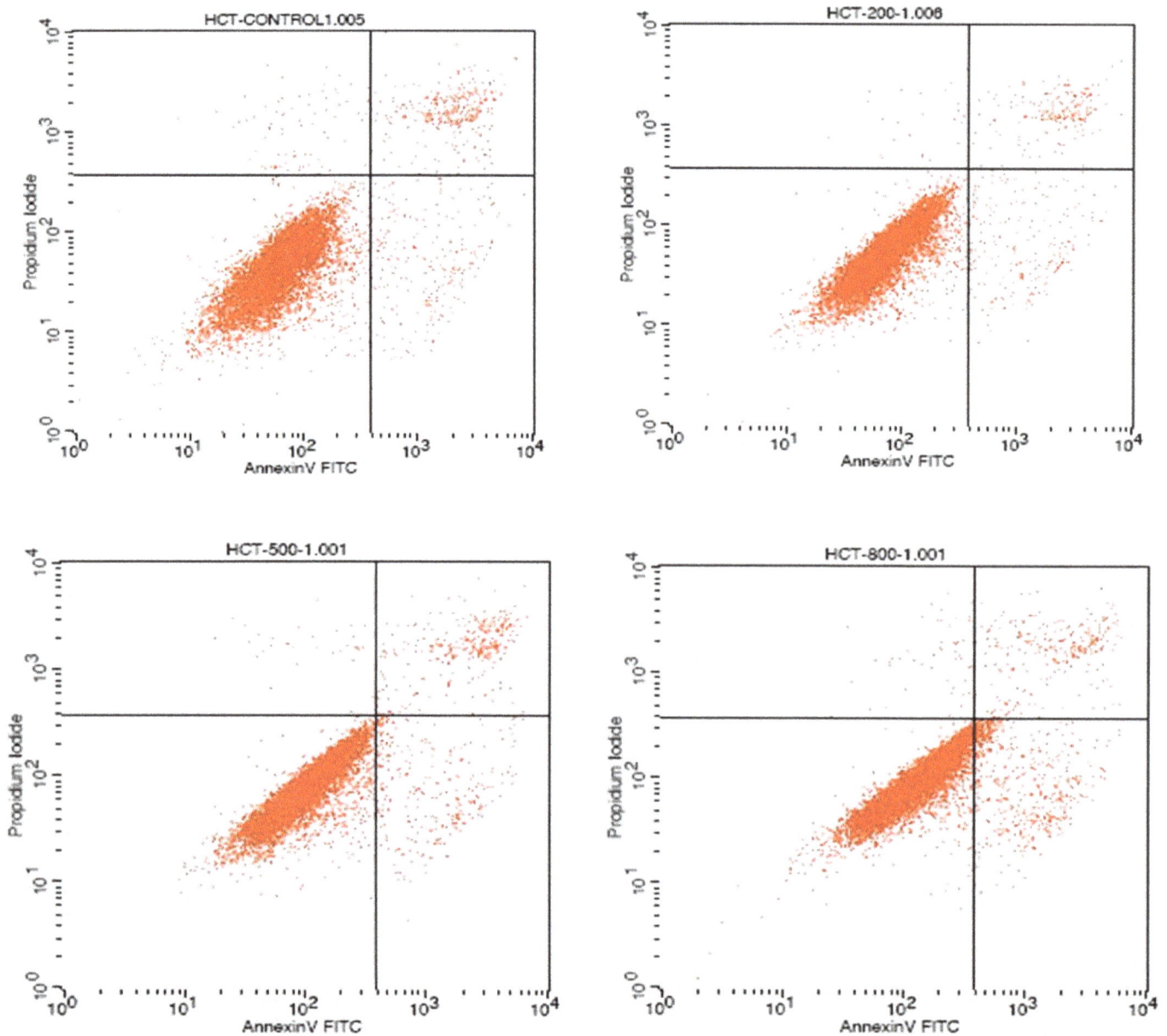

Figure 3b. Flow cytometric analysis of apoptosis induction by ginger in HCT 116 cells. Representative cytograms of apoptosis in HCT 116 cells incubated for 24 hours in the absence (A) or presence of 200 (B), 500 (C) and 800 μg (D) ginger. Within a cytogram, a quadrants 2 and 4 represent late and early apoptotic cells, respectively; quadrant 1, necrotic cells; quadrant 3, viable cells.

primary characteristics of cancer cells. Cell cycle progression is tightly controlled by the regulation of expression and activity of cyclin/cyclin-dependent kinase (CDK) complexes (Taraphdar et al., 2001). Dysregulation of the cell cycle check points and overexpression of growth-promoting cell cycle factors such as cyclin D1 and cyclin-dependent kinase (CDK) are associated with tumorigenesis (Deihl, 2002). Several dietary agents including curcumin (Mukhopadhyay et al., 2002), resveratrol (Estrov et al., 2003), genistein (Li et al., 2005), dietary isothiocyanates (Jakubikova et al., 2005), apigenin (Takagaki et al., 2005) and silibinin (Tyagi et al., 2002)

have been shown to block the deregulated cell cycle in cancers. In our study, ginger managed to arrest cell cycle at G_0/G_1-phase for both colon cancer cells at concentration below IC_{50}, whereas at higher concentrations, both cells were arrested by ginger extract at G_2/M phase. The cell cycle arrest observed could be due to inhibitory effect of ginger on cyclin dependent kinases and activation of cell cycle check points. Palozza et al. (2002) showed that arrest of the cell cycle and accumulation of cells in G_2/M phase in colon cancer cells was followed by the decreased expression of cyclin A, a protein known to regulate cdc2 kinase activity in G_2/M phase after

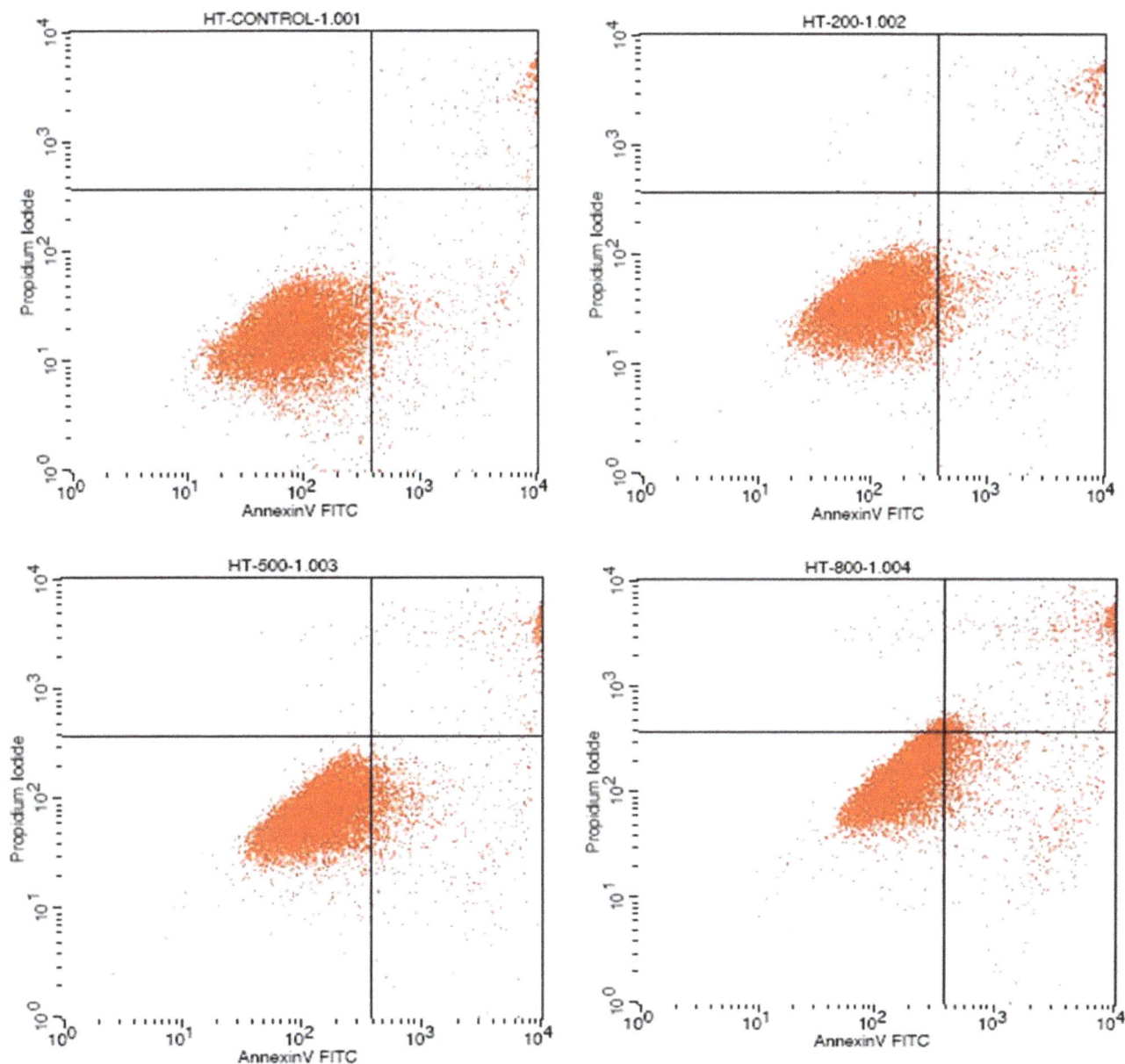

Figure 3c. Flow cytometric analysis of apoptosis induction by ginger in HT 29 cells. Representative cytograms of apoptosis in HT 29 cells incubated for 24 hours in the absence (A) or presence of 200 (B), 500 (C) and 800 µg/ml (D) ginger.

treatment with β-carotene. β-carotene also increased apoptosis through the downregulation of this protein.

In addition to regulation of the cell cycle, apoptosis plays an important role in the maintenance of tissue homeostasis whereby damaged cells are removed since impaired apoptosis contributes to development of cancer (Taraphdar et al., 2001). Compound that can elicit apoptosis is a good chemopreventive agent. In our study, we have shown that ginger at increasing concentration was not only able to inhibit DNA synthesis but also induced apoptosis especially at higher concentration and the percentage of cells that underwent apoptosis, increased dose dependently for both cells. Our

observation of apoptosis in mutant p53- expressing HT 29 is similar to the study by Park et al. (2006) who found that [6]-gingerol induced apoptotic death in pancreatic cells. The percentage of late apoptotic cells was low compared to early apoptotic cells as detected by annexin V staining, suggesting that apoptosis occurred rather slow in colon cancer cells after treatment (Huang and Pardee, 1999). This is consistent with the observation by other researchers whereby colon cancer cells (HCT 116 and HT 29 cells) did not undergo apoptosis rapidly (Goldwasser et al., 1996). Apoptosis that occurred could be due to the down-regulation of NF-κB. Expression of several NF-κB regulated genes including Bcl-2, Bcl-XL, cIAP, surviving, TRAF1 and TRAF2 have been reported to function primarily by blocking the apoptosis pathway (Aggarwal, 2004). Other than that, ginger also could induced apoptosis through mitochondrial pathway involving caspase-8, BID cleavage, cytochrome *c* release and caspase-3 activation like curcumin (Shishodia et al., 2005).

Conclusion

In this from our study, we have demonstrated that ginger extract inhibited the proliferation and induced apoptosis in HCT 116 and HT 29 colon cancer cells. The results suggest that ginger stimulates apoptosis through GoG_1 cell cycle arrest. Future our study will may thus be further investigate the possible use of ginger extract as a new alternative chemotherapeutic agent for human colon cancer.

Competing interests

To our knowledge, previous studies have shown the chemopreventive effect of *Z. officinale* in other cancer cells but less report evidence on colon cancer cells. Thus, our study represent among the first to show the mechanism of anticancer effect of *Z. officinale* in colon cancer cells HCT 116 and HT 29.

ACKNOWLEDGEMENTS

This research was funded by Fundamental Research Grant Science (FRGS), Universiti Kebangsaan Malaysia (UKM), Malaysia through UKM-FF-03-FRGS0007-2007. We also thank Dr. Noor Azian Murad from Center for Lipids Engineering Applied Research (CLEAR), Universiti Teknologi Malaysia, who has kindly provided the ginger extract.

REFERENCES

Aggarwal BB (2004). Nuclear factor-kappaB: the enemy within. Cancer Cell. growth factor-induced cell transformation and activator protein 16(3): 203-208..

Bode AM, Ma WY, Surh YJ, Dong Z (2001). Inhibition of epidermal activation by [6]-gingerol. Cancer Res. 61: 850-853.

Bode A (2003). Ginger is an effective inhibitor of HCT 116 human colorectal carcinoma *in vivo*. Paper presented at the Frontier in Cancer Prevention Research Conference, Phoenix, AZ, October: pp. 26-30.

Deihl JA (2002). Cycling to cancer within cyclin D1. Cancer Biol. Ther. 1(3): 226-231.

Estrov Z, Shishodia S, Faderl S, Harris D, Van Q, Kantarjian HM (2003). Resveratrol blocks interleukin-1beta-induced activation of the nuclear transcription factor NF-kappaB, inhibits proliferation, causes S-phase arrest and induces apoptosis of acute myeloid leukemia cells. Blood 102(3): 987-995.

Evan GI, Vousden KH (2001). Proliferation, cell cycle and apoptosis in cancer. Nature 411: 342-348.

Gerard LCC, Halimah Y (2003). Second Report of the National Cancer Incidence in Malaysia National Cancer Registry, Ministry of Health Malaysia.

Goldwasser F, Bae I, Fornace AJ, Pommier Y (1996). Differential GADD45, p21CIP1/WAF1, MCL-1and topoisomerase II gene induction and secondary DNA fragmentation after camptothecin-induced DNA damage in two mutant p53 human colon cancer cell lines. Oncol. Res. 8: 317-323.

Gruenwald J, Brendler T, Jaenicke C (2000). PDR for Herbal Medicines. Medical Economics Company, Montvale, New Jersey (USA).

Hanif R, Qiao L, Shiff SJ (1997). Curcumin, a natural plant phenolic food additive, inhibits cell proliferation and induces cell cycle changes in colon adenocarcinoma cell lines by a prostaglandin-independent pathway. J. Lab. Clin. Med. 130: 576-584.

Harliansyah NAM, Wan ZWN, Yasmin AMY (2007). Antiproliferative, antioxidant and apoptosis effects of *Zingiber officinale* and 6-Gingerol on HepG$_2$ cells. Asian J. Biochem. 2 (6): 421-426.

Huang L, Pardee AB (1999). β-Lapachone induces cell cycle arrest and apoptosis in human colon cancer cells. Mole. Med. 5: 711-720.

Jakubinova.J, Bao Y, Sedlak J (2005). Isothiocyanates induce cell cycle arrest, apoptosis and mitochondrial potential depolarization in HL-60 and multidrug-resistant cell lines.Anticancer Res. 25(5): 3375-3386.

Katiyar SK, Agarwal R, Mukhtar H (1996). Inhibition of tumor promotion in SENCAR mouse skin by ethanol extract of Zingiber officinale rhizome. Cancer Res. 56: 1023-1030.

Kazuhiro I, Ando T, Maeda O, Ohmiya N, Niwa Y, Kadomatsu K, Goto H (2007). Ginger ingredients reduce viability of gastric cancer cells via mechanisms. Biochem. Biophy. Res. Commun. 362: 218-223.

Kellof GJ, Crowell JA, Steele VE, Lubet RA, Malone WA, Boone CW, Kopelovich L, Hawk ET, Lieberman R, Lawrence JA, Ali I, Viner JL, Sigman CC (2000). Progress in cancer chemoprevention: Development of diet-derived chemopreventive agents. Nutri. Sci. pp. 467-471.

Keum YS, Kim J, Lee KH, Park KK, Surh YJ, Lee JM, Lee SS, Yoon JH, Joo SY, Cha IH, Yook JI (2002). Induction of apoptosis and caspse-3 activation by chemopreventive [6]-paradol and structurally realted compounds in KB cells. Cancer Lett. 177: 41-47.

Kim EJ, Lee YJ, Shin HK, Park JH (2005). Induction of apoptosis by the aqueous of Rubus coreanum in HT-29 human colon cancer cells. Nutr. 21: 1141-1148.

Koshimizu K, Ohigashi H, Tokuda H, Kondo A, Yamaguchi K (1988). Screening of edible plants against possible anti-tumor promoting activity. Cancer Lett. 39: 247-257.

Lee E, Surh YJ (1998). Induction of apoptosis in HL-60 cells by pungent vanilloids, [6]-gingerol and [6]-paradol. Cancer Lett. 134: 163-168.

Lee JH, Yang HS, Park KW, Kim JY, Lee MK, Jeong IY, Shim KH, Kim YS, Yamada K, Seo KI (2009). Mechanisms of thiosulfinates from Allium tuberosum L.-induced apoptosis in HT-29 human colon cancer cells. Toxicology Lett. 188: 142-147.

Li M, Zhang Z, Hill DL, Chen X, Wang H, Zhan R (2005). Genistein, a dietary isoflavane, down-regulates the MDM2 oncogene at both transcriptional levels. Cancer Res. 65(18): 8200-8208.

Manju V, Nalini N (2005). Chemopreventive efficacy of ginger, a naturally occurring anticarcinogen during the initiation, post-initiation stages of 1,2 dimethylhydrazine-induced colon cancer. Clinica Chimica Acta. 358: 60–67.

Miyoshi N, Nakamura Y, Ueda Y, Abe M, Ozawa Y, Uchida K, Osawa T (2003). Dietary ginger constituents, galanals A and B, are potent apoptosis inducers in Human T lymphoma Jurkat cells 199: 113-119.

Mukhopadhyay A, Banarjee S, Stafford LJ, Xia C, Liu M, Aggarwal BB (2002). Curcumin-induced suppression of cell proliferation correlates with down-regulation of cyclin D1 expression and CDK4 –mediated retinoblastoma protein phosphorylation. Oncogene 21(57): 8852-8861.

Murillo G, Naithani R, Mehta RG (2008). Efficacy of Herbal Products in Colorectal Cancer Prevention. Current Colorectal Cancer Reports 4: 34-42.

Nagasawa H, Watanabe K, Inatomi H (2002). Effects of bitter melon (Momordica charantia I) or ginger rhizome (Zingiber officinale rocs) on spontaneous mammary tumorigenesis in SHN mice. Am. J. Chin. Med. 30: 195-205.

Palozza P, Serini S, Maggiano N, Angelini M, Boninsegna A, Fiorella

DN, Ranelletti FO, Calviello G (2002). Induction of cell cycle arrest and apoptosis in human colon adenocarcinoma cell lines by β-carotene through regulation of cyclin A and Bcl-2 family proteins. Carcinogenesis 23: 11-18.

Park YJ, Wen J, Bang S, Park SW, Song SY (2006). [6]-Gingerol induces cell cycle arrest and cell death of mutant p53-expressing pancreatic cancer cells. Yonsei Med. J. 47: 688-697.

Sakamoto S, Kudo H, Kuwa K, Suzuki S, Kato T, Kawasaki T, Nakayama T, Kasahara N, Okamoto R (1991). Anticancer effects of a Chinese herbal medicine, juzen-taiho-to, in combination with or without 5- fluorouracil derivative on DNA-synthesizing enzymes in 1,2-dimethylhydrazine induced colonic cancer in rats. Am. J. Chin. Med. 19(3-4): 233-241.

Shukla Y, Singh M (2007). Cancer preventive properties of ginger: A brief review. Food Chem. Toxico. 45: 683-690.

Sporn MB, Dunlop NM, Newton DL, Smith JM (1976). Prevention of chemical carcinogenesis by vitamin A and its synthetic analogs (retinoids). Federation Proceedings 35: 1332-1338.

Surh Y (1999). Molecular mechanisms of chemopreventive effects of selected dietary and medicinal phenolic substances. Mutation Res. 428: 305-327.

Surh YJ (2002). Anti-tumor promoting potential of selected spice ingredients with antioxidative and anti-inflammatory activities: a short review. Food Chem. Toxico. 40: 1091-1097.

Surh YJ, Lee E, Lee JM (1998). Chemopreventive properties of some pungent ingredients present in red pepper and ginger. Mutation Res. 402: 259-267.

Takagaki N, Sowa Y, Oki T, Nakanishi R, Yogosawa S, Sakai T (2005). 26(1): 185-189.

Taraphdar AK, Roy M, Bhattacharya RK (2001). Natural products as inducers of apoptosis: Implication for cancer therapy and prevention. Current Sci. 80: 10-11.

Tyagi AK, Singh RP, Agarwal C, Chan DC, Agarwal R (2002). Silibinin strongly synergizes human prostate carcinoma DU145 cells to doxorubicin-induced growth inhibition, G2-M arrest and apoptosis.Clin. Cancer Res. 8(11): 3512-3519.

Virginie DM, Pocard M, Richon S, Weiswald LB, Assayag F, Saulnier P, Judde JG, Janneau JL, Auger N, Validire P, Dutrillaux B, Praz FO, Bellet D, Poupon MF (2007). Establishment of Human Colon Cancer Cell Lines from Fresh Tumors versus Xenografts: Comparison of Success Rate and Cell Line Features. Cancer Res. 67(1): 398-407.

Wang G, Li X, Huang F, Zhao J, Ding H, Cunningham C, Coad JE, Flynn DC, Reed E, Li QQ (2005). Antitumor effect of β-elemene in non-small-cell lung cancer cells is mediated via induction of cell cycle arrest and apoptotic cell death. Cell. Mole. Life Sci. 62: 881-893.

Wei QY, Ma JP, Cai YJ, Yang L, Liu ZL (2005). Cytotoxic and apoptotic activities of diarylheptanoids and gingerol-related compounds from the rhizome of Chinese ginger. J. Ethnopharm. 102: 177-184.

Yasmin AMY, Shahriza ZA, Looi ML, Shafina HMH, Harlianshah H, Noor AAH, Suzana M, Wan ZWN (2008). Ginger extract (Zingiber officinale Roscoe) triggers apoptosis in hepatocarcinogenesis induced rats. Med. Health 3(2): 263-274.

Yoshimi N, Wang A, Morishita Y, Tanaka T, Sugie S, Kawai K, Yamahara J, Mori H (1992). Modifying effects of fungal and herb metabolites on azoxymethane-induced intestinal carcinogenesis in rats. Japanese J. Cancer Res. 83: 1273-1278.

Yoshimi N, Wang A, Morishita Y, Tanaka T, Sugie S, Kawai K, Yamahara J, Mori H (1992). Modifying effects of fungal and herb metabolites on azoxymethane-induced intestinal carcinogenesis in rats. J. Cancer Res. 83: 1273-1278.

4

Changes in liver and serum transaminases and alkaline phosphatase enzyme activities in *Plasmodium berghei* infected mice treated with aqueous extract of *Aframomum sceptrum*

George, B. O.[1], Osioma, E.[2]*, Okpoghono, J.[1] and Aina, O. O.[3]

[1]Department of Biochemistry, Faculty of Science, Delta State University, Abraka, Delta State, Nigeria.
[2]Department of Biochemistry, Faculty of Science, University of Ilorin, Kwara State, Nigeria.
[3]Department of Biochemistry, Nigeria Institute of Medical Research, Yaba, Lagos State, Nigeria.

One of the effects of *Plasmodium* infection which causes malaria is the invasion of the hepatocyte which affects liver and blood transaminases activities. This study was design to examine the hepato-protective effect of *Aframomum sceptrum* in mice infected with *Plasmodium berghei*. Thirty-six abino male mice infected (test) and non infected (control) of 8 weeks old were used for this research. The mice were divided into six groups of six mice per group. Biochemical parameters measured in serum and liver samples include, alanine aminotransferase (ALT), aspartate amniotransferase (AST) and alkaline phosphatase (ALP) activities. Serum and liver ALT, AST and ALP activities were significantly higher in parasitized control mice as compared with all other groups. Parasitized mice receiving 250 and 350 mg/kg body weight of *A. sceptrum* have comparable (p>0.05) serum and liver ALT activities with the normal control mice. However, normal mice receiving both doses of *A. sceptrum* did not show any increase of serum ALT, AST and ALP activities with their respective controls. This result is supported by the histological examination of liver section of parasitized mice treated with 350 mg/kg b. wt of *A. sceptrum* showing moderately brought central vein, hepatic cell with preserved cytoplasm and prominent nucleus. It can therefore be inferred from the study that the administration of *A. sceptrum* to malaria infected mice at 250 and 350 mg/kg b. wt did not damage hepatocyte as expressed by the parasitized mice and also treatment with extract could protect hepatocyte integrity of *Plasmodium* infected mice.

Key words: *Aframomum sceptrum,* malaria, liver enzymes, *Plasmodium berghei.*

INTRODUCTION

Malaria remains a devastating global health problem (Trampuz et al., 2003). It is known as the world's most important tropical parasitic infectious disease to human (Donovan et al., 2007). About 300 to 500 millions of cases and a mean death of 2 million are reported every year (Kochar et al., 2003).

Worldwide, the control of malaria has witnessed a serious deterioration (Fernex, 1985). Severe malaria is almost exclusively caused by *Plasmodium falciparum* in

humans, but other forms of *Plasmodium* include *Plasmodium vivax, Plasmodium ovale* and *Plasmodium malariae.* When the infected anopheline mosquito takes a blood meal, sporozoites are inoculated into the blood stream. Within an hour, sporozoites enter hepatocytes and begin to divide into exoerythrocytic merozoites (tissue schizogony) (Trampuz et al., 2003). Once merozoites leaves the liver, they invade erythrocytes. In effect malaria affects the liver. Alanine and aspartate transminease (ALT and AST) activities are used as indicators of hepatocytes damage (Asagba et al., 2004; Coppo et al., 2002; Dede et al., 2002; Whitehead et al., 1999). Akanji et al. (1993) reported that alkaline phosphatase (ALP) is a marker enzyme for the plasma

*Corresponding author. E-mail: ejoviosioma@yahoo.com.

membrane and endoplasmic reticulum.

Aframomun sceptrum (Family- Zingiberaceae, Local name: Urioma/Alaiko) is a local spice commonly used to enhance flavour, aroma and palatability of cooking particularly by the Urhobos, Itsekiris and Ijaws of Delta State (George et al., 2010). Duker-Eshun et al. (2002) reported the anti-plasmodia activity of *A. sceptrum* while, George et al. (2010) reported the normalization of liver enzymes activities (AST and ALT) of diabetic rats treated with 200 mg/kg b. wt of *A. sceptrum*. However, there is dearth of information on the changes of the liver enzymes activities in mice infected with *Plasmodium berghei* treated with *A. sceptrum*. Liver damage caused by malaria infection may be prevented or reduced by treatment with this spice that forms part of our diet, hence, this study aim to examine the hepato-protective effect of *A. sceptrum* in mice infected with *P. berghei*.

MATERIALS AND METHODS

Experimental animals

Thirty-six albino male mice of 8 weeks old obtained from the animal house, Faculty of Basic Medical Sciences, Delta State University Abraka, Delta State, Nigeria were used for this study. They were fed on growers mash obtained from Top-Feeds, Sapele, Delta State, and were given water *ad libitum*. The animals were housed in metal cages under controlled conditions of 12 h light/ 12 h dark cycle. The animals used in this study were maintained in accordance with the guidelines approved by the Animal Ethical Committee, Delta State University, Abraka.

Inoculation of animals

The mice were infected with parasites (*P. berghei*) by obtaining parasitized blood from the cut-tip of the tail of an infected blood (3 to 4 drops) and diluted in 0.9 ml phosphate buffer, pH 7.4. The mice were inoculated intraperitoneally with 0.1 ml parasitized suspension. Parasitaemia was assessed by thin blood film made by collecting blood from the cut tip of the tail and this was stained with Geimisia stain (WHO, 2000). Inoculation was carried out in the Biochemistry Laboratory of the Nigerian Institute of Medical Research, Yaba Lagos.

Collection of *Aframomum sceptrum*

A. sceptrum was purchased from a local market in Abraka, Delta State, Nigeria. The spice was identified at the Department of Botany, Delta State University Abraka.

Preparation of extract

The spice was sun-dried to a constant weight for two weeks. This was followed by grinding to fine powder using warren blender. 100 g of the ground spice material was then soaked in 400 ml of distilled water and boiled for 5 min, followed by mechanical shaking for 10 min, cooled and filtered. The filtrate was then concentrated using rotary evaporator at 40 to 50°C under reduced pressure. The extract was stored frozen in a deep freezer until required for the experiment (Abukakar et al., 2008).

Experimental design

After the confirmation of parasitemia, the mice infected (parasitized) and non infected (normal) were divided into 6 groups of 6 mice per group treated as follows:

Group 1: Normal control
Group 2: Parasitized control
Group 3: Normal mice + *A. sceptrum* (250 mg/kg b. wt)
Group 4: Normal mice + *A. sceptrum* (350 mg/kg b. wt)
Group 5: Parasitized mice + *A. sceptrum* (250 mg/kg b. wt)
Group 6: Parasitized mice + *A. sceptrum* (350 mg/kg b. wt)

The administration of the extract was carried out using an intragastric tube for a period of four days. On the fourth day mice were starved overnight, sacrificed by decapitation and the blood and tissue (liver) were collected for various biochemical estimations.

Preparation of serum

Fasting blood was collected from each mice into a sterile, plain tube, and then it was centrifuged at 1,200 × g for 5 min at room temperature to obtain the serum sample, which was stored frozen at -20°C until analyzed.

Preparation of tissue homogenate

0.5 g of wet liver tissue was homogenized in 4.5 ml of freshly prepared normal saline. The supernatant obtained was used for this experiment.

Biochemical investigations

The biochemical investigations in serum and liver samples were carried out with the following methods, using commercially available kits as supplied by TECO Diagnostic, Anahein, USA. Alanine aminotransferase (ALT) was determined by method of Reitman and Frankel (1957), Aspartate aminotransferase (AST) by the method of Reitman and Frankel (1957) and while alkaline phosphatase (ALP) activity was estimated by the method of Tietz et al. (1983).

Histology

Histological study on liver tissues obtained from experimental mice was carried out following the method described by Drury and Wallington (1973).

Statistics

The repeat measure analysis of variance (ANOVA) was used to compare similar mean values, and the group means were compared by Duncan's multiple range test (DMRT). The level of statistical significant was established at 5% probability level.

RESULTS

The results in Table 1 indicate that the mice infected with *P. berghei* (Group 2) expressed a significantly ($P<0.05$) higher serum alanine aminotransferase (ALT) activity

Table 1. Changes in serum alanine aminotransferase, aspartate aminotransferase and alkaline phosphatase activities in parasitized mice treated with *Aframomum sceptrum*.

Groups	ALT (U/l)	AST (U/l)	ALP (U/l)
1	20.69±0.66[a]	25.34±0.84[a]	42.61±1.86[a]
2	23.83±1.19[b]	28.14±0.46[b]	51.95±1.09[b]
3	21.44±3.75[a]	25.56±0.80[a]	43.37±2.75[a]
4	20.29±1.10[a]	26.52±0.87[c]	44.07±2.34[a]
5	21.21±1.40[a]	23.19±0.61[d]	51.39±1.08[b]
6	19.94±0.88[a]	22.46±0.50[d]	50.63±1.02[b]

Values are means ± SD with n = 6. Column values with different superscript letters are significantly different (P<0.05).

Table 2. Changes in liver alanine aminotransferase, aspartate, aminotransferase and alkaline phosphatase activites in parasitized mice treated with *Aframomum sceptrum*.

Groups	ALT (U/l)	AST (U/l)	ALP (U/l)
1	29.99±0.74[ad]	34.20±0.86[a]	66.18±6.58[ad]
2	32.44±.0.88[b]	37.19±0.50[b]	93.03±4.91[b]
3	29.26±0.60[a]	33.20±1.29[ac]	67.78±7.63[a]
4	28.29±0.94[c]	32.33±1.22[cd]	69.16±7.24[a]
5	30.20±0.67[d]	33.49±1.41[ad]	62.10±4.55[dc]
6	29.36±0.64[ad]	32.70±1.53[cd]	55.29±2.53[c]

Values are means ± SD with n = 6. Column values with different superscript letters are significantly different (P<0.05).

compared with all other groups of experimental animals. The ALT activity of the control mice (Group 1) is comparable (P>0.05) to the normal and parasitized mice treated with 250 and 350 mg/kg b. wt (that is, Groups 3, 4, 5 and 6) respectively. However, the Group 6 animals showed the lowest ALT activity.

Serum aspartate aminotransferase activity is significantly lower (P<0.05) in the parasitized mice (22.46±0.50 U/l) administered with 350 mg/kg b. wt of *A. sceptrum* compared to all other groups except parasitized mice (Group 5) treated with 250 mg/kg b. wt of *A. sceptrum*. Serum AST activity of the control parasitized mice (Group 2) is statistically higher (p<0.05) than the AST activity of all the other groups.

The table also revealed that serum alkaline phosphatase activity significantly (P<0.05) increased in the serum of parasitized control mice (that is, Group 2) as compared with Groups 1, 2 and 3 respectively, but comparable (P>0.05) with parastized mice administered with 250 mg/kg b. wt (51.39±1.08) and 350 mg/kg b. wt (50.63±1.02).

The liver marker enzymes (ALT, AST and ALP) activities according to Table 2 were significantly increased (P<0.05) in the liver of the parasitized mice as compared with all other groups. Alanine aminotransferase activities of parasitize mice in Groups 1 and 2 were statistically comparable (P>0.05) to ALT activity of the control mice (Group 1). According to Table 2 also, no significant difference was observed between aspartate aminotransferase activity of parasitize mice in Groups 5,

6 and control mice (Group 4) receiving 350 mg/kg b. wt of *A. sceptrum*.

Table 2 also reveal that a statistical (P<0.05) reduced alkaline phosphatase activity was observed between parasitized mice receiving 250 and 350 mg/kg b. wt of *A. sceptrum* (Groups 5 and 6) compared with the corresponding normal control mice receiving an equivalent dose of the spice material (Groups 3 and 4). The experimental mice in Group 1 showed a lower level or alkaline phosphatase activity as compared with the control mice in Groups 3 and 4. However, the observed increased was not significant. The histopathological section of mouse liver is presented in Figures 1 to 6.

DISCUSSION

Liver destruction can affect the metabolic processes in the body due to the role of liver in general metabolism of the organism. Enzymes are necessary for normal cellular metabolism including that of the liver (Rajamanickam and Muthuswamy, 2008). Within an hour of which an infected anopheline mosquito takes a blood meal, sporozites enter the hepatocytes and divide into exoerythrocytic merozoites. The degenerative changes in the hepatocytes due to the infection of *Plasmodium* may alter the activities its enzymes. This study investigated the possible effect of *A. sceptrum* amelioration on this cellular damage. Alanine aminotransferase (ALT), aspartate aminotranferase (AST) and alkaline phosphatase (ALP)

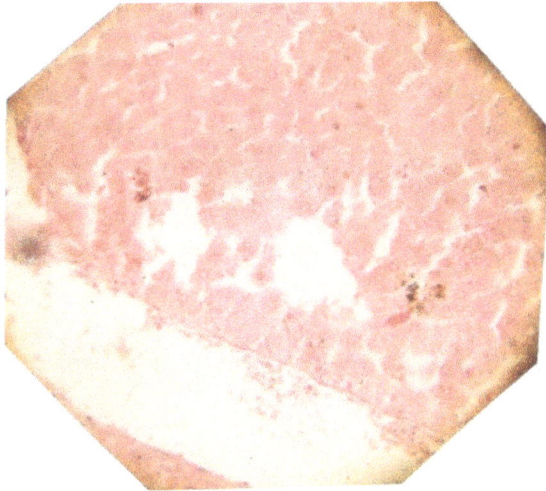

Figure 1. Photomicrograph of Group 1 mouse liver section showing well brought central vein and hepatic cells.

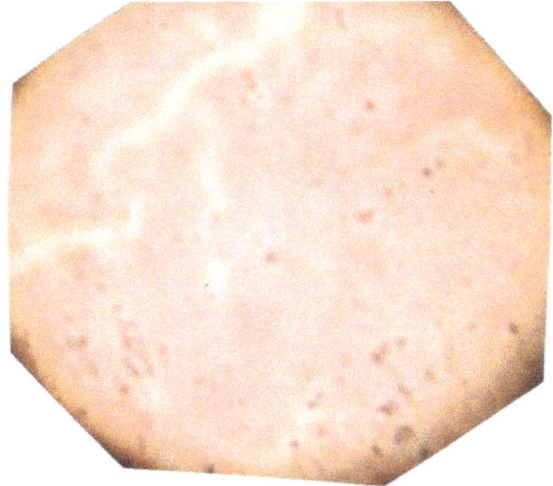

Figure 3. Photomicrograph of mouse liver section of Group 3 showing moderately ballooning degeneration.

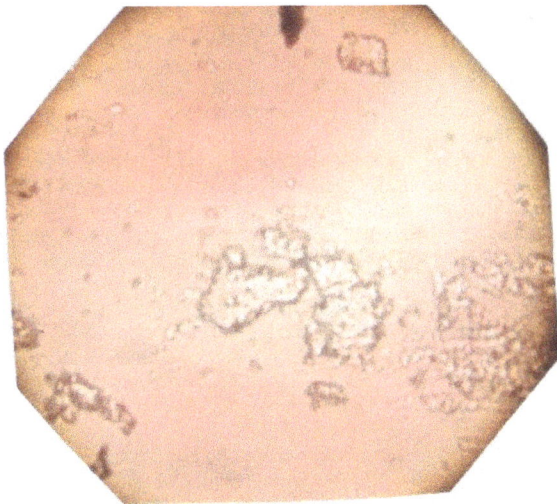

Figure 2. Photomicrograph of Group 2 liver section showing marked steatosis of the hepatocytes with ballooning degeneration, necrosis and mild periportal fibrosis.

Figure 4. Photomicrograph of mouse liver section of Group 4 showing degeneration brought central vein and hepatic cells.

are considered indicators of hepatocellular health (Yang and Chen, 2003; Vozarova et al., 2002).

A significant increase in the activities of ALT, AST and ALP in the blood and liver of control parasitized mice as compared with all the other groups was observed. The observed increase in enzyme activities may be as a result of liver injury and altered hepatocyte integrity caused by the *Plasmodium* infection and the consequent released of the enzymes into the blood stream. This is shown in the histopathological examination of the liver of the parasitized mouse which possesses a marked steatosis of the hepatocytes with ballooning dege-neration, necrosis and mild periportal fibrosis (Figure 2)

as compared with the liver examination of the control mouse that showed a well brought central vein and hepatic cell (Figure 1).

The administration of *A. sceptrum* tends to normalize these enzymes (ALT, AST and ALP) activities which is in agreement with previously reported investigation by George et al. (2010). Although, the aforementioned liver enzymes normalization was more pronounced at a higher dose of 350 mg/kg b. wt of *A. sceptrum*. This spice is extensively used in cooking different types of soups and vegetables. It is very popular in cooking 'pepper soup' commonly prepared for individuals' suffering from malaria fever. The observation in the foregoing is supported by

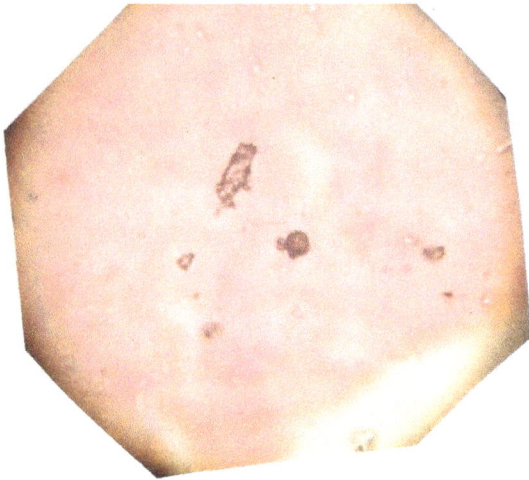

Figure 5. Photomicrograph of mouse liver section of Group 5 showing considerable reduction in necrosis moderately regeneration in hepatocellular architecture.

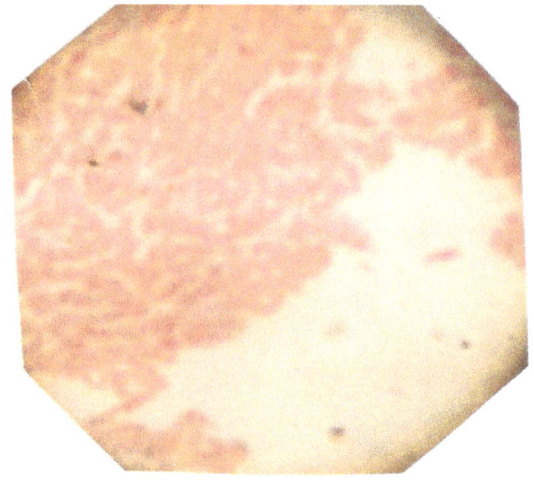

Figure 6. Photomicrograph of mouse liver section of Group 6 showing moderately brought central vein, hepatic cell with preserved cytoplasm and prominent nucleus.

the histological study as seen in Figure 6. Photomicrograph of liver section of parasitized mice treated with 350 mg/kg b. wt of *A. sceptrum* showing moderately brought central vein, hepatic cell with preserved cytoplasm and prominent nucleus.

It could therefore be concluded that the administration of *A. sceptrum* to parasitized mice at 250 and 350 mg/kg b. wt did not contribute to the hepatic injury expressed by the parasitized (Group 2) mice and may not be toxic. Treatment with extract from this study has shown to protect hepatocyte intergrity of parasitized mice.

REFERENCES

Abukakar MG, Uwani AN, Shehu RA (2008). Phytochemical screening and antibacterial activity of Tamaniadus Indica pulp extract. Asia J. Biochem., 3: 134-138.

Akanji MA, Olagoke OA, Oloyede OB (1993). Effect of chronic consumption of metabisulplute on the integrity of rate liver cellular system. Toxicology, 81: 173-179.

Asagba SO, Owhe-Ureghe BU, Falodun A, Okoko F (2004). Evaluation of the toxic effect of *Mansonia altisima* extract after short term oral adminsitartion to rats. West Afr. J. Pharmacol. Drug Res., 20(1/2): 53-57.

Coppo JA, Mussart NB, Fioranelli SA (2002). Physiological variation of enzymatic activites in blood of bull frog, Ranacetesbeiana (Shaw, 1802). Rev. Vet., 12/13: 22-27.

Dede EB, Igboh NM, Ayalogu OA (2002). Chronic toxicity study of the effect of crude petroleum (Bonny light) kerosene and gasoline on rats using hacmotological parameters. J. Appl. Sci. Environ. Manage., 6(1): 60-63.

Donovan MJ, Messmore AS, Scrafford DA, Sacks DL, Kamhawi S, McDowell MA (2007). Uninfected mosquito bites confer protection against infection with malaria parasites. Infect. Immum., 75(5): 2523-2530.

Drury RAB, Wallington EA (1973). Tissue histology. In: Carleton's Histological Technique 4th ed. Oxford University Press, New York. p. 58.

Duker-Eshun G, Jaroszewski JW, Asomaning WA, Oppon-Baclue F, Olisen CE, Christensen SB (2002). Antiplasmodial activity of labdenes from Aframomum latifoluim and Aframomum Sceptrum. Plant Med., 68(7): 642-644.

Fernex M (1985). Mefloquine and its allies. World Health. May 6-7.

George BO, Osioma E, Falodun A (2010). Effect of Atiko (Aframomum Sceptrum) and African Nutmeg (Monodora Myristicca) on reduced glutathione, Uric acid levels and liver marker enzymes in Streptozotocin-induce diabetic rats. Egyptian J. Biochem., 28(2): 67-78.

Kochar DK, Singh P, Agawal P, Kochar SK, Pokharna R, Sareen PK (2003). Malaria Hepatitis. J. Assoc. Phys. India, 51: 1069-1072.

Reitman S, Frankel SA (1957). A colorimetric method for the determination of serum glutamic oxaloacetic and pyruvic transaminases. Am. J. Clin. Pathol., 28(1): 56-63.

Rajamanickam V, Muthuswamy N (2008). Effect of heavy metals induced toxicity on metabolic biomarkers in common carp (Cyprinus carpio L.). Maejo Int. J. Sci. Tech., 2(01): 192-200.

Tietz NW, Burtis CA, Ducan P, Ervin K, Petitclera CJ, Rinker AD, Shney D, Zygowicz ER (1983). A reference method for measurement of alkaline phosphatase activity in human serum. Clin. Chem., 29: 751-761.

Trampuz A, Matjaz J, Igomuzloric RP (2003). Clinical review: Severe malaria. Critical Care, 7(4): 315-323.

Vozarova B, Stefan N, Lindsay RS, Saremi A, Pratley RE, Bogardus C, Tataranni PA (2002). High alanine aminotransferase is associated with decreased hepatic insulin sensitivity and predicts the development of type 2 diabetes. Diabetes, 51: 1889-1895.

Whitehead MW, Hawkes ND, Hainsworth I, Kingham JG (1999). A prospective study of the cause of notably raised aspartate transaminase in liver origin. Gut, 45: 129-133.

World Health Organization (WHO) (2000). Severa falciparum malaria. Trans R. Soc. Trop. Med. Hyg., 94: 1-90.

Yang JL, Chen HC (2003). Serum metabolic enzyme activities and hepatocyte ultra structure of common carp after gallium exposure. Zoological Studies, 42: 455-461.

Nerve growth factor (NGF) combined with oxygen glucose deprivation OGD induces neural ischemia tolerance in PC12 cells

Chunli Mei[1,2], Jinting He[1], Jing Mang[1], GuihuaXu[1], Zhongshu Li[1], Wenzhao Liang[1] and Zhongxin Xu[1]*

[1]Department of Neurology, China - Japan Friendship Hospital, Jilin University, Changchun, 130012, China.
[2]Beihua University, Jilin 132013, China.

Ischemic cerebrovascular disease is a major disease in humans. To better study this disease, a good ischemia model of nerve cells is needed. Nerve growth factor (NGF) could induce PC12 cells to become neurons. Oxygen glucose deprivation (OGD) could lead to hypoxia and neuronal ischemia. In this study, we used NGF and OGD to stimulate PC12 cells and converted them into neurons in order to establish an ischemia model. After stimulation with NGF (100 ng/ml for 6 days), PC12 cells show a neuron - like function as measured by physiology and biochemistry. After 6 days of NGF stimulation, we performed OGD treatment for 16 h to establish an oxygen glucose deprivation model. The results showed that PC12 cells transformed into cells that looked like neurons and that MAP2 was up-regulated in NGF - treated PC12 cells. Cell apoptosis was found to be up-regulated after NGF stimulation and OGD (5% CO_2 and 95% N_2, 1 mmol/l NaS_2O_4 in sugar - free DMEM for 16 h). A western blot analysis showed that OGD treatment increased the expression of HIF - 1. The apoptosis rate after 16 h of OGD was 19.44%. These results postulate that NGF treatment can be combined with OGD to establish an *in vitro* model of acute ischemic brain damage.

Key words: Nerve growth factor, oxygen glucose deprivation, PC12 cells, ischemia tolerance model.

INTRODUCTION

Ischemic cerebrovascular disease, such as cerebral thrombosis (CT), cerebral infaction (CI) and so on, are the most important causes of morbidity, and the third most common cause of death in elderly patients in the world. Effective methods of preventing and controlling ischemic cerebrovascular disease have been a topic of great interest. Ischemic damage of nerve cells leads to a series of complex signaling pathways that produce corresponding biological functions. A better under-standing of the role of the signal transduction mechanisms that underlie brain ischemic injury could identify key targets for neuroprotective substances. Thus, it is very important to establish a stable *in vitro* neuronal ischemia model. Nerve growth factor (NGF) could induce the differentiation of PC12 cells into neuron-like cells. The

sympathetic neuron-like cells are characterized by electrical excitability, expression of neuron - specific genes and neurite outgrowth (Dichter, 1977; Greenberg, 1985). PC12 cells were cultured in different oxygen conditions. The metabolic activity of PC12 cells was measured in the final 4 h prior to cellular characterization using an alamar blue assay (Serotec) following the manufacturer's instructions (Hamid, 2004). A neuronal ischemia model is needed to conduct research on the molecular level. PC12 cells can be induced with NGF to become neuron - like in appearance. Neural ischemia tolerance in PC12 cells via the combination of NGF with OGD was little in references. In this study, we combined NGF stimulation and OGD to establish a neuronal ischemia model. We examined the role of ischemic injury and signal transduction mechanisms.

The model might help to establish NGF treatment followed by OGD as an *in vitro* model of acute ischemic brain damage. The model could provide a new tool for the identification of pathways that were involved in cerebral

*Corresponding author. E-mail: meixiaoqing2007@126.com, xuzhongxin999@yahoo.com.cn.

ischemia. It provide the reference for further study ActA/smads signaling pathways on acute ischemic brain damage. This cerebral ischemic model could be one example of the cell's broader general-stress response.

Therefore, the present model could be applied to the study of mechanisms that were involved in tolerance to other stressful stimuli, and it is the important step to study reperfusion/reoxygenation after OGD. Cell-based assays with high - throughput capacity could be used as direct screens and models to explore molecular mechanisms that were involved in cellular function and pathology. The model presented in this study will be a new tool to study ischemic cerebrovascular disease.

MATERIALS AND METHODS

Cell culture

PC12 cells were purchased from the Cell Bank of the Chinese Academy of Sciences. The cell line was maintained in DMEM medium supplemented with 10% (v/v) fetal bovine serum and 5% horse serum (FBS, GIBCO), 100 IU/ml streptomycin, 100 IU/ml penicillin, pH 7.0, and detached with 0.25% trypsin (Sigma, USA). PC12 cells were grown at 37°C in 5% CO_2.

Differentiation of PC12 cells by NGF

Cells were grown in 5% horse serum containing media on collagen - coated tissue culture dishes before differentiation. After the cells got attached, they were treated with 100 ng/ml nerve growth factor (NGF 2.5S; Promega, Madison, WI) and cultured with serum - free DMEM for 6 days (Michiyoshi, 2010; Jaehoon, 2010), observed and photographed.

MAP2 immunocytochemical analysis

The cells were fixed with 4% paraformaldehyde/PBS and were permeabilized with 0.1% Triton X - 100 in PBS for 10 min. The cells were then incubated in 5% goat serum/PBS for 1 h at room temperature (20 to 25°C). Cells were washed again then incubated at 4°C overnight (14 to 16 h) in the presence of anti- MAP2 (1:1,000 dilution, Santa SC - 20172). After washing twice with PBS, the cells were incubated with fluorescently labeled secondary FITC-goat anti-rabbit (Santa SC - 3839) for 1 h at room temperature (Kumar, 2006). The results were observed by a fluorescence microscope equipped with a photomicrograph system.

OGD model of PC12 cells after NGF treatment

PC12 cells were treated with NGF (100 ng/ml) for 6 days. Cells were then washed 3 times with DMEM, and the cells were cultured with DMEM without sugar and 1 mmol/L NaS_2O_4 in hypoxic conditions (37°C, 5% CO_2 and 95% N_2) for 3 , 6 , 9 , 12 , 16 , or 24 h respectively, (Larsena, 2007; Damian, 2010).

Flow cytometry analysis

Cytometry was used to quantitatively assess the apoptosis detection kit, Annexin V - FITC and PI double-staining followed. Cells (1×10^5) were harvested and stained with Annexin V-FITC and

PI using a double staining kit (Kaiji Bio Co., Nanjing, China) according to the manufacturer's instructions(Beckman coulter, USA). The cells were then immediately analyzed by flow cytometry. Signals from apoptotic cells were localized in the lower right quadrant of the resulting dot-plot graph.

MAP2 and HIT-1 western blot detection

PC12 cells were treated with NGF for 1, 2, 3, 4, 5, or 6 days. Samples were washed twice with cold PBS and then 1×10^6 cells were lysed with RIPA buffer (50 mmol/l Tris (pH 8.0), 150 mmol/L NaCl, 0.1% SDS, 1% NP40 and 0.5% sodium deoxycholate) containing protease inhibitors (1% Cocktail and 1 mmol/L PMSF). Total proteins were separated with 15% SDS-PAGE and transferred to PVDF membranes. The membrane was blocked for with Tris-buffered saline with 0.1% Tween 20 (pH 7.6, TBST) for 1 h at room temperature and then immunoblotted with the primary antibody (1:1000) at 4°C overnight. After washing twice with TBST, the membrane was incubated with HRP-labeled secondary antibody (Santa SC-2073) for 1 h at room temperature and was washed three times with TBST. Final detection was performed with enhanced chemiluminescence (ECL) western blotting reagents (Amersham Biosciences, Piscataway, NJ), and the membranes were exposed to Lumi-Film Chemiluminescent Detection Film (Roche). Differences of loading were normalized using a monoclonal ß-actin antibody. The antibodies used in the study included anti-MAP2 (1:1000 dilution, Santa SC-20172) and anti-HIT-1 (1:1000 dilution, Santa SC-101907). After the PC12 cells were treated with NGF for 6 days, all the samples were treated with OGD for 3, 6, 9, 12, 16, or 24 h,respectively. HIT-1 protein expression was assessed at various times as described previously.

Statistical analysis

SPSS softwareseries and origin were used for statistical analyses, and values are presented as means ± SD. An ANOVA was used to compare the mean valueswithin or between samples. P values less than 0.05 indicated that the results showed statistically significant differences.

RESULTS

Morphological changes of PC12 cells

The results show that samples treated with NGF (100 ng/ml) stimulates neuron-like differentiation of PC12 cells as seen under the microscope. PC12 cells changed into neurons after one day of NGF treatment and followed by the formation of synapses. Synapses extended up the length of the cell after 3 days of treatment. The synaptic length increased 6 to 8 fold after 6 days of treatment (Figure 1). PC12 cells obtained from four cell depositories yielded essentially the same results. PC12 cells were exposed to the indicated concentrations of NGF for 3 days, and the differentiated cells were counted. Significant differences were detected. PC12 cells were treated NGF 6 days. The cultures were photographed under a phase contrast microscope (Olmpus, Japan) (scale bars = 20 um), Values are means ± S.D. of results from four microscopic fields. Asterisks indicate the

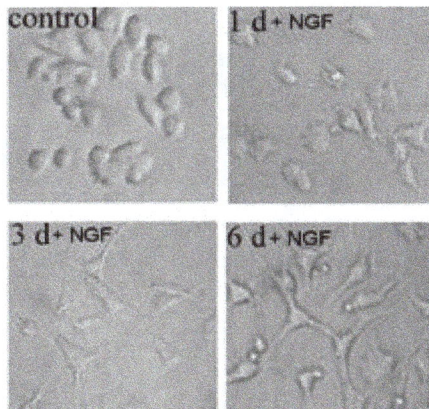

Figure 1. The morphological changes of PC12 cells(×200).PC12 cells were pretreated with 100 ng/mlNGF for 1, 3 or 6 days. The differentiated cells were photographed under a phase contrast microscope.

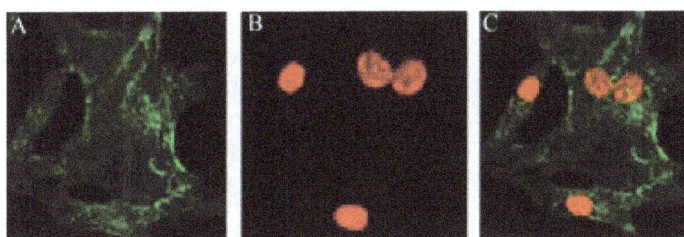

Figure 2. Morphological changes captured with fluorescence microscopy following immunofluorescence staining (×200). PC12 cells were cultured with NGF for 6 days and were then assessed with MAP2 immunofluorescence staining. (A) Anti- MAP2 and FITC-labeled IgG staining under the blue excitation, (B) nuclear counter staining of PI under purple excitation, (C) merged image of the FITC and PI staining.

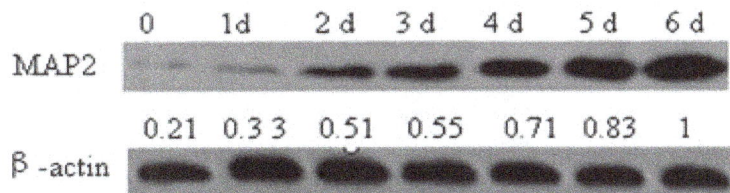

Figure 3. Analysis of western blot with samples treated with NGF. The results shown are representative of three repeated experiments.

statistical significance as determined by unpaired t-tests (P < 0.01).

Immunofluorescence analysis

The results showed that PC12 cells cultured with NGF for 6 days showed characteristic MAP2 immunofluorescence staining (Figure 2). Application plus pro 6.0 software to add image fusion after confirm green fluorescent were for

PC12 cells transformation of neurons appearance cell. MAP2 immunofluorescence stain was strong positive expression. PBS control was negative fluorescence.

MAP2 protein expression increased in PC12 cells after NGF treatment

Expression of MAP2 protein of NGF treatment after 1, 2,

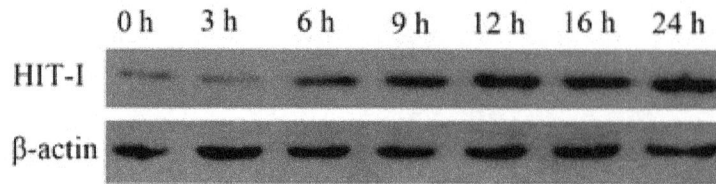

Figure 4. Analysis of western blot with samples treated with OGD after NGF stimulation. The results shown are representative of three repeated experiments.

3, 4, 5, or 6 days (Figure 3) showed that NGF couldincrease MAP2 protein expression. MAP2 expression
wasdetermined by western blotting after different treatment time periods. The results shown are representative of three repeated experiments. NIH imaging indicated that the protein signal densities were increased in cells treated with NGF for 3, 4, 5, or 6 days compared with control cells (no treatment).

OGD increased expression of HIT-1 in PC12 cells

After the PC12 cells were treated with NGF for 6 days, samples were treated with OGD. Western blot analysis was used to study the effect of OGD treatment on HIT-1 expression after 0, 3, 6, 9, 12, 16, or 24 h (Figure 4). The results showed that OGD could increase HIT-1 protein expression. The expression of HIT-1 protein was significantly increased by OGD. The results shown are representative of three repeated experiments. NIH imaging indicated that the protein signal densities were increased in cells treated with OGD for 6 h compared with control cells (no treatment).

OGD induced apoptosis in PC12 cells

To assess the apoptosis of differentiated PC12 cells treated with OGD, the cells were analyzed using a dual-laser FACSVantage SE flow cytometer (Becton Dickinson, Mountain View, CA, USA). Annexin V-FITC and PI signals were excited using a 488 nm laser light and their emissions captured using bandpass filters set at 530 ± 30 and 613 ± 20 nm, respectively. OGD induced apoptosis in PC12 cells. Cells were treated with OGD and cultured for 3, 6, 9, 12, 16, or 24 h. Signals from each group of cells were located in the lower right quadrant of the dot-plot graph. The cells were then analyzed with flow cytometry following AnnexinV-FITC/PI staining. The results are shown in Figure 5. Compared with the control transfected but untreated cells, the proportions of apoptotic cells treated with OGD after 3, 6, 9, 12, 16 and 24 h were 0.25, 0.38, 4.62, 10.54, 19.44 and 23.18%, respectively. After 9 h of OGD, most cells

(95%) were viable with the remaining being either early apoptotic (2%) or necrotic (3%). After 16 h of OGD, a pattern ofincreased AnnexinV-positive, PI-negative apoptotic cells. Quantitative analysis showed that 16 h of OGD produced a 20-fold increase in the number of apoptotic cells compared to controls. Necrotic cells also increased significantly after 16 h of OGD, but less so (2 to 3-fold). Furthermore, preexposure of PC12 cells to 9 h of OGD, attenuated cell death induced by 16 h of OGD 1 d later by significantly decreasing the number of apoptotic, but not necrotic cells (Figure 5). These resultscould help to provide the reference for further study ActA/smads signaling pathways on acute ischemic brain damage by the model.

DISCUSSION

Rat adrenal pheochromocytomas have been made into PC12 cell lines. They have been the object of intense study in neurobiology for the investigation of signal transduction mechanisms (Tischler, 2002). For example, studies of cell differentiation and survival (Agell et al., 2002; Michiyoshi, 2010; Jaehoon, 2010), Ca^{2+} signaling (Ghosh et al., 1994), apoptosis (Macdonald et al., 2003), Parkinson's disease (Ryu et al., 2002) and Huntington's disease (Peters et al., 2002) have been conducted in PC12 cell model systems. Primary neuronal cells came from the animals, which could express neurons tissue damage directly. But it was limited for the purity and number to the experiments. And primary neuronal cells were non-regeneration cells. Because of spontaneous apoptosis *in vitro* the survival time was short. So it was hard to uniform the initial condition of the experiment. In this study, we chose PC12 cells as the experiment model.

MAP2 is a neuron specific protein. It is present in PC12 cells during differentiation. There is a direct correlation between the rate of microtubule assembly and the increase of neurite length (Koji, 2006; Kumar, 2006). Microtubules stabilizing are critical for dendrite development and neurite outgrowth. MAP2 plays a critical role in neurite outgrowth (Dehmelt, 2003). It is a helpful diagnostic and prognostic feature in various neurological disorders. In the present study, we examined MAP2 expression by immunoblotting and western blot analysis.

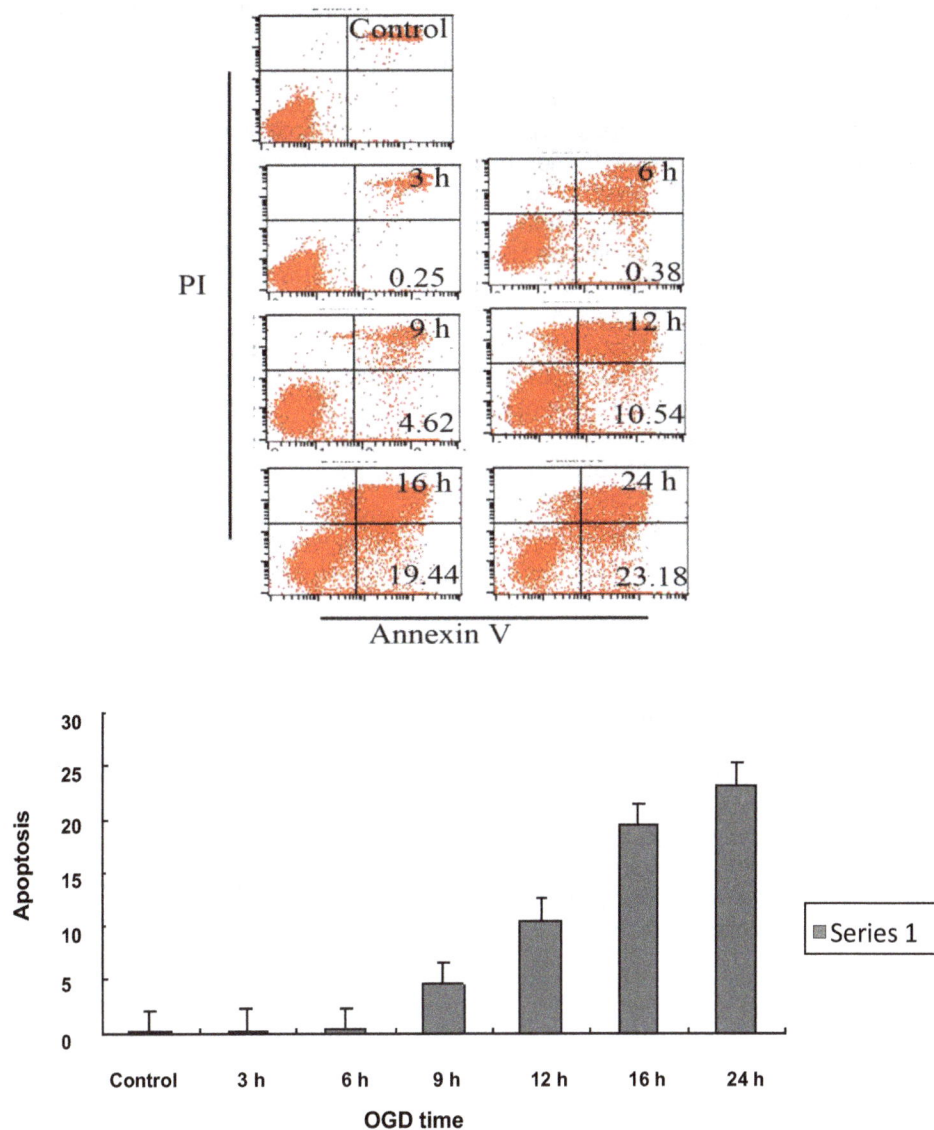

Figure 5. Apoptosis of differentiated PC12 cells treated with OGD was analyzed with Annexin V-FITC and PI double-staining flow cytometry.

Our results showed that MAP2 is differentially expressed in PC12 cells exposed to NGF for different time periods.

Protein signal densities were increased in cells treated with NGF for 3, 4, 5, or 6 days compared with control cells. HIF-1(Hypoxia-inducible factor-1 alpha) is critical mediators of physiological responses to acute and chronic hypoxia. HIF-1 plays an essential role in cellular and systemic homeostatic responses to hypoxia (Semenza, 2000). HIF-1α protein is stabilized and translocation into the nucleus is increased under hypoxia. Tolerance to ischemia and hypoxia can be modeled *in vitro* and has been described in cultured PC12 cells (Larsena, 2007; Damian, 2010). Our results showed thatHIF-1 is differentially expressed in PC12 cells that

have been exposed to OGD for different time periods. These resultscould help to provide the reference for furtherstudy ActA/smads signaling pathways on acute ischemic brain damage by the model.

In this study, physical and chemical methods were usedto establishOGD conditions, and sugar-free culture medium and NaS_2O_4 were used to establish a liquid environment lacking oxygen and sugar. HIF-1 protein expression was assessed to confirm the effects of OGD (Damian, 2010). NGF-stimulated PC12 cells differentiate into sympathetic neurons *in vitro* and have been shown to have neuronal characteristics both physiologically and biochemically. The circulation of oxygen and glucose is necessary to maintain neurons' normal physiologyfunction and survival. During the ischemic damage process, neurons' blood supply is disrupted. Oxygen glucose deprivation (OGD) is a model of oxygen and glucose shortage. We used the OGD on neurons to simulate the ischemic brain damage process that occurs in the body. NGF combined with OGD to set up an ischemia tolerance model. Our study shows that PC12 cells treated with NGF form cells that are neuron-like in appearance. These resultssuggested establishing NGF treatment followed by OGD as an *in vitro* model of acute ischemic brain damage. The model provides a new tool for the identification of pathways that are involved in cerebral ischemia. This model is one example of the cerebral ischemic response. Therefore, the present model could be applied to the study of mechanisms involving in tolerance to other stressful stimuli. Cell-based assays with high-throughput capacity can be used as models and direct screens to explore molecular mechanisms that are involved in pathology and cellular function.

Conclusion

In conclusion, we describe here an *in vitro* model of acute ischemic brain damage in the PC12 cell line established NGF treatment followed by OGD. The model provides a new tool for the identification of pathways involved in ischemic brain damage. The model can be one example of a more widely, general-stress response of the cell. In neurology, the model presented in this study will be a new tool to study ischemic cerebrovascular disease.

ACKNOWLEDGMENT

The project supported by natural science fund of Jilin province in China (Number 201015181).

REFERENCES

AgellN,Bachs O, Rocamora N, Villalonga P (2002). Modulation of the Ras/Raf/MEK/ERK pathway by Ca($^{2+}$), and calmodulin. Cell Signal, 14: 649–54.

Damian C, Genetos, Whitney K (2010). Oxygen tension modulates neurite outgrowth in PC12 cells through a mechanism involving HIF and VEGF. J. Mol. Neurosci., 40: 360-366.

Dehmelt L, Smart FM, Ozer RS (2003).The role of microtubule-associated protein 2c in the reorganization of microtubules and lamellipodia during neurite initiation. J. Neurosci., 23: 9479-9490.

Dichter MA, Tischler AS, Greene LA (1977). Nerve growth factor-induced increase in electrical excitability and acetylcholine sensitivity of a rat pheochromocytoma cell line. Nature, 268: 501-504.

Katarzyna M, David P (2010).Inhibition by anandamide of 6-hydroxydopamine-induced cell death in PC12 cells. Int. J. cell Biol., Article ID 818497: 10.

Koji I, Masahiro M (2006). TTLL7 is a mammalian -tubulin polyglutamylase required for growth of MAP2-positive neurites. J. Biol. Chem., 281(41): 30707-30716.

Kumar MR, Bhat, NityanandMaddodi (2006). Transcriptional regulation of human MAP2 gene in melanoma: role of neuronal bHLH factors and Notch1 signaling. Nucleic acids Res., 34(13): 3819-3832.

Larsena EC, Hatchera JF (2007).Effect of D609 on phospholipid metabolism and cell death during oxygen-glucose deprivation in PC12 cells.Neuroscience, 146(3): 946-961.

Macdonald NJ, Delderfield SM, Zhang W, Taglialatela G (2003).Tumour necrosis factor-alpha- vs. growth factor deprivation-promoted cell death: distinct converging pathways. Aging Cell, 2: 245-256.

Michiyoshi H, Norihiro S (2010). 15d-prostaglandin J2 Enhancement of nerve growth factor–Induced neurite outgrowth is blocked by the chemoattractant receptor–homologous molecule expressed on T-Helper Type 2 cells (CRTH2) Antagonist CAY10471 in PC12 cells. J. Pharmacol. Sci., 113: 89-93.

Peters PJ, Ning K, Palacios F, Boshans RL, Kazantsev A, Thompson LM, Woodman B, Bates GP,Ryu EJ, Harding HP, Angelastro JM, Vitolo OV, Ron D, Greene LA (2002). Endoplasmic reticulum stress and the unfolded protein response in cellular models of Parkinson's disease. J Neurosci., 22: 10690-10698.

Tischler AS (2002).Chromaffin cells as models of endocrine cells and neurons. Ann NY Acad Sci., 971: 366-370.

6

Effect of 50 Hz electromagnetic fields on acid phosphatase activity

K. S. Prashanth[1]*, T. R. S. Chouhan[2] and Snehalatha Nadiger[3]

[1]Research Scholar, Dr M.G.R University, Chennai Tamil Nadu, India.
[2]Ethica Matrix CRO Pvt Ltd, Hyderabad, India.
[3]Department of Biotechnology, New Horizon College of Engineering, Outer Ring Road, Panathur post Bangalore-560087, Karnataka(s) India.

The effect of extremely low frequency (ELF) electromagnetic field (EMF) (50 Hz 0.5 mT) on the activity of acid phosphatase (EC 3.1.3.2) was studied. In addition the factors affecting the enzyme activity such as the temperature, pH and substrate concentration were also investigated. The results show that ELF EMF have significant influence on enzyme activity. Upon EMF exposure K_m increased from 0014 ± 0.005 to 0.040 ± 0.008 mM whereas V_{max} increased from 0.991 ± 0.254 to 1.638 ± 0.345 µmol/min. Further studies can probably help in finding suitable applications for ELF based modulation of enzyme activity.

Key words: ELF EMF, acid phosphatase, enzyme activity.

INTRODUCTION

Daily exposure to electromagnetic field is unavoidable as a consequence of living in a society that depends heavily on the use of the electricity. Over the past several years there has been a growing concern in the general public of a perceived health risk associated with exposure to EMF. The EMF generated by the 50 Hz alternating current traveling through electrical power lines has been of special concern. The first indication of a possible health risk developed from epidemiological studies (Wertheimer and Leeper, 1979). Further investigation in the laboratory has shown that a variety of biological processes can be Influenced by 50 Hz EMF (Shang et al., 2004; Lupke et al., 2006; Mehri et al., 2008). Although EMFs are usually associated with high voltage power lines and power stations, they are also provided by any electrically powered device, typical of those found in households or the workplace. Appliances such as video display terminals, TV's, hair dryers and cellular phones emit EMF's (Goodman et al., 1993).

Human exposures are normally to extremely low frequency ELF EMF's (defined as less than 300 Hz). Heightened public awareness has led to the inclusion of exposure to ELF EMF's as a part of a growing series of environmental conditions related to the "quality of life" in the industrial world. In recent years studies investigating the interaction of extremely low frequency EMF with human subjects, laboratory animals, organ cultures and individual cells have become substantial (Zmyslony et al., 2000; Claudio et al.,2004; Thomas et al., 2006).

There have been reports in the literature that this ELF EMF affects the various biochemical processes. Various surveys (Brix et al., 2001; Kelsh et al., 2003; Henderson et al., 2006) and epidemiological studies (Kuane et al., 2000; Wartenberg, 2001; Szabo et al., 2006) have been carried out to find the effects of these low frequency electromagnetic fields. Several studies have been carried out to investigate the effects on DNA (Ivancsits et al., 2002; Nikolai et al., 2004), enzyme activity (Blank et al., 1995, 1998a, b; Farrell et al., 1997; Morelli et al., 2005; Blank, 2005; Manoliu et al., 2006) and cells (Jin et al., 1997; Chang et al., 2005).

Enzymes play a vital role in the biological processes; also cell communication is facilitated by these biocatalysts. Any alteration in the activity of the enzyme may affect these biological processes. Elevated prostrate acid phosphatase may indicate the presence of prostrate cancer (Bull et al., 2002). Acid phosphatase is a phosphatase, a type of enzyme, used to freely attached phosphate groups from other molecules during digestion. Different forms of acid phosphatase are found in different organs, and their serum levels are used as a diagnostic for disease in the corresponding organs (Dattoli et al.,

---
*Corresponding author. E-mail: rashanth.kallambadi@gmail.com.

Table 1. Effect of EMF on the activity of the acid phosphatase from *Ipomoea batatas*.

Trial	Enzyme activity in the presence of EMF (µmoles/min)	Enzyme activity without EMF (Control) (µmoles/min)
1	0.042	0.022
2	0.050	0.023
3	0.053	0.029
4	0.045	0.026
5	0.042	0.025
Mean	0.0458 ± 0.005^a	0.0232 ± 0.002^a

Acid phosphatase activity measured after 60 min exposure to a field of 0.53 mT, p-value = 0.0054 (paired t-test). [a] mean value with standard deviation of 5 measurements.

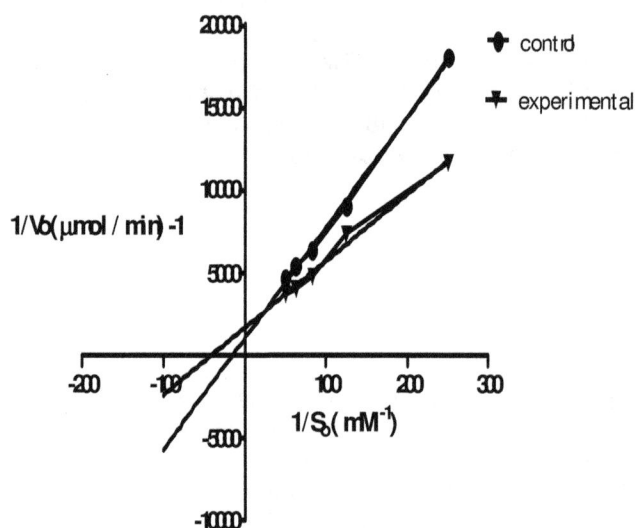

Figure 1. Lineweavear-Burk plot for acid phosphatase in the absence and the presence of EMF. Each value represents the mean ± SD of five measurements.

1999; Giraldo et al., 2000; Tsu et al., 2004). In this paper we have chosen to study the effect of ELF EMF on the activity of acid phosphatase enzyme.

MATERIALS AND METHODS

Exposure system and EMF characteristics

The exposure system consisted of a Helmholtz coil pair 17 cm in diameter, mounted on a wooden frame. Each coil had 500 turns of 0.25 mm diameter copper wire. The inner radius of each Helmholtz coil was 7 cm, while the outer radius was 10 cm. The system arrangement generated a uniform magnetic field of 0.53 mT ac rms. The distance between the two coils was 7 cm. At the centre of the arrangement a shelf was placed to hold the samples to be exposed. The signal was provided using step down ac transformer 6 V, 50 Hz duty cycle and the field intensity 0.53 mT rms measured by a gauss meter GM O5 (Hirst magnetic instruments UK, range 0 mT -3 T,

frequency 15 Hz to 10 KHz and with an accuracy of ± 1%) the gauss meter was connected to a laptop with a RS 232 interface and using Microsoft Visual Basic software programming tool the real time data was captured and stored in the system. The magnetic field of 0.53 mT was provided within Helmholtz coils supplied by electric generator able to deliver 50 Hz electric current. A PVC test tube stand to hold the samples was positioned in the middle of the Helmholtz coils. This arrangement allowed a uniform magnetic field of ac rms 0.53 mT.

Assays for enzyme activity from Ipomoea batatas

Acid phosphatase activity was determined by the method of Peter Bernfield (1955). Sodium β-glycerophosphate was used as the substrate. Acid phosphatase reactions were performed at 37℃ in 0.1 M citrate buffer containing 0.1 M magnesium acetate, pH 5.6. Kinetic measurements were determined by activity assays at pH 5.6 with various substrate concentrations. Acid phosphatase activity was measured before and after 60 min exposure to a field of 0.53 mT based on the absorbance at 660 nm (Systronics 117 type spectrophotometer). The kinetic parameters V_{max} and K_m were evaluated by Lineweavear Burk method. Acid phosphatase was assayed for its pH optimum and temperature optimum with the same substrate to check the effect of EMF over a range of pH and temperature.

Statistical analysis

Statistical analysis was done using Graph pad prism version 5.0. Two tailed paired t-tests were applied to compare the enzyme activities which were exposed to electromagnetic fields and the control samples which were not exposed to electromagnetic fields and results were considered to be statistically significant with $P<0.05$.

RESULTS AND DISCUSSION

Table 1 shows that the acid phosphatase activity increased significantly ($P<0.05$) when it was exposed to EMF. The exposure to electromagnetic field induced about 50% increase in the enzyme activity (0.0468 ± 0.005) when compared to control group (0.0232 ± 0.002). Figure 1 represents a Lineweavear Burk plot with control and exposed samples. Upon EMF exposure K_m increased from 0014 ± 0.005 to 0.040 ± 0.008 mM whereas V_{max} in

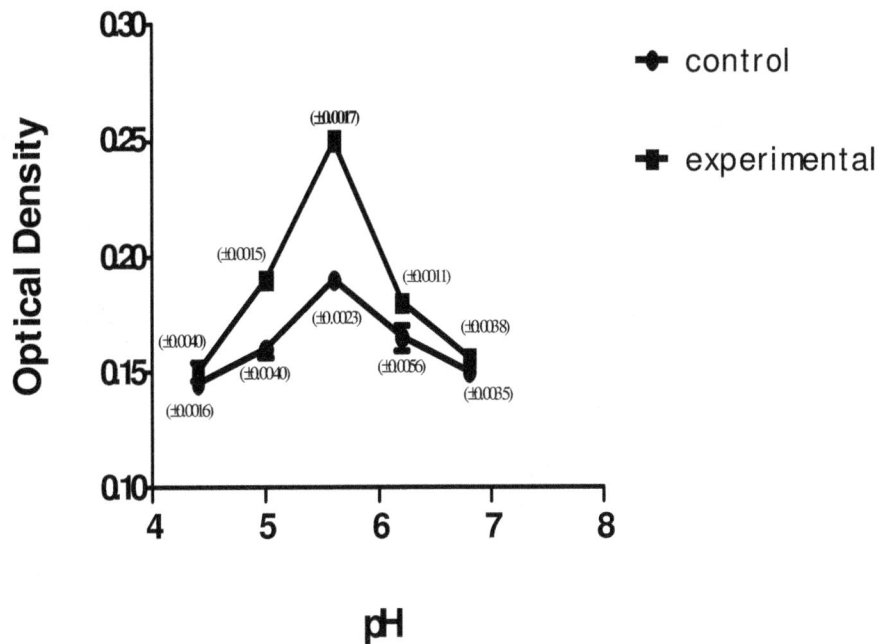

Figure 2. Effect of pH in the absence and the presence of EMF. Each value represents the mean ± SD of five mea-surements.

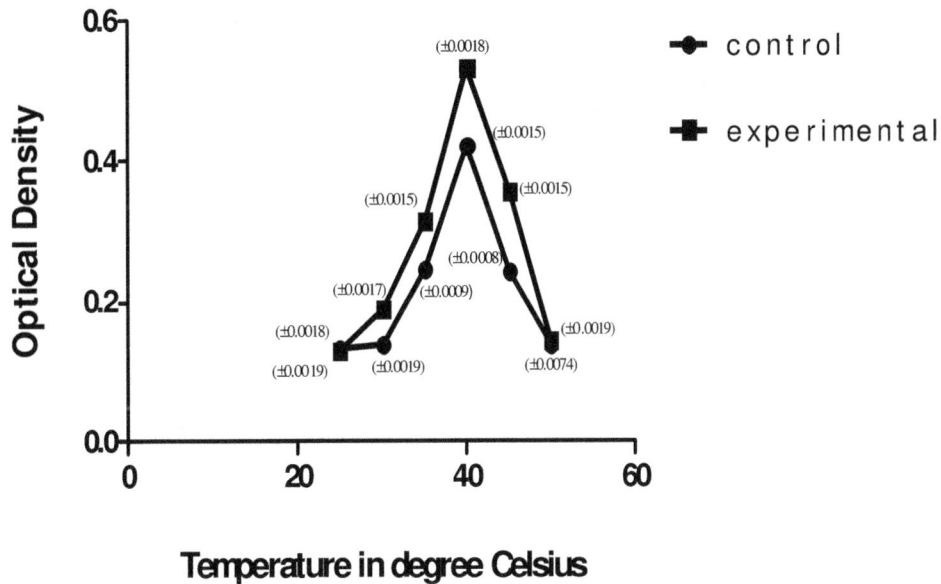

Figure 3. Effect of temperature in the absence and the presence of EMF. Each value represents the mean ± SD of five measurements.

increased from 0.991 ± 0.254 to 1.638 ± 0.345 μmol/min. Figure 2 represents the effect of pH on enzyme activity in the absence and the presence of EMF. No significant differences were detected for the optimum pH between control and exposed samples when acid phosphatase was assayed over a range of pH. Figure 3 represents the effect of temperature on enzyme activity in the absence and the presence of EMF. No significant differences were detected for the optimum temperature between control and exposed samples when acid phosphatase was assayed over a range of temperature. However the curves in Figures 2 and 3 for the exposed samples show a different trend with an optimal pH and temperature where maximal activity is seen in the presence of ELF EMF.

Dick et al. (1983) showed that the optimum pH for corn roots' acid phosphatase was 4, while Ullah and Gibson (1988) showed that the optimum pH for acid phosphatase was pH 5. In soil, the pH optimum for acid phosphatase ranges from pH 4 to 6.5 (Eivazi and Tabatabai, 1976). Figure 3 represents the effect of temperature on enzyme activity which is similar to that reported by Panaral et al. (1990) for acid phosphatase from barley and by Basha (1984) for peanut.

Based on the experimental work, and theoretical calculations (Laurence et al., 2000; Pickard and Moros, 2001; Laurence et al., 2003; Stuerga and Gaillard, 1996; Saxena et al., 2003) several mechanisms have been proposed of how ELF EMF is likely to interact with tissue, cells and proteins in solutions. The Moving Charge Interaction (MCI) model proposes that moving electrons affect enzyme activity. The activation of gene expressions (Olivares et al., 2004) and synthesis of stress proteins can be initiated by an EM field (both electric and magnetic fields) by moving electrons. The optimal frequency dependence is related to the turnover numbers (Blank and Soo, 1998b; 2001; Blank and Goodman, 1997, 1999, 2000, 2002) of the enzyme reaction. It is interesting to notice this effect of ELF EMF on enzyme functions inspite of small amount of energy carried by applied field; which is well below relevant ionizing or binding energy.

Balcavage et al. (1996) have shown that EMFs indeed affect metabolic processes by regulating flow of cellular cations. Wolf et al. (2005) have reported that EMF, in the presence of a transition metal, generates reactive oxygen species to cause DNA damage. Low frequency EMF, and pulsed EMF affect biological systems via information transfer; this information transfer can trigger biochemical processes such as ion binding and signal transduction (Rosch and Markov, 2004). Theoretical models have depicted existence of low frequency (ELF) magnetic field interactions with biosystems at ion cyclotron resonance frequencies, that is, at frequencies corresponding to charge to mass ratios of ions such as Ca^{2+}, Mg^{2+} and K^+ (Bruce et al., 1992). Experimental investigations reveal that through calcium signaling pathways and cystolic calcium oscillator low frequency electromagnetic fields effects living systems (Galvanovskis and Sandblom, 1998). The models on the effects of ELF-EMFs on enzyme activities refer to changes of motion of ions at the active site (Edmonds, 1993).

Structural investigations indicate that the catalytically active form of sweet potato purple acid phosphatase contains binuclear Fe-Mn clusters (Gerhard et al., 2001).

Susan and Bruce, (1987) predicted iron in sweet potato purple acid phosphatase is expected to be most likely in the Fe (III) state. The structural study by Gerhard et al. (1999) revealed the presence of a strongly antiferromagnetically coupled binuclear Fe(III) - Mn (II) center in sweet

potato purple acid phosphatase which is responsible for the enzyme's biological activity. The presence of thioether bond would put more structural constraints at the active site and henceforth increases the electron density (Atila et al., 1999). Therefore one may speculate EMF would affect the electron density which may in turn cause deprotonation which may lead to the enhancement of the enzyme activity.

More important results have been reported for enzymes exposed to ELF EMF. Venkatachalam et al. (2005) have shown a reduction in the activities of lysosomal enzymes; cathepsin D, acid phosphatase and myeloperoxidase in the liver and serum of arthritic rats subjected to a field of 4 µT. Zhang and Liu (1992) have shown that magnetic field effect on acid phosphatase is physically reversible by studying the effects of a strong constant magnetic field on the activity and localization pattern of acid phosphatase in Blepharisma. The magnetic fields in the range 0 - 70 Hz and 0 - 2 G increased Na, K-ATPase activity by 5 - 10% (Blank et al., 1995).

It was reported that 60 Hz magnetic fields accelerate the oxidation of cytochrome C, a reaction catalyzed by cytochrome oxidase, an electron transport enzyme of the mitochondrial redox chain; the acceleration varied with the magnetic field strength, the increase in the oxidation rate constant was 20 – 30% at field strengths below 3 mT (Blank et al., 1998). Farrell et al. (1997) have demonstrated that 4 pT, 60 Hz magnetic field enhances ornithine decarboxylase activity during gastrulation while studying the biochemistry of developing chicken embryos. Morelli et al. (2005) have shown that ELF EMF of 75 Hz, 2.5 mT above a threshold produces a decrease of about 54 – 61% of the enzymatic activities of three membrane-bound enzymes: alkaline phosphatase, phosphoglycerate kinase and acetylcholinesterase from blood cell or from synaptosomes. A more recent study by Manoliu et al. (2006) shows that the exposure to the field strength of 10 mT induced a 55% increase in the catalase activity and a 50% increase in the peroxidase activity in a fungi culture medium. The data reported in this article show that ELF EMF of 50 Hz, 0.53 mT induced about 50% increase in the enzyme acid phosphatase activity.

Blank and Soo (2001) while explaining the EMF interaction mechanism with Na^+-K^+ ATPase suggests the electrons move regularly and the threshold force producing the effect on the enzyme is due to acceleration of the electron regardless of direction. Therefore, it is very likely that the effects of the field on the acid phosphatase could involve electron density at the active site and the antiferromagnetically coupled binuclear Fe (III) - Mn (II) center. The results of the study clearly indicate that ELF EMF influence enzyme activity irrespective of temperature, pH and substrate concentration.

Further studies are required, however it is possible that a whole new process of controlling enzyme activity using ELF EMF with commercial implications in industrial enzymology.

REFERENCES

Atila D, Christoph E, Friedrich S, Bernt K (1999). Cloning and comparative protein modeling of two purple acid phosphatase isozymes from sweet potatoes. Arch. Biochem. Biophys 1434: 202-209.

Balcavage WX, Alvager T, Swez J, Goff CW, Fox MT, Abdullyava S, King MW (1996). A Mechanism for Action of Extremely Low frequency electromagnetic fields on biological systems. Biochem. Biophys. Res. Comm. 222: 374–378.

Basha SM (1984). Purification and characterization of an acid phosphatase from peanut seed. Can. J. Bot. 62:385–391.

Bernfield P (1955). Methods of Enzymology, Academic Press, New York, pp. 148-155

Blank M, Soo L, Papstein V (1995). Effects of low frequency magnetic fields on Na, K-ATPase activity. J. Bioelectrochem. Bioenerg 38: 267-273.

Blank M, Goodman R (1997). Do electromagnetic fields interact directly with DNA? J. Bioelectromagnetics 18:111-115.

Blank M, Soo L (1998a). Enhancement of cytochrome oxidase activity in 60 Hz magnetic fields. J. Bioelectrochem Bioenergetics 45:253-259.

Blank M, Soo L (1998b). Frequency dependence of cytochrome oxidase activity in magnetic fields. J. Bioelectrochem. Bioenerg 46: 139-143.

Blank M, Goodman R (1999). Electromagnetic fields may act directly on DNA. J.Cell. Biochem 75: 369-374.

Blank M, Goodman R (2000). Stimulation of the stress response by low-frequency electromagnetic fields: Possibility of direct interaction with DNA. IEEE Transactions on Plasma Science 28: 168-172.

Blank M, Soo L (2001). Electromagnetic acceleration of electron transfer reactions. J. Cell. Biochem 81:278-283.

Blank M (2005). Do Electromagnetic Fields Interact With Electrons in the Na, K-ATPase? Bioelectromagnetics 26: 677-683

Brix J, Wettemann H, Scheel O, Feiner F, Matthes R (2001). Measurement of the individual exposure to 50 and 16 2/3 Hz magnetic fields within the Bavarian population. J. Bioelectromagnetics 22: 323-332.

Bruce M, Abraham R, Liboff A, Stephen DS (1992). Electromagnetic gating in ion channels. J. Theor. Biol. 158:15-31.

Bull H, Murray PG, Thomas D, Fraser AM, Nelson PN (2002). Acid Phosphatases. J. Clini. Pathol: Mol. Pathol 55:65-72.

Chang K, Chang WHG, Huang S, Shih C (2005). Pulsed electromagnetic fields stimulation affects osteoclast formation by modulation of osteoprotegerin, rank ligand and macrophage colony-stimulating factor. J. Ortho. Res 23:1308–1314.

Claudio G , Marcello DA, Angela T , Giovanni M ,Federica W , Achille C, Gian BA (2004). Effects of 50 Hz electromagnetic fields on voltage-gated Ca2+ channels and their role in modulation of neuroendocrine cell proliferation and death. J. Cell Calcium 35:307–315.

Dattoli M, Wallner K, True L, Sorace R, Koval J, Cash J, Acosta R, Biswas M, Binder M, Sullivan B, Lastarria E, Kirwan N, Stein D(1999). Prognostic role of serum prostatic acid phosphatase for 103Pd-based radiation for prostatic carcinoma. Int.J. Radiat. Oncol. Biol. Phys 45: 853–856.

Dick WA, Juma NG, Tabatabai MA (1983). Effects of soils on acid phosphatase and inorganic pyrophosphatase of corn roots. J.Soil. Sci 136: 19–25.

Edmonds DT (1993). Larmor precession as a mechanism for the detection of static and alternating magnetic fields. J. Bioelectrochem. Bioenerg 30:3-12.

Eivazi F, Tabatabai TA (1976). Phosphatase in soil. J. Soil. Biol. Biochem 9: 167–172

Farrell JM, Barber M, Krause D, Litovitz TA (1997).Effects of low frequency electromagnetic fields on the activity of ODC in Chicken embryo. J. Bioelectrochemistry and Bioenergetics 43: 91-96.

Galvanovskis J, Sandblom J (1998). Periodic forcing of intracellular calcium oscillators theoretical studies of the effects of low frequency fields on the magnitude of oscillations. J. Bioelectrochem. Bioenerg 46:161-174.

Gerhard S, Yubin G, Lyle EC, Ceridwen JW, Iain RS, Bernard JC, Susan H, John J (1999). Binuclear metal centers in plant purple acid phosphatases: Fe±Mn in sweet potato and Fe±Zn in soybean. Arch. Biochem. Biophy 370:183–189.

Gerhard S, Clare LB, Lyle EC, Christopher JN, Boujemaa M, Keith SM, John J, Graeme RH, Susan H, (2001). A purple acid phosphatase from sweet potato contains antiferromagnetically coupled binuclear Fe-Mn center. J. Biological Chemistry 276:19084–19088.

Giraldo P, Pocovi M, Perez CJ, Rubio FD, Giralt M (2000). Report of the Spanish Gaucher's disease registry: clinical and genetic characteristics. Haematologica 85:792–799.

Goodman R, Chizmadhev Y, Henderson AS (1993). Electromagnetic fields and the cells. J.Cell. Biochem 51:436-441.

Goodman R, Blank M (2002). Insights into electromagnetic interaction mechanisms. J. Cell. Physio192:16-22.

Henderson SI, Bangay MJ (2006). Survey of RF exposure levels from mobile telephone base stations in Australia. J. Bioelectromagnetics 27:73-76.

Ivancsits S, Diem E, Pilger A, Rüdiger HW, Jahn O (2002). Induction of DNA strand breaks by intermittent exposure to extremely-low-frequency electromagnetic fields in human diploid fibroblasts. J. Mut. Res 519: 1–13.

Jin, M, Lin H, Han LI, Opler M, Maurer S, Blank M, Goodman R (1997). Biological and technical variables in myc expression in HL60 cells exposed to 60 Hz electromagnetic fields. J. Bioelectrochem. Bioenerg. 44:111 -120.

Kaune WT, Bracken D, Rankin RF, Niple JC, Kavet R (2000). Rate of Occurrence of Transient Magnetic Field Events in U.S.Residences. J. Bioelectromagnetics 21:197-213.

Kelsh MA, Bracken TD, Sahl JD, Shum M, Ebi KL (2003). Occupational Magnetic Field Exposures of Garment Workers: Results of Personal and Survey Measurements. J. Bioelectromagnetics 24:316-326.

Laurence JA, French PW, Lindner RA, McKenzie DR (2000). Biological effects of electromagnetic fields-mechanisms for the effects of pulsed microwave radiation on protein conformation. J. Theo. Biol. 206: 291-298.

Laurence JA, McKenzie DR, Foster KR (2003). Application of the heat equation to the calculation of temperature rises from pulsed microwave exposure. J. Theo. Biol. 22:403-405.

Lupke M, Frahm J, Lantow M, Maercker C, Remondini D, Bersani F, Simko M (2006). Gene expression analysis of ELF-MF exposed human monocytes indicating the involvement of the alternative activation pathway. Biochim Biophys Acta 2006; 1763 (4): 402 – 412.

Manoliu, AI, Oprica L, Olteanu Z, Neacsu I, Artenie V, Creanga DE, Rusu I, Bodale I (2006). Peroxidase activity in magnetically exposed cellulolytic fungi. J. Magnetism and Magnetic Materials 300: e323–e326.

Mehri KM, Saied MF, Mahyar J (2008.) 50 Hz alternating extremely low frequency magnetic fields affect excitability, firing and action potential shape through interaction with ionic channels in snail neurons. J. Environ. 28:341-347.

Morelli A, Ravera S, Panjolis I, Pepe IM (2005). Effects of extremely low frequency electromagnetic fields on membrane-associated enzymes. Arch. Biochem. Biophy 441:191–198.

Nikolai KC, Gapeyev AB, Sirota NP, Gudkov OY, Kornienko NV, Tankanag AV, Konovalov IV, Buzoverya ME, Suvorov VG, Logunov VA (2004). DNA damage in frog erythrocytes after in vitro exposure to a high peak-power pulsed electromagnetic field. J. Mut. Res 558: 27–34.

Olivares BT, Navarro L, Gonzalez A, Drucker CR (2004). Differentiation of chromaffin cells elicited by ELF MF modifies gene expression pattern. J. Cell. Bio. Inter. 28: 273–279.

Panara F, Pasqualini S, Antonielli M (1990). Multiple forms of barley root acid phosphatase: purification and some characteristics of the major cytoplasmic isoenzyme. Biochim. Biophys. Acta 1073:73–80.

Pickard WF, Moros EG (2001). Energy deposition processes in biological tissue: nonthermal biohazards seem unlikely in the ultra-high frequency range. J. Bioelectromagnetics 22: 97-105.

Rosch PJ, Markov MS (2004). Bioelectromagnetic Medicine, Marcel Dekker, New York. pp 850 -858.

Saxena A, Jacobson J, Yamanashi W, Scherlag B, Lamberth J, Saxena B (2003). A hypothetical mathematical construct explaining the mechanism of biological amplification in an experimental model utilizing picoTesla (PT) electromagnetic fields. J. Med. Hypoth 60: 821-839.

Shang GM, Wu JC, Yuan YJ (2004). Improved cell growth and Taxol production of suspension cultured Taxus Chinensis var.mairei in alternating and direct current magnetic fields. Biotechnol. Lett. 26: 875-878.

Stuerga DAC, Gaillard P (1996). Microwave athermal effects in chemistry - a myths autopsy, orienting effects and thermodynamic consequences of electric field. J. Microwave Power & Electromagnetic Energy 31: 101-113.

Susan KH, Bruce AA (1987). The "Manganese (III)-Containing" Purple Acid Phosphatase from sweet potatoes is an iron enzyme. Biochem. Biophys. Res. Comm. 1173-1177.

Szabo J, Mezei K, Czy GT, Mezei G (2006). Occupational 50 Hz Magnetic Field Exposure Measurements Among Female Sewing Machine Operators in Hungary. J. Bioelectromagnetics 27:451-457.

Thomas EP, Yoshitada S, Mark DG, Michael I, Ronald JM, Maciej Z, Alan W (2006) . Exposure of murine cells to pulsed electromagnetic fields rapidly activates the mTOR signaling pathway.J. Bioelectromagnetics 27: 535 – 544.

Tsu YC, Ching LH, Su HL, Mei JMC, Janckila A, Yam LT (2004).Tartrate-Resistant Acid Phosphatase 5b as a Serum Marker of Bone Metastasis in Breast Cancer Patients. J. Biomed. Sci. 11:511-516.

Ullah AHJ, Gibson DM (1988). Purification and characterization of acid phosphatase from cotyledons of germinating soybean seeds. Arch. Biochem. Biophys 260: 514–520.

Venkatachalam SK, Dilly AK, Kalaivani K, Gangadharan AC, Narayanaraju KVS, Pammi T, Manohar BM, Puvanakrishnan R (2005). Optimization of Pulsed Electromagnetic Field therapy for Management of Arthritis in Rats. J. Bioelectromagnetics 26:431-439.

Wartenberg D (2001). Residential EMF Exposure and Childhood Leukemia Meta-Analysis and Population Attributable Risk.   J. Bioelectromagnetics Suppl 5:S86-S104.

Wertheimer N, Leeper E (1979). Electrical wiring configurations and childhood cancer. Am.J.Epidemiol.109:273-284.

Wolf FI, Torsello A, Tedesco B, Fasanella S, Boninsegna A, D'ascenzo M, Grassi C, Azzena GB, Cittadini A (2005). 50-Hz extremely low frequency electromagnetic fields enhance cell proliferation and DNA damage: possible involvement of a redox mechanism. Biochim Biophys Acta 22:1743(1-2):120-9.

Zhang X, Liu D (1992). The effects of a strong constant magnetic field on the activity and localization pattern of acid phosphatase in Blepharisma. J. Bioelectrochem. Bioenerg 27: 513-517.

Zmyslony M, Palus J, Jajte J, Dziubaltowska E, Rajkowska E (2000). DNA damage in rat lymphocytes treated in vitro with iron cations and exposed to 7 mT magnetic fields (static or 50 Hz). Mutation Research 453: 89–96.

Thermogenic response of guinea pig adipocytes to noradrenaline and β3-AR agonists

7

Thermogenic response of guinea pig adipocytes to noradrenaline and β3-AR agonists

Ghorbani Masoud[1]* and Mehdi Shafiee Ardestani[2]

[1]Department of Research and Development, Pasteur Institute of Iran (Research and Production complex), 25th km Tehran Karaj Highway, Tehran, Iran.
[2]Department of Hepatitis and Aids, Pasteur Institute of Iran, Tehran, Iran.

Brown adipocytes isolated from warm acclimated guinea pig do not respond to noradrenaline (NA) in spite of their marked thermogenic response *in vivo*. In contrast, in cold-acclimated guinea pigs, isolated brown adipocytes show substantial increase in capacity for a thermogenic response to NA. Chronic stimulation with a β3-AR agonist increased insulin-sensitivity of brown adipose tissue (BAT) in the guinea pig. To investigate the responsiveness of glucose transport to insulin, to NA, and to β3-AR agonists, we used BAT of guinea pigs as an animal known to be insulin-resistant. Results of this study showed that no β3-adrenergic agonist used, such as CL 316,243, BRL 37344, was able to stimulate oxygen uptake in BAT cells from cold-acclimated or new born guinea pigs. However, noradrenaline (NA), adrenaline (A) and isoproterenol (ISO) had a marked thermogenic effect on these cells. In contrast, in a comparative study in warm-acclimated rat BAT cells, CL 316,243 was even more potent than NA in stimulating oxygen uptake. We concluded that guinea pigs lack β3-ARs in their BAT. These results were interesting and noteworthy, suggesting that guinea pigs can be a natural model for β3-ARs knockout.

Key words: Brown adipose tissue (BAT), White adipose tissue (WAT), β3-ARs, guinea pig.

INTRODUCTION

Brown adipose tissue (BAT) cells from cold-adapted guinea pigs show a greater response to NA compared to cells from animals acclimated to room temperature (Hamilton and Doods, 2008). In most mammalian adipose tissues, the induction of lipolysis by catecholamines is mediated by β-adrenergic receptor (β-AR) subtypes such as β1- and β3-ARs (Rial and Nicholls, 1984). The guinea pig, however, shows low expression with lack of function of β3-ARs (Atgié et al., 1996). During cold-acclimation, there is a marked increase in cell number, mitochondrial content of each adipocyte, and the response of each mitochondrion to NA stimulation of adipose tissue. The increase of thermogenesis per mitochondrion is not because of an increase of respiratory chain, since cytochrome c oxidase remains unchanged. Instead, it is due to a marked increase in the uncoupling protein per mitochondrion (Himms-Hagen, 1995).

In these experiments, we prepared and incubated brown adipocytes from guinea pigs during the stages of cold adaptation and warm adaptation, and from new born guinea pigs.

The objective of this study was to assess whether noradrenaline (NA) and β3-AR agonists such as BRL and CL 316,243 could increase oxygen uptake in brown adipocytes of warm-acclimated guinea pigs, cold-acclimated guinea pigs and newborn guinea pigs.

MATERIALS AND METHODS

Female Dunkin-Hartley guinea pigs were obtained at 3 weeks of age. They were housed individually at 28°C, with free access to guinea pig chow and water. After one week, some guinea pigs were acclimated to 4°C for up to 4 weeks while other remained at 28°C during the same period. Pregnant guinea pigs were also obtained at

*Corresponding author. E-mail: mghorbani@irimc.org.

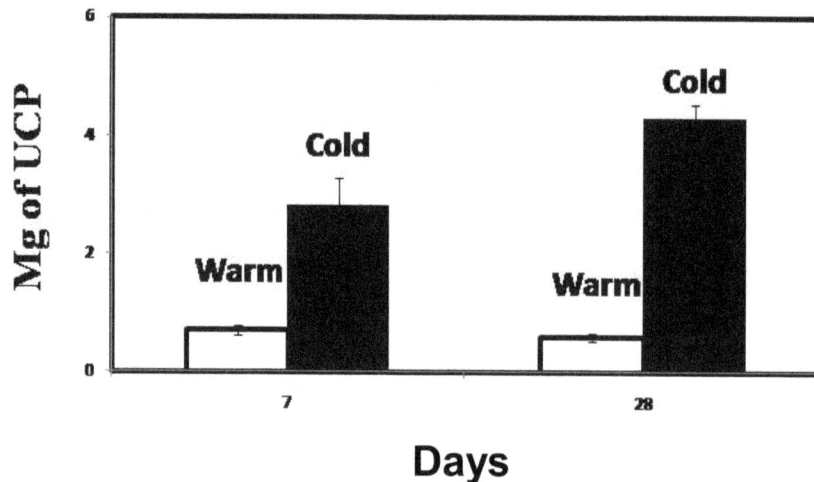

Figure 1. Total UCP of interscapular BAT of warm-acclimated guinea pigs and cold-acclimated (7 to 28 days) guinea pigs. At 7 days, it was 224 ± 35.6 for cold vs. 24.5 ± 4.3 for warm ($P < 0.001$); At 28 days, it was 150.0 ± 19.9 for cold vs. 15.4 ± 1.2 for warm ($P < 0.001$). Total mg protein changed slightly at 7 days, 84.5 ± 10.46 for cold vs. 42.0 ± 2.11 for warm ($P < 0.001$).

advanced stage of gestation. They were housed at room temperature with free access to food and water. The new born guinea pigs were used at the ages of 1, 2 or 3 days after delivery.

For measurement of metabolic rate, conscious guinea pigs were placed in a water-jacketed respiration chamber at 28°C with circulating air at the same temperature attached to an oxygen analyzer (Bechman Indistrial Oxygen Analyzer model 755) to monitor the oxygen uptake.

In all cases, BAT was isolated and transferred into Krebs-Ringer buffer (0.25 M sucrose, 0.2 mM dipotassium EDTA, and 1.0 mM HEPES, in distilled water, pH 7.2) and brown adipocytes were isolated by collagenase digestion of interscapular BAT and washed by flotation using a bicarbonate-buffered culture medium (Gibco DMEM 380-2320 AG) supplemented with 5.5 mM glucose, 1 mM ascorbic acid and 40 mg/ml fatty acid-free bovine serum albumin and equilibrated with 95% oxygen, 5% carbon dioxide, pH 7.4. Oxygen uptake of cells was measured polarographically (YSI Oxygen monitor) in the same medium.

Uncoupling protein (UCP) was measured in BAT homogenates by solid phase radioimmunoassay using antibody to purified hamster UCP and [^{125}I] protein A for detection.

The β-AR agonists we used were L-noradrenaline-D-bitartrate (Arterenol) (Sigma A 9512), L-isoproterenol-D-bitartrate (Sigma I2760), L-epinephrine-D-bitartrate (Sigma E4375), CL 316,243 (American Cyanamid Co), and BRL 37,344 (SmithKline Beecham).

Statistical analysis

Statistical analysis used Instat software to do one-way ANOVA, followed by Student-Newman-Keuls post hoc test. Significant differences are based on $P < 0.05$.

RESULTS

UCP content of interscapular BAT was markedly increased in cold acclimated guinea pigs (Figure 1). Cold acclimation induced a significant increase in UCP content

in both times ($P < 0.001$). The increase is based on the mitochondrial concentration of UCP in μg per mg mitochondrial protein. Since UCP is a specific marker for BAT activity, the increase of UCP content confirms that cold-acclimation has a stimulatory effect on BAT and activates the tissue to increase the UCP concentration in mitochondria to generate heat to fight against cold exposure. In another experiment, we demonstrated the substantial thermogenic response of brown adipocytes isolated from BAT of cold-acclimated but not of warm-acclimated guinea pigs to NA. In warm-acclimated guinea pigs, there was no response to NA by brown adipocytes (Table 1), whereas there was a substantial increase in oxygen consumption and thermogenesis in brown adipocytes isolated from cold-adapted guinea pigs (Table 1 and Figure 2). The experiment was continued with adrenaline (A) and isoproterenol (ISO) (Figure 2). None of these agonists had any effect on isolated brown adipocytes from warm-acclimated guinea pigs. Because of variation in the rate of oxygen consumption from one preparation to another, data are expressed as % of maximum for each agonist in each experiment. Isolated brown adipocytes were then incubated with two selective β3-agonists, BRL 37,344 and CL 316,243 in the same situation and preparations that responded to NA. Surprisingly, there was no response to these β3-AR selective compounds (Figure 3).

Thinking that there might have been a loss of β3-ARs during the isolation of brown fat cells, in another experiment, intact conscious cold-acclimated guinea pigs received a subcutaneous injection containing CL 316,243 and their oxygen uptake measured. Results showed that injection of 10 mg CL 316,243 per kg body weight had no effect on oxygen uptake (Figure 4). In contrast, injection

Table 1. Relative potencies (EC$_{50}$) of different β-adrenergic agonists in stimulating the respiration of brown adipocytes isolated from BAT of guinea pigs.

Agonists	Warm-acclimated guinea pig	Cold acclimated guinea pig
	Dose response (EC$_{50}$)	Dose response (EC$_{50}$)
Isoproterenol	No response	5.0×10^{-7}*
Adrenaline	No response	2.0×10^{-6}*
Noradrenaline	No response	1.5×10^{-6}*
CL 316,243 – β3-agonist	No response	No response
BRL 26810 - β3-agonist	No response	No response

Iso > A = NA are considered as β1-adrenergic agonists.* Concentration of agonist in M.

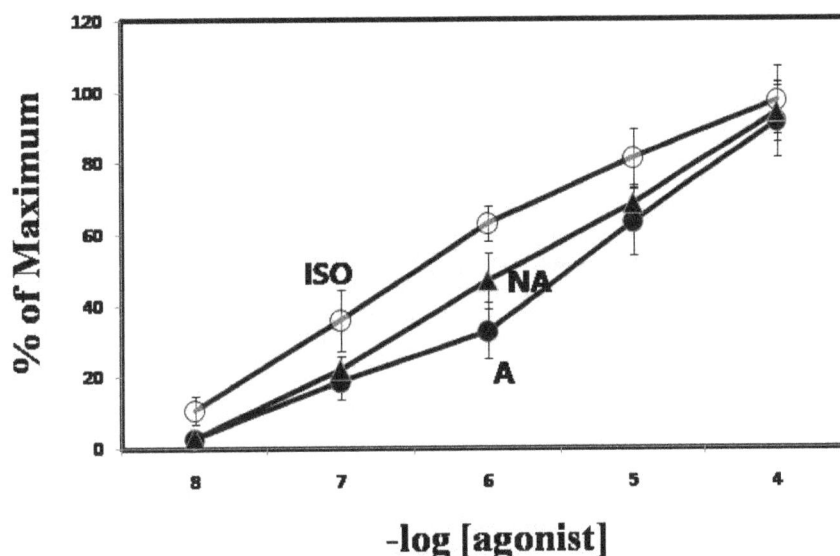

Figure 2. Thermogenic effect of adrenergic agonists on brown adipocytes isolated from cold-acclimated guinea pigs. Symbols: Isoproterenol (ISO) ○, adrenaline (A) ●, noradrenaline (NA) ▲. Actual maximum rates (in nmol per 10^6 cells, n = 6 cells preparations) were 247 ± 46.8 for ISO, 212 ± 34.0 for A, and 191 ± 43.7 for NA. Calculated EC$_{50}$ was 5.0×10^{-7} M for A, and 1.5×10^{-6} for NA.

of NA or cold exposure increased oxygen uptake in guinea pigs by 87 and 108%, respectively (Figure 4).

Since brown adipocytes of warm-acclimated guinea pigs had no thermogenic response to NA, due to very low level of UCP, it was not feasible to assess their response to β3-AR agonists. It has previously been shown that the stimulation of adenylate cyclase activity in crude membranes isolated from BAT of both warm- and cold-acclimated guinea pigs had no response to various β-adrenergic agonists including CL 316,243, whereas it was stimulated by ISO, A, NA and salbutamol in membranes of both warm- and cold-acclimated guinea pigs (Duffaut et al., 2006; Carpéné et al., 1994). Because β3-ARs might have been present in young guinea pigs and lost with aging, as occurs in other precocial mammals such as bovine (Rafael et al., 1986), we also studied the effects of NA and CL 316,243 on brown adipocytes isolated from BAT of newborn guinea pigs. β3-AR agonists also failed to stimulate oxygen uptake in BAT cells from cold acclimated or new born guinea pigs, while NA had a marked effect on increasing oxygen uptake as well as stimulation of adenylate cyclase activity in membrane of these cells. Therefore, it is suggested that the stimulation of thermogenesis by ISO, A, and NA in both warm- and cold-acclimated guinea pigs is mediated by β1-ARs and not by β3-receptors, and these animals do not express β3-AR.

DISCUSSION

Present results show chronic exposure to cold increases mitochondrial uncoupling protein synthesis in BAT of guinea pigs and induces thermogenesis in isolated BAT

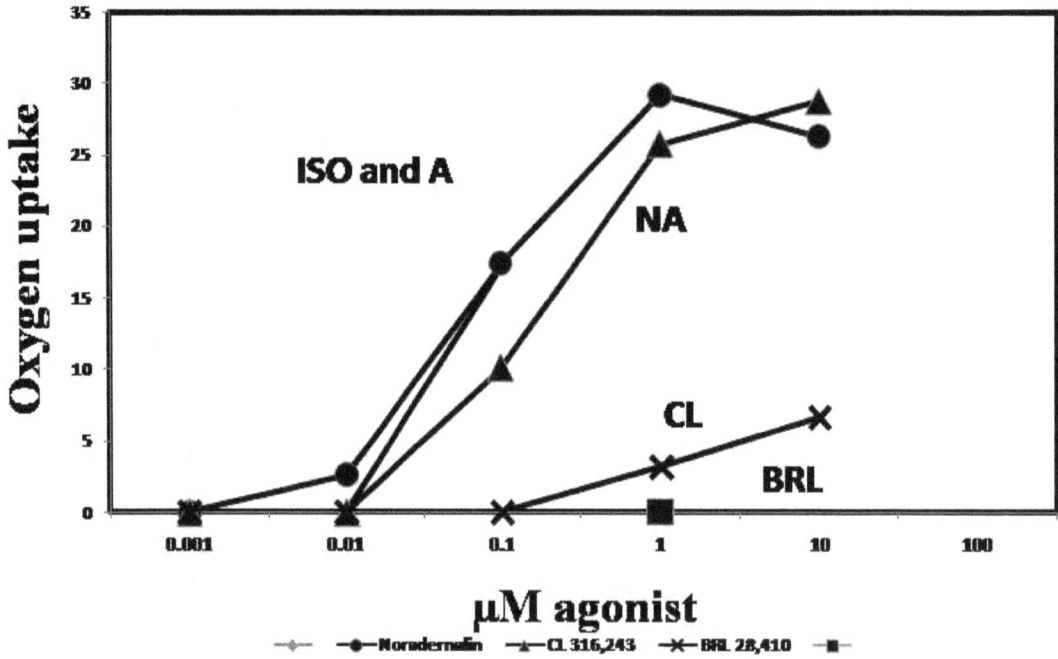

Figure 3. Effect of β-adrenergic agonists on respiration of isolated brown adipocytes from cold-acclimated guinea pigs. Symbols: Isoproterenol (ISO) ○, adrenaline (A) ●, noradrenaline (NA) ▲, CL 316,243 (CL) ×, BRL 28,410 (BRL) ■, Oxygen uptake is in nmol per 10^6 cells per minute.

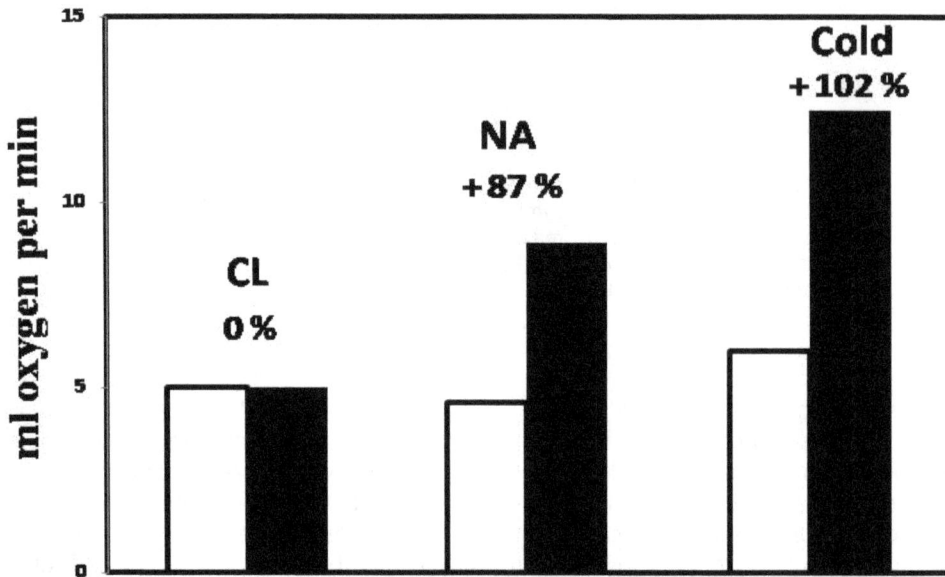

Figure 4. Effect of cold-acclimation and β-adrenergic agonists on energy expenditure of conscious cold-acclimated guinea pigs. Animals were either exposed to cold, infused with CL 316,243 (CL) or infused with noradrenaline (NA). Open bars show the oxygen uptake before injection or cold exposure and solid bars show the oxygen uptake after injection or cold exposure.

cells. Our data also confirmed the remarkable increase in thermogenic responsiveness to NA of brown fat cells isolated from BAT of cold-acclimated guinea pigs (Hamilton and Doods, 2008). Since, NA is able to stimulate lipolysis in brown fat cells of both warm- and cold-acclimated guinea pigs (Cunningham et al., 1986;

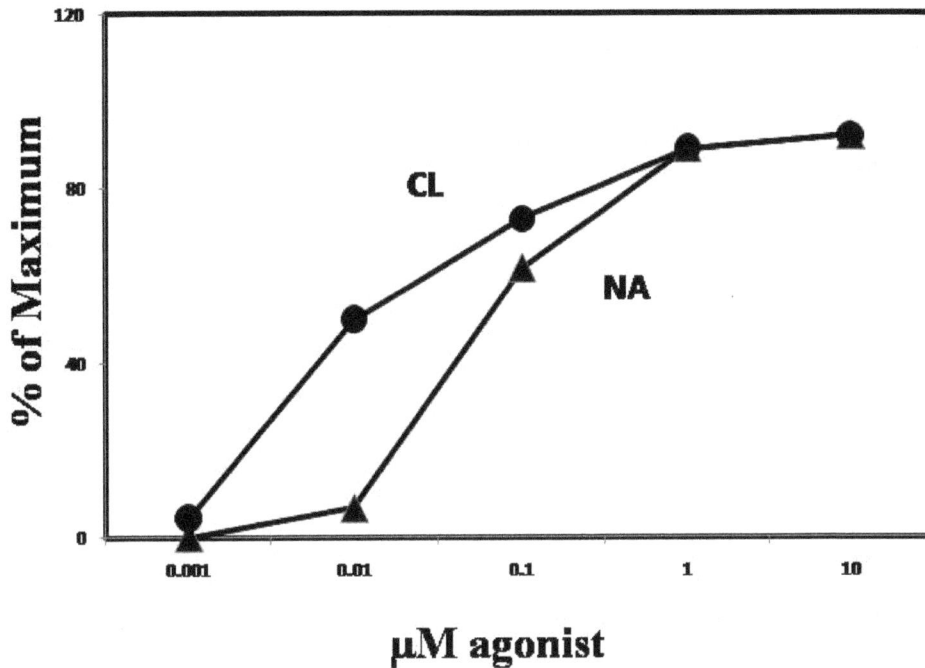

Figure 5. Effect of β-adrenergic agonists on oxygen uptake of brown adipocytes isolated from rats. Symbols: noradrenaline (NA) ▲, CL 316,243 (CL) ●.

Pond and Mattocks, 1991) and they exhibit no responses to all the β3-agonists tested, we conclude that the β-adrenergic responsiveness of guinea pig fat cells essentially involves β1- and β2-ARs and not β3-receptors (Carpéné et al., 1999).

It is also known that β3-adrenergic agonists are inactive on lipolysis in white adipose tissue (WAT) from guinea pigs (Lafontan et al., 1992), in contrast to their lipolytic effect on rat WAT cells. Differences in the adrenergic regulation of lipolysis have also been reported between rat and guinea pigs (Rafael et al., 1986).

However, there is evidence of existence of β3-AR mediating relaxation of various segments of the gastro-intestinal tract of guinea pigs (Bond and Vanhoutte, 1992).

Since, β3-ARs are present in intestinal tissue and absent in adipose tissues of guinea pigs, we conclude that the regulation of β3-ARs expression is different in guinea pigs and rodents, in which they are present in both adipose tissues and intestine. In contrast to guinea pig, in a comparative study in rat BAT cells reported previously (Himms-Hagen et al., 1994), the capability of CL 316,243 was even more potent than NA in stimulating of oxygen uptake (Figure 5). The guinea pig thus differs from rodents by the absence of β3-adrenergic effects in its adipose tissue and thus resembles the human and bovine species (Strosberg and Pietri-Rouxel, 1996; Himms-Hagen et al., 1996). More recent studies on cloning and sequencing of the guinea pig β3-AR gene have revealed a slightly higher amino acid sequence

similarity with the human than with the rodent β3-ARs (Atgié et al., 1996). β3-ARs in guinea pig ileum have the pharmacological characteristics of human β3-ARs (Emorine et al., 1989). It is now known that the pharmacological characteristics of human β3-ARs differs from that in rodents (Strosberg and Pietri-Rouxel, 1996; Emorine et al., 1992) in which β3-adrenergic signalling acutely down regulates adipose triglyceride lipase in brown adipocytes (Deiuliis et al., 2010). It is noteworthy that the guinea pig is not a rodent, thus, it is not surprising that it does not possess rodent-type β3-ARs.

ACKNOWLEDGEMENT

This work has been performed in part at the University of Ottawa, Canada.

REFERENCES

Atgié C, Tavernier G, Allaire F, Bengtsson T, Marti L, Carpéné C, Lafontan M, Bukowiecki LJ, Langin D (1996). β3-adrenoceptor in guinea pig brown and white adipocytes: low expression and lack of function. Am. J. Physiol., 271: R1729-R1738.

Bond RA, Vanhoutte PM (1992). Interaction of tetratolol at the atypical or beta3-adrenoceptor in guinea-pig ileum. Gen. Pharmacol., 23: 171-176.

Carpéné C, Castan I, Collon P, Galitzky J, Moratinos J, Lafontan M (1994). Adrenergic lipolysis in guinea pig is not a β3-adrenergic response: comparison with human adipocytes. Am. J. Physiol. 266: R905-R913.

Carpéné C, Galitzky J, Fontana E, Atgié C, Lafontan M, Berlan M (1999). Selective activation of beta3-adrenoceptors by octopamine:

comparative studies in mammalian fat cells. Arch. Pharmacol., 359(4):310-321.

Cunningham SA, Wiesinger H, Nicholls DG (1986). Quantification of fatty acids activation of the uncoupling protein in brown adipocytes and mitochondria from the guinea-pig. Eur. J. Biochem., 157: 415-420.

Deiuliis JA, Liu LF, Belury MA, Rim JS, Shin S, Lee K (2010). Beta(3)-adrenergic signaling acutely down regulates adipose triglyceride lipase in brown adipocytes. Lipids, 45(6): 479-489.

Duffaut C, Bour S, Prévot D, Marti L, Testar X, Zorzano A, Carpéné C (2006). Prolonged treatment with the beta3-adrenergic agonist CL 316243 induces adipose tissue remodeling in rat but not in guinea pig: 2) modulation of glucose uptake and monoamine oxidase activity. J. Physiol. Biochem., 62(2):101-11.

Emorine LJ, Féve B, Pairault J, Briend-Sutren MM, Nahmias C, Marullo S, Delavier-Klutchko C, Strosberg AD (1992). The human β_3-adrenergic receptor: relationship with atypical receptors. Am. J. Clin. Nutr., 55: 215S-218S.

Emorine LJ, Marullo S, Briend-Sutren MM, Patey G, Tate K, Delavier-Klutchko C, Strosberg AD (1989). Molecular characterization of the human β_3-adrenergic receptor. Science, 245: 1118-1121.

Hamilton BS, Doods HN (2008). Identification of potent agonists acting at an endogenous atypical beta3-adrenoceptor state that modulate lipolysis in rodent fat cells. Eur. J. Pharmacol., 580(1-2):55-62.

Himms-Hagen J (1995). Role of brown adipose tissue thermogenesis in control of thermoregulatory feeding in rats: a new hypothesis that links thermostatic and glucostatic hypotheses for control of food intake. Proc. Soc. Exp. Biol. Med., 208: 159-169.

Himms-Hagen J, Danforth E (1996). The potential role of β_3 adrenoceptor agonists in the treatment of obesity and diabetes. Curr. Opin. Endocrinol. Diabetes, 3: 59-65.

Himms-Hagen J, Cui J, Danforth E, Taayjes DJ, Lang SS, Waters BL, Claus TH (1994). Effect of CL-316,243, a thermogenic β3-agonist, on energy balance and brown and white adipose tissues in rats. Am. J. Physiol., 266: R1371-R1382.

Lafontan M, Saulnier-Blache JS, Carpéné C, Langin D, Galitzky J, Portillo M, Larrouy D, Berlan M (1992). Fat cell adrenergic receptors: from molecular approaches to trapeutic strategies. Obes. Eur., 141-153.

Pond CM, Mattocks CA (1991). The effects of noradrenaline and insulin on lipolysis in adipocytes isolated from nine different adipose depots of guinea pigs. Int. J. Obes., 15: 609-618.

Rafael J, Fesser W, Nicholls DG (1986). Cold adaptation in guinea pig at level of isolated brown adipocytes. Am. J. Physiol., 250: C228-C235.

Rial E, Nicholls DG (1984). The mitochondrial uncoupling protein from guinea-pig brown adipose tissue: Synchronous increase in structural and functional parameters during cold adaptation. Biochem. J., 222: 685-6.

Strosberg AD, Pietri-Rouxel F (1996). Function and regulation of the β_3-adrenoceptor. Trends Biochem. Sci., 17: 373-381.

8

Deficiency of vitamin B12 among Jordanian people with psychological and biological activity

Mohammad A. Al-Fararjeh[1], Nader Jaradat[2] and Abdulrahim Aljamal[2]*

[1]Department of the Medical Laboratory Technology, Hashemite University. Zarqa, Jordan.
[2]Department of Medical Technology, Faculty of Allied Medical Sciences, Zarqa University, P. O. Box 132222-Zarqa 13110- Jordan.

This study was conducted to evaluate vitamin B12 level among Jordanians healthy people and its effects on psychological parameters. One hundred and eighty six subjects were recruited in the study. The subjects were chosen to be healthy, aged between 16 to 71 years. Participants were asked to fill a detailed questionnaire that covered medical data then they were divided into three age groups. Serum vitamin B12 levels were done for all volunteers. Results showed the mean values of B12 concentrations in males and females. There were lower mean B12 concentration in females (264.02 pmol/L) than males (281.56 pmol/L) and a total mean concentration was 265.79 pmol/L that is near the cutoff value which was 258 pmol/liter. The mean value of the B12 distributed in different age ranges for the mean concentration of the 16 to 25 years group was 249.03 pmol/L significantly lower than the other group. There were highly significant difference between the groups in terms of low sexual activity, Memory loss, Loss of appetite and depression and deficiency of vitamin B12.

Key words: Vitamin B12, B12 deficiency and psychological parameters.

INTRODUCTION

The study was conducted to assess if objective evidence for sexual activity, heart palpitation, loss of appetite, memory loss and depression was associated with low vitamin B12 concentrations. In Jordan vitamin B12 deficiency has increased in an alerting rate in the last decade. Several studies done in Jordan reported prevalence between 16 to 50% (Barghouti et al., 2009; Abu-Samak et al., 2008). Serum vitamin B12 test measures total amount of circulating cobalamin, either bound to transcobalamin, which represents the functionally important fraction of plasma vitamin B12 (Miller et al., 2006) or haptocorrin, which makes the vitamin unavailable for uptake by cells (Holleland et al., 1999). In a primary care setting, the indications for the measurement of serum vitamin B12 concentrations typically include a history suggestive of pernicious anaemia, poor nutritional status, cognitive impairment, dementia, neuropathy or malabsorption. But although vitamin B12 supplementation is clearly indicated for the treatment of pernicious anemia or malabsorption associated with low vitamin B12, the relevance of supplementation with vitamin B12 (and other B vitamins) for the treatment of cognitive impairment is uncertain. Because poor nutritional status is common among older people with dementia, as is age-associated decline in the intestinal absorption of vitamin B12, it is not surprising that a substantial proportion of older people with dementia have low serum vitamin B12 concentrations. Nevertheless, it is widely believed that vitamin B12 deficiency accounts for only a small fraction of the reversible causes of dementia in older people (Malouf and Areosa, 2003; Magri et al., 2003).

More commonly, such individuals present with non-specific symptoms of fatigue and cognitive impairment that can be attributed to 'old age'. Some of the

*Corresponding author. E-mail: aljamalzpu@yahoo.com.

Table 1. The mean value of B12 levels of the study participants.

Sex of volunteers	No	Mean B12 concentration pmol/liter	S.D.
males	60	281.56	99.6
females	126	264.02	98.65
Totals	186	265.79	91.85

uncertainty about the importance of vitamin B12 deficiency relates to the limitations of the standard vitamin B12 assays. Low serum vitamin B12 concentrations do not accurately reflect intracellular vitamin B12 concentrations, and blood levels of homocysteine or methylmalonic acid are believed to be more reliable indicators of intracellular vitamin B12 status (Clarke et al., 2004). About 80% of vitamin B12 circulating in blood is biologically unavailable for most cells; the rest comprises holotranscobalamin (holoTC), which is the part of serum vitamin B12 bound to transcobalamin, the protein that delivers the vitamin to cells in the body and is easily measured (Morkbak et al., 2005).

Data collection

One hundred and ninety nine people were selected for the study, 13 were excluded; 6 had a history of systemic illness (e.g. diabetes mellitus, liver cirrhosis), 5 were on vitamin and/or supplement therapy and 2 had serum creatinine level ≥ 1.4 mg/dL. Finally, 186 (60 men and 126 women) adult Jordanian were randomly selected from all regions of Zarka city. All were in the age range 16 to 71 years. Those who were vegetarians and those who had any systemic illness, serious organ diseases, renal failure (creatinine ≥ 1.4 mg/dl), liver failure, alcoholism, taking food or multivitamin supplements or taking anticonvulsants, antimetabolites and antiviral drugs were excluded. Pregnant and lactating women were also excluded. The local ethics committee in Zarka University approved the study. Informed written consent was obtained from all respondents before being admitted into the study.

Data were collected using a questionnaire including items related to sexual activity, heart palpitation, loss of appetite, memory loss, and depression. Blood samples were taken in the morning. The serum was separated within 1 h of sampling by centrifugation (10 min, room temperature, at 2000 rpm) and was stored at −70°C until analysis.

Methods

Serum vitamin B12 level was measured by competitive protein binding assays in the Jerusalem consulting laboratory (Zarka-Jordan) by using Chemiluminescence imunoassay; Immulite 2000 (Siemens Medical Solutions Diagnostics, Deerfield, IL). As in an earlier report (Stabler et. al., 1999), vitamin B12 deficiency was defined according to two classifications, using a high and a low cutoff, in order to explore the existence of a dose-response relationship. High-cutoff vitamin B12 deficiency is present when the serum vitamin B12 level is less than 258 pmol/liter. A low-cutoff vitamin B12 deficiency is present when the serum vitamin B12 level is less than 148 pmol/liter.

Statistical analysis

Data were presented as mean ±SD (standard deviation), of the indicated number of experiments, students t-test and one-way ANOVA. All analysis was calculated utilizing the computer data processing (SPSS, version 14) a probability value (*P < 0.05*) was considered to be statistically significant.

RESULTS

One hundred and ninety nine people were selected for the study, 13 were excluded; 6 had a history of systemic illness (e.g. diabetes mellitus, liver cirrhosis), 5 were on vitamin and/or supplement therapy and 2 had serum creatinine level ≥ 1.4 mg/dL. Finally, 186 respondents (60 men and 126 women) were assessed. Table 1 shows the mean values of B12 concentrations in males and females in the selected group of population. There was lower mean B12 concentration in females (264.02 pmol/L) than males (281.56 pmol/L) and a total mean concentration (265.79 pmol/L) that is near the selected cutoff value which was 258 pmol/liter.

Table 2 demonstrates the mean as well as the range of concentration of B12, distributed in different ranges of age groups. The participants were divided into three age groups: 16 to 25 years; in this group the participants were 60 and the mean concentration was low in this group sample according to the serum vitamin B12 low high cutoff (249.03±78.7). Forty six of the participants in this group (76.6%) showed vitamin B12 deficiency according to high cutoff, while eight of them (13.3%) showed deficiency at low cutoff.

The total mean values of the other groups selected, showed no vitamin B12 deficiency, while 38 (46.3%) and 24 (54.5%) showed deficiency at high cutoff in the age groups 25 to 45 and 45 to 71, respectively. Furthermore, only 8 (9.7%) and 2 (4.5%) showed deficiency at the selected low cutoff.

Table 3 shows relationship between B12 levels and psychological parameters and biological activity, 90 volunteers with a mean vitamin B12 level of 260.7±95.1, 52 of them (58%) are low in their vitamin B12 level are suffering from low sexual activity according to high cutoff, while only 3 (7%) are affected according to low cutoff. 42 (52%) out of 82, 44 (50%) out of 88, 54 (52%) out of 104

Table 2. The mean value of B12 levels according to age groups.

Age (years)	No.	Vitamin B12 (pmol/liter) Mean ± (SD)	Low vitamin B12 level			
			High cutoff (258 pmol/liter)		Low cutoff (148 pmol/liter)	
			No.	%	No.	%
16 to 25	60	249.03±78.7	46	76.6	8	13.3
25 to 45	82	284.3±112.6	38	46.3	8	9.7
45 to 71	44	289.19±113.85	24	54.5	2	4.5

Table 3. The relationship between the mean value of B12 levels with psychological and biological activity.

Say yes for	No.	Vitamin B12 (pmol/liter) Mean ± (SD)	Low at high cutoff		Low at low cutoff	
			No.	%	No.	%
Low sexual activity	90	260.7±95.1	52 *	58	3	7
Depression	82	271.1±92.2	42 *	52	8	10
Heart palpitation	88	290.4±105.3	44 *	50	12	14
Memory loss	104	268.2±90.7	54 *	52	8	8
Loss of appetite	88	262.5±90.4	46*	53	10	12

* $P < 0.01$.

and 46 (53%) out of 88, are answered yes for suffering of depression, heart palpitations, memory loss and loss of appetite, respectively.

DISCUSSION

The results of our study showed that the mean serum B12 level in the younger age group (16 to 25years) was significantly lower than the older group (25 to 45years and 45 to 70years) (Table 2). These results are consistent with those of Barghouti et al. (2009) who found that high frequency of vitamin B12 deficiency (serum level <180 pg/ml) was observed in the age group between 18 to 24 years (51.8%) as compared with other age groups. This prevalence of B12 deficiency in younger adults can be explained by their eating habits that are usually highly dependent on consumption of junk foods by the younger group. High frequencies of vitamin B12 deficiency among young adults were also reported in studies conducted in several countries. In India, Yajnik et al. (2006) found that 67% of the participating healthy men (aged 30 to 50 years) had low vitamin B12 levels (<150 pmole/L). In Iran, Shams et al. (2009) studied vitamin B12 deficiency in 984 healthy subjects aged 20 to 80 years. Their study was conducted on 408 subjects who were between 20 to 40 years. Shams and his colleagues found that among this young age group, 28.3% (156 of 408) had serum B12 level less than 200 pg/ml and this percentage increased to reach 53.9% in those who had

serum B12 <250 pg/ml (220 of 408). All these results agree with our results. However, in a study conducted in Spain, serum vitamin B12 was reported to be significantly higher in younger age (25 to 39 years) than in age groups (40 to 49 years) (Planells et al., 2003).

Gender was shown to have no effect on the occurrence of low vitamin B12 level. There was no significant difference in the mean serum vitamin B12 level between males and females (p = 0.05). Similar results were found in the studies done by other authors (Fora and Mohammad, 2005; Planells et al., 2003). These authors reported no significant differences in the mean serum B12 between both sexes. However, Barghouti et al. (2009) and Hakooz et al. (2006) found that males had significantly lower vitamin B12 level than females (p<0.05). In contrast, Habib (2008) found that males had marginally higher serum B12 concentration than females. The difference on results could be related to eating habits of the study samples.

We found that low vitamin B12 level was more common in our sample than has been previously reported in patients with major depression. Mischoulon et al. (2000) found correlation between the severity of depression and the level of vitamin B12 at baseline, displayed a weak positive correlation. Engström and Träskman-Bendz, (1999) found no correlation between the levels of folate or B12 and the severity of depression, but an inverse relationship between the level of folate and severity of depression has been reported in some other studies (Carney et al., 1990; Wesson et al., 1994). A low B12 level was relatively common among our samples with depressive disorder, which is in accordance with previous studies (Alpert and Fava, 1997). Moreover, our study included only young and middle-aged respondents with depression, which may have influenced the results. The association between B12 and depression may be more prominent in elderly subjects, among whom folate deficiency has been relatively common in some studies (Quinn and Basu, 1996). Poor appetite and inappropriate food intake as symptoms of depression could result in low levels of vitamin B12. These results agree with our study. It might reflect to lower intake of vitamins from food or assimilation from the gastrointestinal track, or a higher rate of metabolism of these vitamins. Depression may also affect the quality of food in the diet. Morris et al. (2003) found that levels of blood folate had decreased after an episode of depression. However, loss of appetite, weight loss and being underweight were not related to folate levels. There are several theories concerning potential associations between depression and levels of vitamin B12 and folate. Vitamin B12 and folate are connected with the synthesis of monoamines and are involved in single carbon transfer methylation reactions connected with the production of monoamine neurotransmitters (Bottiglieri, 1996). However, Vitamin B12 deficiency may also result in the accumulation of homocysteine, which has been suggested to lead to exito-toxic reactions and may enhance depression (Parnetti et al., 1997). Bottiglieri et al. (2000) found raised levels of homocysteine in 52% of depressed inpatients. Vitamin B12 is also required in the synthesis of S-adenosylmethionine (SAM), which is needed as a methyl donor in many methylation reactions in the brain.

This study showed that low serum vitamin B12 concentrations were associated with symptoms of memory impairment. These associations provide support for the suggestion that biologically active fraction of vitamin B12, are a more reliable indicator of intracellular vitamin B12 status than the standard vitamin B12 assay and agree with study of Lewerin et al. (2005).

Our sample included more women than men; all these variables may influence the results. Men and women may have different dietary habits; a low vitamin B12 level and B12 deficiency have been found to be common among younger women (Penninx et al., 2000). Vitamin B12 is an important vital coenzymes that could interfere with the production of many enzymes and hormones in humans. One of these hormones could be testosterone that affect the sexual habits and performance, that's why nearly two third of the volunteers who have deficiency are suffering from low sexual activates.

Finally, patients with psychotic depression may have lower B12 levels than non-psychotic depressives (Bell et al., 1991). Psychotic depression is a common indication for inpatient treatment. For these entire reasons one should be careful about generalizing these findings to all groups of depressive patients. Another explanation is that depression may also affect the quality of food in the diet which lowers levels of blood vitamins during an episode of depression (Morris et al., 2003).

Conclusions

The results of this study revealed that there are relation between deficiency of vitamin B12 and Low sexual activity, depression, Heart palpitation, memory loss and loss of appetite. Further intensive research, that studies the concentration of sex hormones and using international criteria for proper assessment of depression and other psychological parameters are proposed.

ACKNOWLEDGMENTS

The authors thank Hassan al Zeben, Bushra al-Reemawi, Muna faiez, Mamoun Abu Baker and Dr Wafa Abu hatab for their participation.

REFERENCES

Abu-Samak MR, Khuzaie M, Abu-Hasheesh M, Jaradeh MF (2008). Relationship of vitamin B12 deficiency with overweight in male Jordanian youth. J. Appl. Sci., 8: 3060-3063.

Alpert JE, Fava M (1997). Nutrition and depression: the role of folate. Nutr. Rev., 55: 145-149.

Barghouti FF, Younes LJ, Halaseh TT, Said SM (2009). High frequency of low serum levels of vitamin B12 among patients attending Jordan University Hospital. EMRO, 15: 853-860.

Bell IR, Edman JS, Morrow FD, Marby DW, Mirages S, Rerrone G, Kayne HL, Cole JO (1991). B complex vitamin patterns in geriatric and young adult in patients with major depression. J. Am. Geriatr. Soc., 39: 252-257.

Bottiglieri T, Laundy M, Crellin R, Toone BK, Carney MWP, Reynolds EH (2000). Homocysteine, folate, methylation, and monoamine metabolism in depression. J. Neurol. Neurosurg. Psychiatry, 69: 228-232.

Bottiglieri T (1996). Folate, vitamin B12, and neuropsychiatric disorders. Nutr. Rev., 54: 382-390.

Carney M, Chary T, Loundy M, Bottiglieri T, Chanarin I, Reynolds EH, Toone T (1990). Red cell folate concentrations in psychiatric patients. J. Affect. Disord., 19: 207-213.

Clarke R, Grimley Evans J, Refsum H (2004). Vitamin B12 and folate deficiency in older people. Age Ageing, 33: 34-41.

Engström G, Träskman-Bendz L (1999). Blood folate, vitamin B12 and their relationship with cerebrospinal fluid monoamine metabolites, depression and personality in suicide attempters. Nord. J. Psychiatry, 53: 131-137.

Fora MA, Mohammad MA (2005). High frequency of suboptimal serum vitamin B12 level in adults in Jordan. Saudi Med. J., 26: 1591-1595.

Habib ABA (2008). Factors Affecting Vitamin B12 Deficiency among College Students. Master's Thesis, Jordan University of Science and Technology, Jordan.

Hakooz N, Abu-Dahab R, Arafat T, Hamad M (2006). A trend of low serum vitamin B12 in Jordanian adults from two ethnic groups in Amman. J. Med. J., 40: 80-87.

Holleland G, Schneede J, Ueland PM, Lund PK, Refsum H, Sadberg S (1999). Cobalamin deficiency in general practice assessment of the diagnostic utility and cost-benefit analysis of methylmalonic acid determination in relation to current diagnostic strategies. Clin. Chem.,

45: 189-198.

Lewerin C, Matousek M, Steen G, Johansson B, Steen B, Nilsson-Ehle H (2005). Significant correlations of serum homocysteine and serum methylmalonic acid with movement and cognitive performance in elderly subjects but no improvement from short-term vitamin therapy: a placebo-controlled randomized study. Am. J. Clin. Nutr., 81: 1155-62.

Magri F, Borza A, del Vecchio S (2003). Nutritional assessment of demented patients: a descriptive study. Aging Clin. Exp. Res., 15: 148-153.

Malouf R, Areosa SA (2003). Vitamin B12 for Cognition. The Cochrane Database of Systematic Reviews, Issue 3. Art. No.: CD004394. DOI: 10.1002/14651858. CD004394.

Miller JW, Garrod MG, Rockwood AL, Kushnir MM, Allen LH, Haan MN, Green R (2006). Measurement of total vitamin B12 and holotranscobalamin, singly and in combination, in screening for metabolic vitamin B12 deficiency. Clin. Chem., 52: 278-285.

Mischoulon D, Burger JK, Spillmann MK, Worthington JJ, Fava M, Alpert JE (2000). Anaemia and macrosytosis in the prediction of serum folate and vitamin B12 status, and treatment outcome in major depression. J. Psychosom. Res., 49: 183-187.

Morkbak AL, Heimdal RM, Emmens K (2005). Evaluation of the technical performance of novel holotranscobalamin (holoTC) assays in a multicenter European demonstration project. Clin. Chem. Lab. Med., 43: 1058-1064.

Morris MS, Fava M, Jacques PF, Selhub J, Rosenberg IH (2003). Depression and folate status in the US population. Psychother. Psychosom., 72: 80-87.

Parnetti L, Bottiglieri T, Lowenthal D (1997). Role of homocysteine in age-related vascular and non-vascular diseases. Ageing, 9: 241-257.

Penninx BH, Guralnik JM, Ferucci L, Fried LP, Allen RH, Stabler SP (2000). Vitamin B12 deficiency and depression in physically disabled older women: Epidemiologic evidence from the women's health and aging study. Am. J. Psychiatry, 157: 715-721.

Planells, E, Sanchez C, Montellano MA, Mataix J, Llopis J (2003). Vitamin B6 and B12 and folate status in an adult Mediterranean population. Eur. J. Clin. Nutr., 57: 777-785.

Quinn K, Basu TK (1996). Folate and vitamin B12 status of the elderly. Eur. J. Clin. Nutr., 50: 340-342.

Shams M, Homayouni K Omrani GR (2009). Serum folate and vitamin B12 status in healthy Iranian adults. EMRO, 15: 1285-1292.

Stabler SP, Allen RH, Fried LP, Pahor M, Kittner SJ, Penninx BW, Guralnik JM (1999). Racial differences in prevalence of cobalamin and folate deficiencies in disabled elderly women. Am. J. Clin. Nutr., 70: 911-919

Wesson VA, Levitt AJ, Joffe RT (1994). Change in folate status with antidepressant treatment. Psychiatry Res, 53: 313-322.

Yajnik, SCS, Deshpande SS, Lubree HG, Naik SS, Bhat DS, Uradey BS, Deshpande JA, Refusm H (2006). Vitamin B12 deficiency and hyperhomocysteinemia in rural and urban Indians. J. Assoc. Physicians India, 54: 775-782.

9

Measurement of cholesterol sub-fractions, high density lipoprotein 2 and high density lipoprotein 3, in HIV infected patients treated with highly active antiretroviral therapy in Burkina Faso

Jean Sakandé[1]*, Josiane B Kaboré[1], Elie Kabré[1], Boblwendé Sakandé[2] and Mamadou Sawadogo[1]

[1]Laboratory of Biochemistry, University of Ouagadougou, Burkina Faso.
[2]Laboratory of Philadelphie Clinic of Ouagadougou, 09 BP 863 Ouagadougou, Burkina Faso.

This paper described dyslipidemia caused by the metabolic effects of HIV and exposure to antiretroviral therapy. The aim of this study was to appreciate the interest of the measurement of HDL sub-fractions in the monitoring of human immunodeficiency virus (HIV) infected patients on antiretroviral treatment. A case-control study was carried out on 31 HIV infected before (naive) highly active antiviral therapy (HAART naive), 33 HIV infected on HAART during one year (HAART 1 Year), 47 HIV infected on HAART between one year to five years (HAART 1 to 5 Years) and control group (43 HIV negative). Total HDL (high density lipoprotein), HDL2 and HDL3 were determined by using dextran sulfate - $MgCl_2$ precipitation technique in human serum. The lipid profile showed that HDL2 significantly decreased according to the rate of CD4 cells. Significant decreases of total cholesterol, HDL, HDL3 and HDL2 were observed among patients at advanced WHO CDC clinical stage. Correlation analysis showed decrease of HDL2 with age in HIV negative group ($r = -0.401$, $p = 0.008$). This correlation was not significant in HIV infected group ($r = -0.134$, $p = 0.162$). Obviously, HDL2 was inversely correlated than total HDL with BMI ($r = -0.229$, $p = 0.001$ versus $r = -0.292$, $p = 0.001$). HDL2 was positively correlated with CD4 lymphocytes ($r = 0.322$, $p = 0.004$) and duration of antiretroviral treatment ($r = 0.347$, $p = 0.002$). As a result, we have found a significant increase of HDL2 associated with a lower risk of cardiovascular diseases in patients infected with HIV-1 which were treated with a regimen including Nevirapine. If confirmed in larger studies, this finding may influence the initial choice of therapy for HIV-1 infection, and might also lead to novel approaches targeted at raising HDL2 cholesterol for cardiovascular diseases prevention.

Key words: Acquired immunodeficiency syndrome, nevirapine, dyslipidemia, cardiovascular disease.

INTRODUCTION

HIV-infected patients are at increased risk of cardiovascular disease (CVD), reflecting interaction of risk associated with host, virus, and antiretroviral therapy factors (Currier, 2009). Among the causes, dyslipidemia led by the metabolic effects due to the cytopathogenicity of HIV was described (Kanjanavanit et al., 2011). Then, after introduction of highly active antiretroviral therapy (HAART), cohort studies showed associations between exposure to antiretroviral therapy and an increased risk of myocardial infarction (Friis-Moller et al., 2003). Effects of HAART such as dyslipidemia and insulin resistance on increased coronary diseases in HIV are well established (Rossi et al., 2009). Indeed, HIV infection is associated with endothelial dysfunction, hypercoagulability, vascular damage and inflammation with higher level of high sensitivity C reactive protein (hsCRP) compared to

---

*Corresponding author. E-mail: jsakande@gmail.com.

negative controls (Huang et al., 2009; Francisci et al., 2009; Soro-Paavonen et al., 2006; van Vonderen et al., 2009). These observations coupled with changes in platelet reactivity contribute to atheromatosis. The invert relation between the cardiovascular accident and the concentration of HDL, particularly its subfraction HDL2, is well established (Maeda et al., 2012). Several methods of measurement of HDL subclasses were proposed and many authors have demonstrated the interest of HDL subclasses in the monitoring of metabolic diseases as diabetes (Superko et al., 2009; Shuhei et al., 2010). The aim of this study is to appreciate the interest of HDL sub-fractions in the monitoring of HIV infected patients on antiretroviral treatment in Burkina Faso.

MATERIALS AND METHODS

Type and period of the study

A case-control study was carried out from March 2010 to May 2011 in Ouagadougou, capital city of Burkina Faso (West Africa). HIV infected subjects were recruited at Pissy Medical Center of Ouagadougou and HIV negative subjects (control group) at the National Center of Blood Transfusion. Biological analyses were performed at the laboratory of Philadelphie Private Clinic of Ouagadougou.

Study population

Study population was 154 subjects (111 HIV infected subjects and 43 HIV negative) with average age of 31 ± 8.6 years in HIV negative group versus 38.8 ± 9.2 in HIV infected group. The 111 HIV infected subjects (41 men and 70 women, sex ratio: 0.58) were classified: 31 HIV infected naive of antiretroviral treatment (HAART naive), 33 HIV infected on antiretroviral treatment for one year (HAART1 Year), and 47 HIV infected on antiretroviral treatment between one year to five years (HAART1-5 Years). The control group comprised 43 healthy subjects (24 men and 19 women, sex ratio: 1.26) who showed no serological evidence for HIV and/or HBV, HCV (HIV negative). All subjects were recruited in accordance with the Helsinki Declaration, and the National Ethics Committee approved the protocol; in which, they had to complete a consent form before their participation in the study. Also, subjects must not be enlisted in another concomitant study. The following criteria of exclusion were defined: diabetes, coronaropathy (cardiac events), treatment with any lipid medication (statines, nicotinic acid, fibrate, resins), smokers and drinkers. The patients who had no good adherence to the treatment and patients with incomplete data were excluded. The patients following an antiretroviral treatment for more than 5 years were also excluded because of the risk of antiretroviral regimen switching in this group.

Methods

After an overnight fast, venous blood was collected on a dry tube for biochemical analyses and an EDTA tube for CD4 lymphocytes count. Serums were separated by centrifugation at 3000 g for 10 min at 4°C, stored at -80°C and analyzed within a week. The CD4 lymphocytes count was carried out with a flow cytometer BD-Fascount (California, USA). Serum total Cholesterol (TC) and triglycerides (TG) were determined with an automated KONELAB 20i analyzer (Thermo Electron Corporation, Vantaa, Finland) by

fully enzymatic methods (Thermo kits 981812 and 981301 respectively). The dual-step precipitation of HDL subfractions was performed according to the procedure described by Hirano (2008) with light modifications. To isolate total HDL-C by precipitation, a combined precipitant consisting of 100 µl (0.02 mmol/L) of dextran sulfate (Mr 500000, SIGMA, France) and 25 µl (200 mmol/L) of $MnCl_2$ ($MgCl_2$-$6H_2O$, MERCK, France) was added to 1 ml of serum. After 15 min of standing at room temperature, the mixture was centrifuged at 3,400 g for 20 min at 4°C. Aliquots of the resulting supernatant (S1) were taken for the assay of the HDL-C and precipitation of the HDL2. The HDL2 was precipitated by a combined precipitant consisting of 100 µl (0.02 mmol/L) of dextran sulfate (Mr 500000, SIGMA, France) and 50 µl (200 mmol/L) of $MnCl_2$ ($MgCl_2$-$6H_2O$, MERCK, France) added to 500 µl of supernatant (S1). After 2 h at room temperature, the mixture was centrifuged at 3,400 g for 20 min at 4°C. Aliquots of the resulting supernatant (S2) were taken for the assay of the HDL3-C. The measured value for total HDL-C was multiplied by 1.125 and that for HDL_3 was multiplied by 2.92 to correct for dilution by the reagents. HDL_3-C was measured by the direct HDL-C homogenous assay instead of the original TC assay. The sub-fraction HDL2 was calculated by the following formula: cholesterol HDL2 = Cholesterol HDL-cholesterol HDL3.

We calculated the LDL-Cholesterol (in mmol/L) by using the Friedewald formula: LDL-C = TC − HDL-C − TG/2.2 (Srisawasdi et al., 2011). Our laboratory is standardized for the determination of TC, TG, HDL-C, HDL2 and HDL3 by the National External Quality Assessment Program. The accuracy and precision of the measurements during the study were within the acceptable criteria of literature (Hirano et al., 2008).

Statistical analysis

The quantitative variables were expressed on means ± standard deviation and the qualitative variables in percentages. The Analysis of Variance (ANOVA) was used to determine quantitative variables with normal distribution, followed by the Bonferonni multiple comparisons test to compare the means between groups. Mann-Whitney test was used to compare the non-parametric distribution. Correlations were evaluated by Spearman's rank correlation analysis. The statistical analysis was performed using the statistical software PASW, version 18 for Windows (SPSS CPSC., Chicago, USA). Probability levels of 0.05 or less were considered significant.

RESULTS

The demographic and clinical characteristics presented in Table 1 showed that study population was hetero-geneous in age (31 ± 8.6 in HIV negative group versus 38.8 ± 9.2 in HIV infected group). High frequency of women was observed among HIV infected subjects (sex ratio: 0.58). The infected men were 44.3 ± 7.2 years old against 35.8 ± 8.8 years old in women ($p < 0.001$). The average body mass index (BMI) was 23.3 ± 3.9 kg/m² in HIV infected group versus 24.5 ± 3.8 in control group. The distribution of infected subjects according to the type of virus showed high frequency of HIV1 with 106 subjects (95.5%) against 3 subjects infected by HIV2 (2.7%) and 2 co-infected subjects by HIV1 and 2 (1.8%). The distribution according to CD4 count reported 18 subjects (16.2%) with CD4 lymphocytes lower than 200 cells/µl, 70 subjects (63.1%) with CD4 lymphocytes included

Table 1. Demographic and clinical characteristics of the population of study.

Demographic characteristics	Variable	HIV negative	HIV positive	HAART naive	HAART ≤ 1 year	HAART1-5 years
Sex (n)	men	24	41	7	8	25
	women	19	70	24	25	22
	Sex Ratio	1.26	0.58	0.29	0.32	1.13
	Total	43	111	31	33	47
Age (years)	men	34 ± 9	44.3 ± 7.2	43 ± 7	42 ± 10	45 ± 6
	women	28 ± 7	35.8 ± 8.8	35 ± 10	37 ± 11	36 ± 5
	Total	31.0 ± 8.6	38.8 ± 9.2	36.8 ± 9.8	37.9 ± 10.4	41 ± 7.6
Types of HIV	HIV 1	-	106 (95.5)	28 (90.3)	32 (97)	46 (97.9)
	HIV 2	-	3 (2.7)	02 (6.5)	-	01 (2.1)
	HIV 1+2	-	2 (1.8)	01 (3.2)	01 (3)	-
BMI	Total	24.5 ± 3.8	23.3 ± 3.9			
CD4 (cell μl^{-1})	<200	-	18 (16.2)	07 (22.6)	06 (18.2)	05 (10.6)
	200 - 499	-	70 (63.1)	20 (64.5)	26 (78.8)	24 (51.1)
	> 500	-	23 (20.7)	04 (12.9)	01 (3)	18 (38.3)
WHO Staging	Clinical stage 1	-	90 (81.1)	22 (71.0)	24 (72.7)	44 (93.6)
	Clinical stage 2	-	14 (12.6)	05 (16.1)	06 (18.2)	03 (6.4)
	Clinical stage 3	-	7 (6.3)	04 (12.9)	03 (9.1)	-
Therapies	AZT/3TC/NVP	-		-	27 (81.8)	37 (78.7)
	AZT/3TC/EFV	-		-	03 (9.1)	02 (4.3)
	D4T/3TC/NVP	-		-	02 (6.1)	05 (10.6)
	TDF/3TC/NVP	-		-	01 (3)	01 (2.1)
	AZT/3TC/LPVRITO	-		-	-	02 (4.3)

AZT: Zidovudine; D4T: Stavudine; 3TC: Lamivudine; EFV: Efavirenz; TDF: Tenofovir; LPV-RITO: Lopinavir-Ritonavir; NVP: Nevirapine.

between 200 and 499 cells/μl, and 23 subjects (20.7%) with CD4 lymphocytes up to 500 cells/μl. According to the WHO-CDC clinical staging, 90 subjects (81.1%) were at stage 1, 14 subjects (12.6%) at stage 2 and 7 subjects (6.3%) at stage 3. Concerning the therapeutic antiretroviral protocol followed, 78 subjects (97.5%) were treated with 2 NRTIs (Nucleoside Reverse Transcriptase Inhibitors) + 1 NNRT (Non Nucleoside Reverse Transcriptase Inhibitor), 2 subjects (2.5%) treated with 2 NRTIs + 1 PI (Protease Inhibitor). The majority (80%) followed a treatment combining Zidovudine + Lamivudine + Nevirapine (AZT + 3TC + NVP).

The total cholesterol level (Table 2) was not significantly different between controls and infected subjects (p = 0.288). However, in the group on HAART between one and five years (HAART1-5Years), total cholesterol level was significantly higher in men (p = 0.004). In this group, values of HDL was significantly higher (P=0.001) and particularly among women (P=0.001). On the contrary, the LDL cholesterol was low in HAART1-5Years group (P=0.003) and the decrease was more remarkable among women (P=0.001). High values of triglycerides were also observed among men of HAART1-5Years group (P=0.007). HDL3 was not significantly increased in patients on HAART compare to HIV negative and HAART naive groups (P=0.052). The HDL2 level in HAART naive group was significantly lower compared to control group and patients on HAART (P=0.001). Particularly, HDL2 level was very lower in men of HAART naive group (0.17±0.08 mmol/L) than men of control group (0.45±0.29 mmol/L). Results showed an increase of HDL2 with HAART. Particularly in women, HDL2 level was twice higher in HAART1-5 Years group (1.24±0.64 mmol/L) as compared to 0.62±0.61 mmol/L in men (P=0.001). The lipid profile recorded in Table 3 showed only HDL2 significant increase according to the

Table 2. Blood lipid profile in HIV patients and controls (mmol/L).

Lipid	Variable	HIV negative (n=43)	HAART naive (n=31)	HAART ≤ Year (n=33)	HAART1-5Years (n=47)	P
Total cholesterol	Population	4.69 ± 0.84	4.72 ± 1.61	5.02 ± 1.3	5.27 ± 1.55	0.288
	Men	4.71 ± 0.94	4.16 ± 0.76	5.35 ± 1.39	5.88 ± 1.60	0.004
	Women	4.67 ± 0.72	4.89 ± 1.76	4.90 ± 1.27	4.56 ± 1.16	0.809
	P	0.883	0.301	0.401	0.003	
LDL	Population	3.23 ± 0.8	3.49 ± 1.51	3.52 ± 1.28	3.22 ± 1.61	0.301
	Men	3.26 ± 0.86	3.10 ± 0.80	3.94 ± 1.61	4.05 ± 1.54	0.569
	Women	3.20 ± 0.74	3.60 ± 1.65	3.38 ± 1.16	2.26 ± 1.10	0.003
	P	0.821	0.448	0.287	0.001	
Triglycerides	Population	1.10 ± 0.43	1.18 ± 0.62	0.97 ± 0.45	1.37 ± 1.28	0.057
	Men	1.09 ± 0.40	0.89 ± 0.24	1.2 ± 0.70	1.84 ± 1.61	0.62
	Women	1.10 ± 0.47	1.27 ± 0.68	0.90 ± 0.30	0.85 ± 0.32	0.012
	P	0.965	0.169	0.102	0.007	
HDL	Population	1.23 ± 0.3	1.00 ± 0.45	1.31 ± 0.5	1.78 ± 0.73	0.001
	Men	1.19 ± 0.27	0.87 ± 0.20	1.20 ± 0.41	1.46 ± 0.67	0.03
	Women	1.28 ± 0.34	1.04 ± 0.50	1.34 ± 0.53	2.21 ± 0.63	0.001
	P	0.331	0.373	0.505	0.001	
HDL 2	Population	0.50 ± 0.28	0.32 ± 0.30	0.44 ± 0.37	0.91 ± 0.69	0.001
	Men	0.45 ± 0.29	0.17 ± 0.08	0.39 ± 0.27	0.62 ± 0.61	0.09
	Women	0.56 ± 0.28	0.37 ± 0.33	0.45 ± 0.40	1.24 ± 0.64	0.001
	P	0.216	0.133	0.671	0.001	
HDL 3	Population	0.76 ± 0.18	0.68 ± 0.32	0.88 ± 0.3	0.87 ± 0.18	0.052
	Men	0.76 ± 0.20	0.69 ± 0.20	0.81 ± 0.34	0.84 ± 0.20	0.368
	Women	0.76 ± 0.15	0.67 ± 0.36	0.89 ± 0.30	0.88 ± 1.15	0.016
	P	0.941	0.89	0.525	0.509	

Table 3. Blood lipids variation with CD4 cells count and clinical staging.

Lipids (mmol/L)	CD4 count (cells/µl)			P
	<200 (n=18)	200–499 (n=70)	>500 (n=23)	
Total cholesterol	4.48 ± 1.32	5.00 ± 1.60	5.15 ± 1.27	0.186
LDL	3.01 ± 1.19	3.21 ± 1.63	3.30 ± 1.23	0.932
HDL	1.25 ± 0.61	1.43 ± 0.65	1.55 ± 0.80	0.171
HDL3	0.79 ± 0.27	0.84 ± 0.30	0.78 ± 0.22	0.803
HDL2	0.46 ± 0.20	0.59 ± 0.55	0.77 ± 0.70	0.045
Triglycerides	1.05 ± 0.29	1.14 ± 0.78	1.54 ± 1.52	0.081

rate of CD4 cells. Significant decrease of total cholesterol, HDL, HDL3 and HDL2 was observed among patients at WHO Clinical stage 3 (Table 4). Correlation analysis showed decrease of HDL2 with age in HIV negative group (r=-0.401, P=0.008). This correlation was not significant in HIV infected group (r=-0.134, P=0.162). HDL2 was inversely correlated than total HDL with BMI (r=-0.229, P=0.001 versus r=-0.292, P=0.001). HDL2 was positively correlated with CD4 lymphocytes (r=0.322, P=0.004) and duration of antiretroviral treatment

Table 4. Blood lipids variation with clinical staging.

Lipids (mmol/L)	WHO CDC clinical staging			P
	Stage 1 (n=90)	Stage 2 (n=14)	Stage 3 (n=07)	
Total cholesterol	5.21 ± 1.55	4.56 ± 0.81	3.90 ± 1.34	0.01
LDL	3.45 ± 1.57	3.17 ± 1,06	2.93 ± 1.19	0.458
HDL	1.51 ± 0.65	1.19 ± 0.81	0.79 ± 0.31	0.002
HDL3	0.85 ± 0.27	0.74 ± 0.32	0.60 ± 0.30	0.01
HDL2	0.66 ± 0.57	0.46 ± 0.20	0.19 ± 0.09	0.01
Triglycerides	1.25 ± 1.03	0.99 ± 0.42	1.06 ± 0.13	0.367

(r=0.347,P=0.002).

DISCUSSION

The most common lipid profile in untreated HIV patients reported by literature (Ngoundi et al., 2007) is an elevated triglycerides level and decreased HDL cholesterol. However, in our study, only HDL2 significant decrease was observed in HAART naive group compared to control group. This rarity of signs of dyslipidemia in untreated subjects in our study could be due to the context of malnutrition in which those patients live. Indeed, those patients are particularly vulnerable, in a Sahelian country like Burkina Faso. This dyslipidemia is increased on the other hand in treated HIV patients in particular in the group treated between one to 5 years (HAART1-5Years). According to the literature, the total cholesterol level observed in this study increased with duration of antiretroviral treatment (Arun et al., 2011; Riddler et al., 2003). Total cholesterol level was higher in men on antiretroviral treatment as reported in Nigeria (Awah et al., 2011). The low level of total cholesterol in women can be explained by the effect of estrogens in lipid metabolism (Ratiani et al., 2011). An increase of triglycerides was also observed in men of HAART1-5 group (p = 0.007). This hypertriglyceridemia in HIV infected subjects can be explained by a decrease of the catabolism of triglyceride (Ware et al., 2009). The increase in triglycerides is usually severe with PIs (Acosta et al., 2002; Haubrich et al., 2009; Hicks et al., 2006; Walmsley et al., 2002).

An increase of HDL was observed with the antiretroviral treatment and with its duration in accordance to literature (Alonso-Villaverde et al., 2005; Ngoundi et al., 2007; Tashim et al., 2000; Troll et al., 2011). This could be accounted for by the fact that the majority of the subjects in this study were treated with NNRT based HAART (AZT / 3TC / NVP). Indeed, previous studies showed that Nevirapine, an NNRT, increased the rate of HDL by 16 to 49% (Murphy et al., 2000; Smith et al., 2009; van der Valk et al., 2001). Nevirapine increases High-Density Lipoprotein Cholesterol concentration by stimulation of Apolipoprotein A-I production (Franssen et al., 2009). HDL2 increase was also observed in antiretroviral treated

group and its level was higher in HAART1-5 group (p = 0.001). HDL2 was positively correlated with CD4 lymphocytes (r=0.322, P=0.004) and duration of antiretroviral treatment (r=0.347, P=0.002). HDL2 was inversely correlated than total HDL-C with BMI (r=-0.229, P=0.001 versus r=-0.292, P=0.001). The increase of HDL3 in treated group was not significant. These results showed that Nevirapine combining treatment increases only HDL2 fraction of High-Density Lipoprotein Cholesterol.

Conclusion

In HIV-1 infected patients treated with regimen including nevirapine, we found more significant increase of HDL2 which is associated with a decrease in risk for cardiovascular diseases. If confirmed in larger studies, this finding may influence the initial choice of therapy for HIV-1 infection and might lead to novel approaches targeted at raising HDL2 cholesterol for cardiovascular diseases prevention.

REFERENCES

Acosta EP (2002). Pharmacokinetic enhancement of protease inhibitors. J. Acquir. Immune Defic. Syndr. 29(1):11–8.

Alonso-Villaverde C, Coll B, Gómez F, Parra S, Camps J, Joven J, Masana L (2005). The efavirenz-induced increase in HDL-cholesterol is influenced by the multidrug resistance gene 1 C3435T polymorphism. AIDS. 19(3):341-2.

Arun K, Brajesh S (2011). Assessment of lipid profile in patients with human immunodeficiency virus (HIV/AIDS) without antiretroviral therapy. Asian Pacific J. Trop. Dis. (5):24-27.

Awah, Francis M, Onyinye A (2011). Effect of highly active anti-retroviral therapy (HAART) on lipid profile in a human immunodeficiency virus (HIV) infected Nigerian population. Afri. J. Biochem. Res. 5(9):282-286.

Currier JS (2009). Update on cardiovascular complications in HIV infection. Top HIV Med. 17(3):98-103.

Francisci D, Giannini S, Baldelli F (2009). HIV type 1 infection, and not short-term HAART, induces endothelial dysfunction. AIDS 23:589–96.

Franssen R, Sankatsing RR, Hassink E, Hutten B, Ackermans MT, Brinkman K, Oesterholt R, Arenas-Pinto A, Storfer SP, Kastelein JJ, Sauerwein HP, Reiss P, Stroes ES (2009).Nevirapine increases high-density lipoprotein cholesterol concentration by stimulation of apolipoprotein A-I production. Arterioscler Thromb. Vasc. Biol. 29(9):1336-41.

Friis-Moller N, Sabin CA, Weber R (2003). Combination antiretroviral therapy and the risk of myocardial infarction. N. Engl. J. Med. 349:1993–2003.

Haubrich RH, Riddler SA, DiRienzo AG (2009). Metabolic outcomes in a randomized trial of nucleoside, nonnucleoside and protease inhibitor-sparing regimens for initial HIV treatment. AIDS 23:1109–18.

Hicks CB, Cahn P, Cooper DA (2006). Durable efficacy of tipranavir-ritonavir in combination with an optimised background regimen of antiretroviral drugs for treatment-experienced HIV-1-infected patients at 48 weeks in the randomized evaluation of strategic intervention in multi-drug resistant patients with Tipranavir (RESIST) studies: An analysis of combined data from two randomised open-label trials. Lancet 368:466–75.

Hirano T, Nohtomi K, Koba S, Muroi A, Ito Y (2008).A simple and precise method for measuring HDL-cholesterol subfractions by a single precipitation followed by homogenous HDL-cholesterol assay. J. Lipid Res. 49(5):1130-6.

Huang RC, Mori TA, Burke V, Newnham J, Stanley FJ, Landau LI, Kendall GE, Oddy WH, Beilin LJ (2009). Synergy between adiposity, insulin resistance, metabolic risk factors and inflammation in adolescents. Diabetes Care 32:695–701.

Kanjanavanit S, Puthanakit T, Vibol U, Kosalaraksa P, Hansudewechakul R, Ngampiyasakul C, Wongsawat J, Luesomboon W, Wongsabut J, Mahanontharit A, Suwanlerk T, Saphonn V, Ananworanich J, Ruxrungtham K (2011).High prevalence of lipid abnormalities among antiretroviral-naive HIV-infected Asian children with mild-to-moderate immunosuppression. Antivir. Ther. 16(8):1351-5.

Maeda S, Nakanishi S, Yoneda M, Awaya T, Yamane K, Hirano T, Kohno N (2012). Associations between Small Dense LDL, HDL Subfractions (HDL2, HDL3) and Risk of Atherosclerosis in Japanese-Americans. J. Atheroscler. Thromb. 25, 19(5):444-52.

Murphy RL, Sanne I, Cahn P (2003). Dose-ranging, randomized, clinical trial of atazanavir with lamivudine and stavudine in antiretroviral-naive subjects: 48-week results. AIDS 17:2603–14.

Ngoundi, JL, Etame SHL, Fonkoua M, Yangoua H and Oben J (2007). Lipid profile of infected patients treated with highly active antiretroviral therapy in cameroon. J. Med. Sci. 7:670-673.

Ratiani L, Parkosadze G, Koptonashvili L, Ormotsadze G, Sulaqvelidze M, Sanikidze T (2011). Correlation of atherogenetic biomarkers and estradiol changes in post-menopause. Georgian Med. News (195):100-5.

Riddler SA, Smit E, Cole SR (2003). Impact of HIV infection and HAART on serum lipids in men. JAMA 289:2978–82.

Rossi R, Nuzzo A, Guaraldi G, Squillace N, Orlando G, Esposito R, Lattanzi A, Modena MG (2009). Metabolic disorders induced by highly active antiretroviral therapy and their relationship with vascular remodeling of the brachial artery in a population of HIV-infected patients. Metabolism 58(7):927-33.

Shuhei N, Söderlund S, Jauhiainen M, Taskinen MR (2010). Effect of HDL composition and particle size on the resistance of HDL to the oxidation. Lipids Health Dis. 9(1):104–114.

Significantly Improved Cardiovascular Risk Profile In Women Who Switch From A Protease Inhibitor To Efavirenz. Antiviral Therapy 5:77-91.

Smith KY, Patel P, Fine D (2009). Randomized, double-blind, placebo-matched, multicenter trial of abacavir/lamivudine or tenofovir/emtricitabine with lopinavir/ritonavir for initial HIV treatment. AIDS 23:1547–56.

Soro-Paavonen A, Westerbacka J, Ehnholm C, Taskinen MR (2006). Metabolic syndrome aggravates the increased endothelial activation and low-grade inflammation in subjects with familial low HDL. Ann. Med. 38:229–238.

Srisawasdi P, Chaloeysup S, Teerajetgul Y, Pocathikorn A, Sukasem C, Vanavanan S, Kroll MH(2011). Estimation of plasma small dense LDL cholesterol from classic lipid measures. Am. J. Clin. Pathol. 136(1):20-9.

Superko H Robert A. (2009). Advanced Lipoprotein Testing and Subfractionation Are Clinically Useful. Circulation 119:2383-2395.

Tashim KT, Flynn MM, Bausserman L, Spigno MG Di and Carpenter CCJ (2000). The efavirenz-induced increase in HDL-cholesterol is influenced by the multidrug resistance gene 1 C3435T polymorphism. AIDS. 2005. 19(3):341-2.

Troll JG (2011). Approach to Dyslipidemia, Lipodystrophy, and Cardiovascular Risk in Patients with HIV Infection. Curr. Atheroscler. Rep. 13:51–56.

van der Valk M, John J, Kasteleina P, Robert L, Murphy B, van Leth F, Katlamac C(2001). Nevirapine-containing antiretroviral therapy in HIV-1 infected patients results in an anti-atherogenic lipid profile. AIDS 15: 2407-2414.

Van Vonderen MG, van Agtmael MA, Hassink EA (2009). Zidovudine/lamivudine for HIV-1 infection contributes to limb fat loss. PLoS One 4:5647–50.

Walmsley S, Bernstein B, King M (2002). Lopinavir-ritonavir versus nelfinavir for the initial treatment of HIV infection. N. Engl. J. Med.346:2039–46.

Ware LJ, Jackson AG, Wootton SA, Burdge GC, Morlese JF, Moyle GJ, Jackson AA, Gazzard BG (2009). Antiretroviral therapy with or without protease inhibitors impairs postprandial TAG hydrolysis in HIV-infected men. Br. J. Nutr.1038-46.

10

Adsorption, metabolism and degradation of erythromycin in giant freshwater prawn and tilapia aquaculture in Mekong River Delta

N. P. Minh*, T. B. Lam and T. T. D. Trang

Department of Food Technology, Ho Chi Minh City University of Technology, 268 Ly Thuong Kiet Street, District 10, Ho Chi Minh City, Vietnam.

Adsorption, metabolism and degradation of erythromycin in freshwater prawn and tilapia aquaculture in Mekong River Delta were monitored and evaluated. They were fed practical diets medicated with erythromycin (50 and 100 mg. kg^{-1} body weight for 7 days). Erythromycin residues in their muscle were determined by the liquid chromatography - mass spectrometry/ mass spectrometry (LC-MS/MS) method. Our study provided preliminary data for a more prudent use of erythromycin in giant freshwater prawn and tilapia, suggesting a possible withdrawal time after treatment as well as clearing away the awareness of forming and accumulating a harmful over-threshold level of derived products from parental drug during veterinary usage in aqua culture.

Key words: Giant freshwater prawn, tilapia, erythromycin, metabolism, degradation, LC-MS/MS.

INTRODUCTION

Erythromycin is a macrolide antibiotic that is produced by the actinomycete species, *Streptomyces erythreus*. The chemical structures of erythromycin A (EA), which is the major component of erythromycin base, and its related substances, are depicted in Figure 1. Erythromycin is a polyhydroxylactone that contains two sugars. The aglycone portion of the molecule, erythranolide, is a 14-membered lactone ring. An amino sugar, desosamine, is attached through a β-glycosidic linkage to the C-5 position of the lactone ring. The tertiary amine of desosamine confers a basic character to erythromycin (pK_a 8.8). Through this group, a number of acid salts of the antibiotic have been prepared. A second sugar, cladinose, which is unique to erythromycin is attached via a β-glycosidic linkage to the C-3 position of the lactone ring.

The fermentation process that produces commercial grade erythromycin is not entirely selective. It results in the production of small quantities of erythromycin B (EB), C (EC), D(ED), E(EE) and F(EF), in addition to EA, which is the major component. EB, EC and EE are the most important impurities found in commercial samples of erythromycin (Table 1).

In addition to the related substances, the metabolite, demethylerythromycin (dMeE), and acidic and basic degradation products are also present in small quantities in commercial samples of erythromycin. These include erythromycin enol ether (EEE), anhydroerythromycin (AE), erythrolosamine (ESM), pseudoerythromycin A hemiketal (psEAHK), pseudoerythromycin A enol ether (psEAEE) and dehydroerythromycin (DE). Other related substances such as erythromycin A *N*-oxide (EANO), erythromycin oxime (EOXM) and erythromycylamine also exist and are structurally very similar, differing by only hydrogen, hydroxyl and/or methoxy groups.

Moreover, erythromycin also exists in forms of erythromy-cin stearate (ES), erythromycin ethylsuccinate (EESC), propionyl erythromycin (PE), erythromycin estolate (EES), erythromycin lactobionate (EL), erythromycin glucoheptonate (EG), erythromycin ethyl carbonate (EEC) and erythromycin acistrate.

Giant freshwater prawn (*Macrobrachium rosenbergii*) and Nile tilapia (*Oreochromis niloticus*) have been considered two of the most important species of freshwater aquaculture in Viet Nam, especially in the Mekong

---

*Corresponding author. E-mail: stapimex@gmail.com.

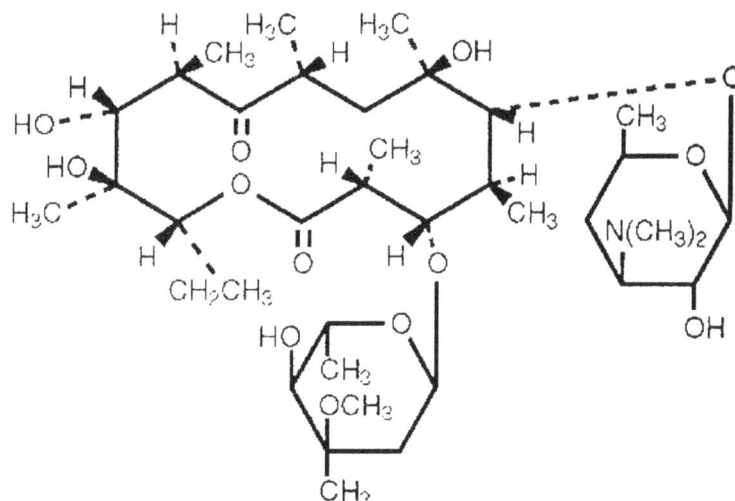

Figure 1. Chemical structures of erythromycin A.

Table 1. Formula of erythromycin A and related substances.

Erythromycin	Formula	Molecular mass	R_1	R_2	R_3	R_4	R_5
A	$C_{37}H_{57}NO_{13}$	734	OH	H	H	OCH_3	CH_3
B	$C_{37}H_{57}NO_{12}$	718	H	H	H	OCH_3	CH_3
C	$C_{38}H_{55}NO_{13}$	720	OH	H	H	OH	CH_3
D	$C_{36}H_{65}NO_{12}$	704	H	H	H	OH	CH_3
E	$C_{37}H_{67}NO_{13}$	748	OH	-O-	OCH_3		CH_3
F	$C_{37}H_{67}NO_{14}$	750	OH	OH	H	CH_3	CH_3

River Delta. Bacterial necrosis is a common disease observed in adult prawns (Winton Cheng and Jiam-Chu Chen, 1998). Bacterial necrosis has variously been termed as 'black spot', 'brown spot', 'shell disease' or chitinolytic bacterial disease. It is caused by the invasion of chitinolytic bacteria, which break down the chitin of the exoskeleton. *Aeromonas hydrophila*, *Aeromonas caviea*, *A. sorbia* and *Aeromonas sp.* were bacterial flora isolated from necrosis prawns (Dat N. T., 2002). *Pseudomonas fluorescens*, *Aeromonas* sp., *Lactococcus garvieae* and *Edwardsiella tarda* were bacteria flora isolated from adult prawns (Ahmed, 2003; Be, 2002; Lalitha and Surendran, 2006; Shih-Chu et al., 2001; Tran et al., 2002). Meanwhile, the most significant diseases in Nile tilapia (*Oreochromis niloticus*) culture are caused by *Streptococcus iniae*, *Aeromonas hydrophila*, *Trichodina sp.*, *Flexibacter*, *Edwardssiella* spp. (Nagla et al., 2005).

The macrolide antibiotic erythromycin has long been the chemotherapeutant of choice to prevent and tackle these pathogenic bacteria. However, there are limited studies being published relating to adsorption, metabolism and degradation of erythromycin in aquatic species. Purpose of this study was to survey the adsorption, metabolism and degradation of erythromycin in giant freshwater prawn muscle (*Macrobrachium rosenbergii*) and Nile tilapia (*Oreochromis niloticus*) after oral administration of the drug given by medicated feed. From that, we can interpolate appropriately tentative withdrawal times. In addition, an evidence of bio-transformative forms of erythromycin, not only in fermentation process but in endo-enzymatic aquaculture pathway as well can be obviously seen.

MATERIALS AND METHODS

Materials

Erythromycin base in white powder and purity 96.5% was purchased from DHG Pharma (Can Tho, Vietnam). Feed and coating agent (squid liver oil) were supplied from Grobest Ltd.

Animals and diet

750 adult giant freshwater prawns (*Macrobrachium rosenbergii*), with an average weight 40 ± 2 g and 120 adult Nile tilapias (*Oreochromis niloticus*) with an average weight 500 ± 5 g were used

for the investigation.

750 adult giant freshwater prawns (*Macrobrachium rosenbergii*) were separated into two groups: Group A (375 prawns) and Group B (375 prawns). Two different diets were prepared for the experimental trial. Group A was treated with 50 mg kg^{-1} prawn body weight day^{-1} for 7 days through medicated feed (water temperature, 28°C). Group B was treated with 100 mg kg^{-1} prawn body weight day^{-1} for 7 days through medicated feed (water temperature, 28°C). 120 adult Nile tilapias (*Oreochromis niloticus*) were divided into Group A (60 tilapias) and Group B (60 tilapia). Two different diets were prepared for the experimental trial. Group A was treated with 50 mg.kg^{-1} tilapia body weight.day^{-1} for 7 days through medicated feed (water temperature, 28°C) while Group B was treated with 100 mg.kg^{-1} tilapia body weight.day^{-1} for 7 days through medicated feed (water temperature, 28°C).

Two groups of medicated feed were conditioned by weighing and mixing feed with erythromycin base at appropriate dosages. Combination between drug and feed was adhesively guaranteed by a coating agent (squid liver oil).

Temperature monitoring

Freshwater prawn and tilapia were poikilothermic species. The optimum metabolic temperature range for them is between 26 and 32°C. Temperature could strongly affect to their survival and enzymatic metabolism, including drug biotransformation. So influence of temperature fluctuation at sampling time was recorded and mentioned in drug metabolism calculation.

Sample collection

Erythromycin base material had been estimated previously to screen and confirm whether other derivatives of erythromycin A, such as erythromycin B, C, D, E and F have been available or not. Sampling times for the prawn and fish in Group A and B were 1, 3, 6, 9 and 23 days after 7 days of the pharmacological treatment. At each sampling time, individuals in each group were sacrificed to confirm erythromycin A residue. Meanwhile, bio-transformation of erythromycin in prawn and tilapia was monitored by screening and confirming derivative forms of erythromycin at the beginning and the end of sampling stage.

Muscle samples in natural proportion were collected, and placed into polyethylene bags, coded and transferred to the laboratory on dry ice, stored at –40°C before analysis.

Analytical procedures

The methodology used for the determination of erythromycin A as well as derivatives of erythromycin in erythromycin base material, in prawn and fish muscle was based on LC-MS/MS. Parameters of measurements: methanol as the extraction solvent; a temperature of 80°C; a pressure of 1500 psi; an extraction time of 15 min; 2 cycles; a flush volume of 150% and a purge time of 300 s.

RESULTS

Adsorption and depletion of erythromycin A

The influence of water temperature on fish metabolism and, consequently, on the drug pharmacokinetics, the time

parameter was also expressed as °C-day. Degree-days were calculated by multiplying the mean daily water temperature by the total number of days at which the temperature was measured to that point.

Prawn

Results of erythromycin depletion at different times in prawn muscle samples treated with 50 mg kg^{-1} (Group A) and 100 mg kg^{-1} (Group B) prawn body weight day^{-1} for 7 days were shown in Table 2 and Figure 2.

Tilapia

Results of erythromycin A depletion at different times in tilapia samples treated with 50 and 100 mg kg^{-1} fish body weight day^{-1} for 7 days were shown in Table 3 and Figure 3.

Determination of withdrawal time and metabolism and degradation of parental drug

The MRL value for erythromycin was set at 30 µg.kg^{-1}, as reported by CFIA (*Canadian Food Inspection Agency*), date 17/11/2009. The regression line and the upper, one-sided tolerance limit (95%) regression line with a confidence of 95% were also traced. This graph had been obtained using the statistical program recommendded by the European Agency for the Evaluation of Medicinal Products (EMEA) and was downloadable from the same EMEA web site (2009).

Prawn

A withdrawal time of 976°C days was interpolated for giant freshwater prawn treatment - Group B (Figure 4). The results for metabolism and degradation of parental drug are shown in Table 4.

Tilapia

A withdrawal time interpolated for tilapia treatment was 908°C-days (Group A) and 1150 °C-days (Group B) (Figures 5 and 6). The results for metabolism and degradation of parental drug are shown in Table 5.

DISCUSSION

Our research was designed in conditions that were quite close to actual aquaculture. The minimum inhibited concentrated of erythromycins A, B, C and D and some of

Table 2. Erythromycin depletion at different times in giant prawn muscles treated with 50 mg kg^{-1} prawn body weight day^{-1} for 7 days.

Time		Erythromycin residue in prawn muscle (μg/kg) [α]	
Day	°C – Day	Group A	Group B
1	28	15.4 ± 3.3	632.4 ± 74.1
3	84	10.6 ± 2.1	199.0 ± 31.2
6	168	5.9 ± 3.1	141.8 ± 3.1
9	252	5.5 ± 4.1	54.2 ± 9.0
23	644	2.8 ± 0.8	31.4 ± 7.5

α Values shown are concentration means ± standard deviations from 5 prawn samples.

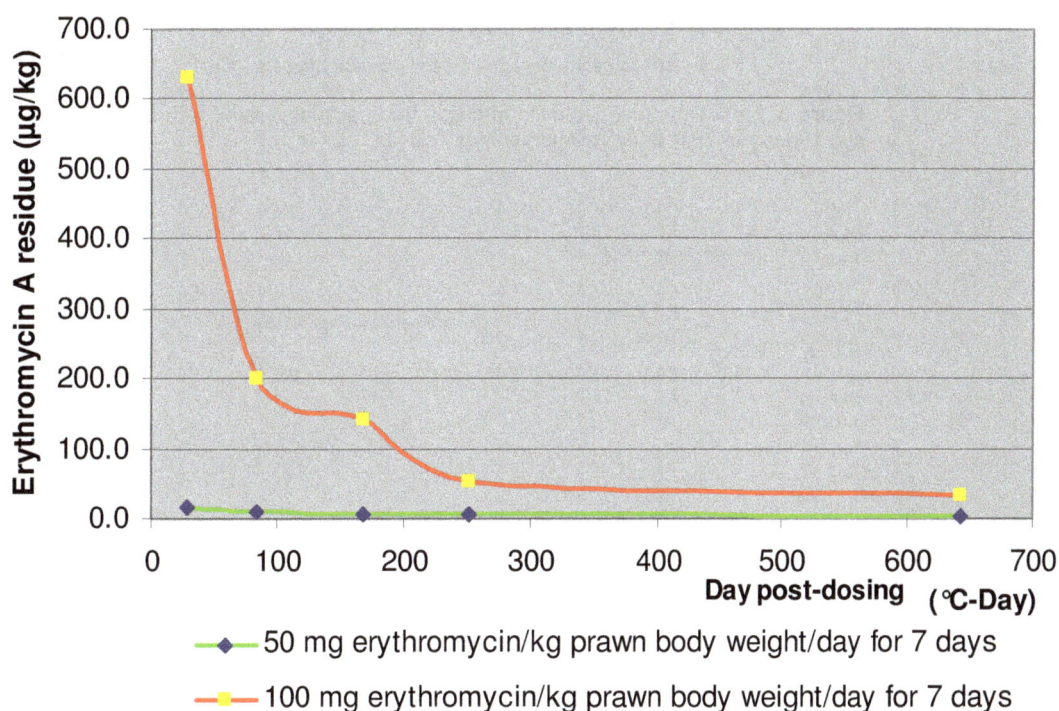

Figure 2. Erythromycin A depletion at different times in giant prawn muscles treated with 50 and 100 mg.kg^{-1} prawn body weight.day^{-1} for 7 day.

Table 3. Erythromycin A depletion at different times in tilapia fillet samples treated with 50 mg.kg^{-1} and 100 mg.kg fish body weight.day^{-1} for 7 days.

Time		Erythromycin A residue in tilapia fillet (μg/kg) [α]	
Day	°C – Day	Group A	Group B
1	28	22,216.0 ± 22,023.0	46,960.0 ± 9,054.7
3	84	13,590.0 ± 14,415.9	14,328.0 ± 18,336.1
6	168	940.8 ± 460.3	6,382.0 ± 5,582.5
9	252	131.4 ± 31.9	379.7 ± 99.3
23	644	34.7 ± 9.6	42.9 ± 17.4

α Values shown are concentration means ± standard deviations from 5 tilapia fillet samples.

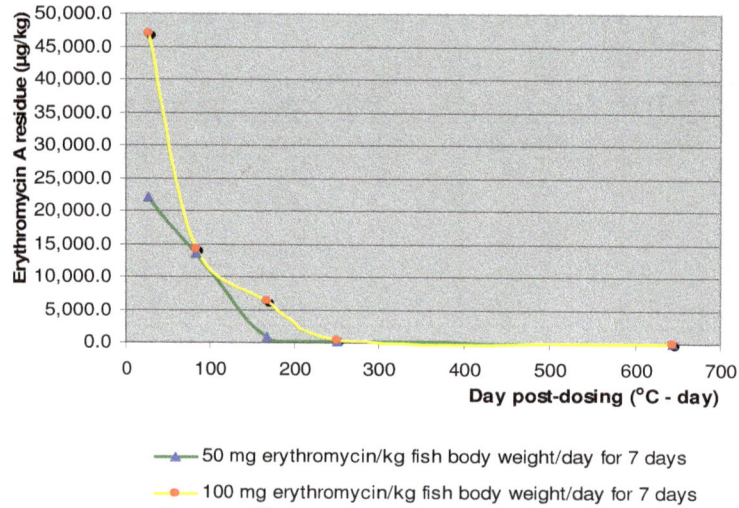

Figure 3. Erythromycin A depletion in tilapia fillet samples treated with 50 and 100 mg.kg^{-1} fish body weight.day^{-1} for 7 days.

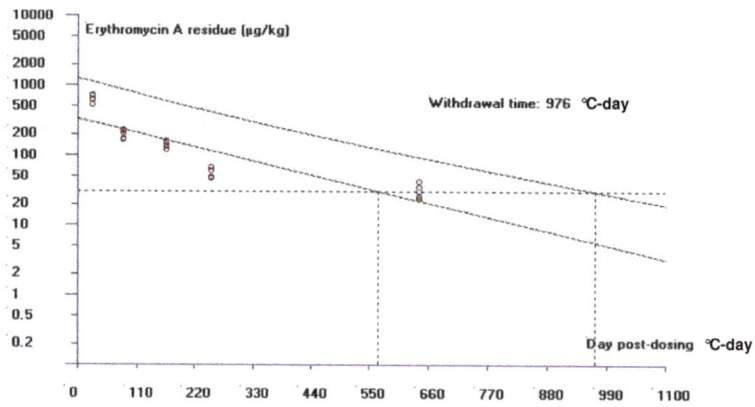

Figure 4. Withdrawal time in muscle prawn treated with erythromycin - Group B.

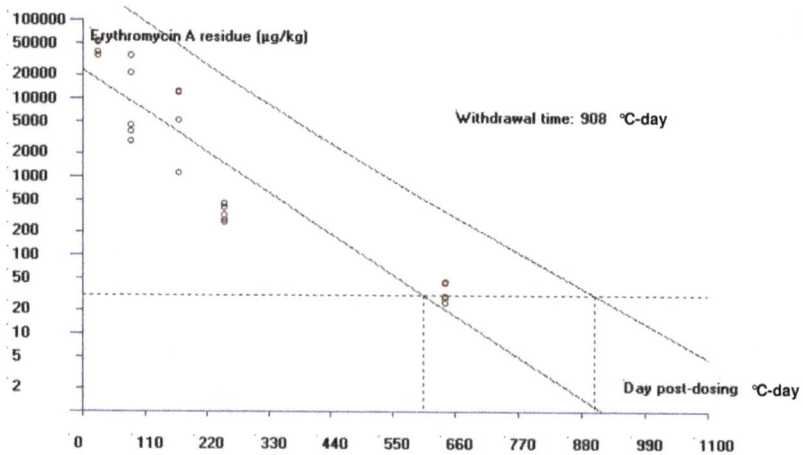

Figure 5. Withdrawal time in muscle tilapia treated with erythromycin - Group A.

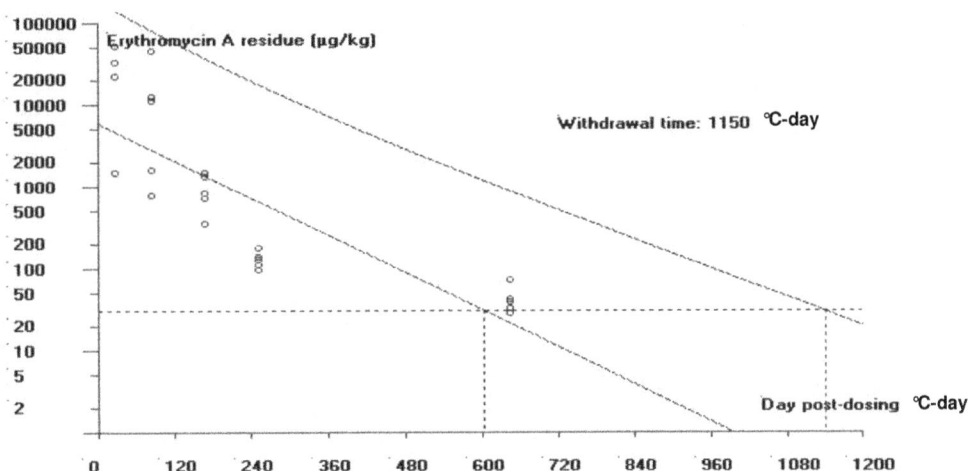

Figure 6. Withdrawal time in muscle tilapia treated with erythromycin – Group B.

Table 4. Degradation of erythromycin at different times in giant prawn muscle samples treated with 50 and 100 mg.kg^{-1} prawn body weight.day^{-1} for 7 days.

Name of Sample	Identification	Test parameter	MDL (µg/kg)	Result (µg/kg)	
				Group A	Group B
Erythromycin base	EBS/0901-RC-002	Erythromycin B	1.0	N.D	N.D
		Erythromycin C	1.0	N.D	N.D
		Erythromycin D	1.0	N.D	N.D
		Erythromycin E	1.0	N.D	N.D
		Erythromycin F	1.0	5.00	5.00
Giant prawn muscle	GP – S1	Erythromycin B	10.0	N.D	N.D
		Erythromycin C	10.0	N.D	N.D
		Erythromycin D	10.0	N.D	N.D
		Erythromycin E	10.0	N.D	2.09
		Erythromycin F	10.0	N.D	N.D
Giant prawn muscle	GP – S5	Erythromycin B	10.0	N.D	N.D
		Erythromycin C	10.0	N.D	N.D
		Erythromycin D	10.0	N.D	N.D
		Erythromycin E	10.0	N.D	5.81
		Erythromycin F	10.0	N.D	3.52

* MDL: Method Detection Limit. ** N. D: Not detected.

concentration of erythromycins A, B, C, and D and some of their derivatives were determined against 21 gram-positive and 15 gram-negative microorganisms. Antibacterial activity was confined to gram-positive and very few gram-negative bacteria. Erythromycin B was somewhat less active than erythromycin A, and erythromycin C and D showed about half that activity or even less. Most other derivatives had negligible activity (Isaac Ongubo Kibwage

et al., 1985).

Prawn

The mean concentration of erythromycin in Group A was lower in comparison with that in Group B. However, the eliminating slope of erythromycin residue in Group B

Table 5. Degradation of erythromycin at different times in tilapia muscle samples treated with 50 and 100 mg.kg^{-1} fish body weight.day^{-1} for 7 days.

Name of sample	Identification	Test parameter	MDL (µg/kg)	Result (µg/kg) Group A	Group B
Erythromycin Base	EBS/0901-RC-002	Erythromycin B	1.0	N.D	N.D
		Erythromycin C	1.0	N.D	N.D
		Erythromycin D	1.0	N.D	N.D
		Erythromycin E	1.0	N.D	N.D
		Erythromycin F	1.0	5.00	5.00
	TL - S1	Erythromycin B	10.0	N.D	N.D
		Erythromycin C	10.0	N.D	131.49
Tilapia Fillet		Erythromycin D	10.0	N.D	N.D
		Erythromycin E	10.0	N.D	258.28
		Erythromycin F	10.0	N.D	N.D
	TL - S5	Erythromycin B	10.0	N.D	N.D
		Erythromycin C	10.0	N.D	N.D
Tilapia Fillet		Erythromycin D	10.0	N.D	N.D
		Erythromycin E	10.0	0.30	6.94
		Erythromycin F	10.0	1.37	5.90

* MDL: Method Detection Limit. ** N. D: Not detected.

was faster than in Group A.

Salmon *Oncorhynchus mykiss*, after its erythromycin administration at 100 mg.kg^{-1} trout body weight.day^{-1} for 21 days through medicated feed (water temperature, 11.5°C) gave a withdrawal time of 255°C days (Annarita Esposito et al., 2007). Salmon *Oncorhynchus tshawytscha* through intraperitoneal injection (William et al., 2006) as well as orally administered erythromycin (Fairgrieve et al., 2005), the mechanism of its retention and depletion was also investigated.

The digestive enzymes of tryptase, pepsin, cellulase, amylase, and metabolic enzymes of alkaline phosphatase (AKP), acid phosphatase (ACP), superoxide dismutase (SOD) and glutathione-S-transferase (GST) were dominated in the hepatopancreas of *M. rosenbergii* (11). Only erythromycin F (5 µg/kg) presented in erythromycin base. During medication at dose 100 mg.kg^{-1} prawn body weight.day^{-1} for 7 days via feed, erythromycin has slightly changed to erythromycin E (2.09 µg/kg) after ceasing drug one day. At day 23 of post-treatment, erythromycin E (5.81 µg/kg) and erythromycin F (3.52 µg/kg) was detected and fortunately was not significant to our concern (Table 4). Considering the fact that biotransformation has been kept erythromycin derivatives at safe residue if we tightly obey recommendation of withdrawal time and drug dosage.

Drug residue levels dropped quickly during the first 3 days after treatment termination, then slowly and steadily until a residue level of < 100 µg/kg, considered a safe

limit by requirements of FDA and the European Community was attained at day 9 of erythromycin withdrawal. However, a longer withdrawal period (35 days of post-treatment) was recommended to ensure complete drug depletion to satisfy CFIA's concern.

Tilapia

The high metabolic rate of furazolidone, AOZ in Nile tilapia was 22 days at least (Weihai Xu et al., 2006). Mean while, a research of accumulation and clearance of florfenicol in tilapia didn't rule out the withdrawal times (P. R. Bowser et al., 2009s).

When tilapias were medicated with erythromycin base at low dose (Group A), none of derivatives of erythromycin was detected in tilapia muscle at day 1 of post-treatment. At day 23 of post-treatment, erythromycin E (0.30 µg/kg) and erythromycin F (1.37 µg/kg) was not significant to our concern (Table 5).

In case tilapias were fed with erythromycin at higher dose (Group B), two derivatives erythromycin C (131.49 µg/kg) and erythromycin E (258.28 µg/kg) appeared right after ceasing drug treatment. At day 23 of post-treatment, erythromycin E (6.94 µg/kg) and erythromycin F (5.90 µg/kg) was detected and fortunately was also not significant to our concern (Table 5). . This phenomenon could be explained by intestinal and hepatic enzymes. Maltase, leucine aminopeptidase, dipeptidyl aminopeptidase IV,

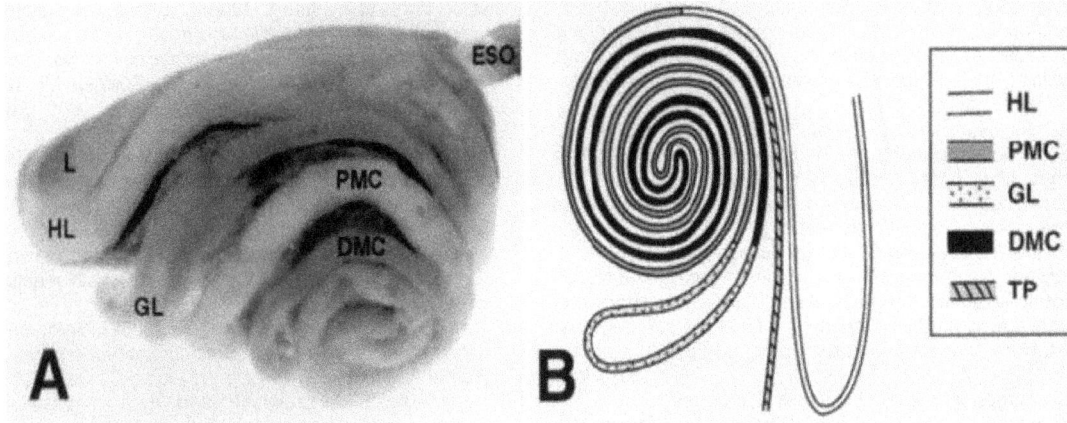

Figure 7. Photomicrograph (A) and schematic drawing (B) of five intestinal segments of tilapia. HL, hepatic loop; PMC, proximal major coil; GL, gastric loop; DMC, distal major coil; TP, terminal portion of the intestine

IV, lipase, non-specific esterases, and alkaline phosphatase were their intestinal enzymes participated in erythromycin metabolism.

Maltase, leucine aminopeptidase, dipeptidyl aminopeptidase IV, lipase, non-specific esterases, and alkaline phosphatase were present at specific sites along the first four intestinal segments (Bundit et al., 2000). Strong reaction for maltase was present in the third intestinal segment, while aminopeptidases and alkaline phosphatase were detected in the first three parts. The most intense activity for lipase was present in the first two parts, while non-specific esterases were observed in the first four portions. Activities of all these enzymes were demonstrated in the brush border. Non-specific esterases were also present in the cytoplasm of the enterocytes. In addition to its brush border localization in the cranial segments, dipeptidylaminopeptidase IV was also observed in the basal lamina of all segments, including the terminal segment. The first four regions played the most important role in both digestion and absorption of erythromycin (Figure 7).

Parallel with intestinal enzymes, hepatic biotransformation enzymes in tilapia such as CYP1A protein, 7-ethoxyresorufin O-deethylase (EROD), glutathione S-transferase (GST), UDP-glucuro-nasyl transferase (UDP-GT) and lipogenic enzyme were also dominated and highly correlated with erythromycin demethylase (Bernard Kwaku-Mensah Gadagbui et al., 1996). They eliminated erythromycin derivatives to minor level at date 23 post-dosing.

Our study provides preliminary data for a prudent use of the antimicrobial drug erythromycin in Nile tilapia, in order to guarantee safety in foods for the consumers and to improve fish farming management. The withdrawal time of erythromycin in Nile tilapia was recommended 33 days or 42 days at least depend on dosage of chemotherapy.

Conclusion

There were lots of researches about toxicity and carcinogenicity of erythromycin A and erythromycin derivatives on mice, rat, dog, and even on human. However, studies about toxicity and carcinogenicity of erythromycin A and erythromycin derivatives in aquatic animals were scarcely investigated. Bio-transformative forms of erythromycin B, C, D, E, F were always in mind of human whether they could be transformed from parental drugs in aquaculture or not. Whether could they create harmful risks to human health? So this research confirmed an obvious evidence that bio-transformative forms of erythromycin appeared through endo-enzymatic mechanism and quickly decomposed to minor level. This would set a basic foundation that there will be no risk of toxicity and carcinogenicity of erythromycin B, C, D, E, and F in aquaculture if farmers strictly follow the recommended veterinary dose.

ACKNOWLEDGEMENT

The technical assistance of Dr. Nguyen Ba Hoai Anh was highly appreciated.

REFERENCES

Ahmed HA (2003). Bacterial flora of freshwater prawn, *Macrobrachium rosenbergii* (de Man), cultured in concrete tanks in Saudi Arabia. J. Appl. Aqua., 14: 113-124.

Annarita E, Laura F, Dario L, Luigi M, Ettore C, Emilio G (2007). Orally administered erythromycin in rainbow trout (*Oncorhynchus mykiss*): Residues in edible tissues and withdrawal time. Anti. I Ag. Chem., 51: 1043-1047.

Be LM (2002). Investigation on diseases of giant freshwater prawn (Macrobrachium rosenbergii) in ponds and rice-prawn farming systems in A Giang province. *Msc. thesis* (in Vietnameses).

Bernard KMG, Marian A, Anders G (1996). Species characteristics of hepatic biotransformation enzymes in two tropical freshwater tilapia (Oreochromis niloticus) and mudfish (Clarias anguillaris). Comp. Biochem. Physiol. Part C: Pharmacol., Toxicol. Endocrinol., 114: 201-211.

Bundit T, Bonnie JS, Thomas C, Stephen AS (2000). Distribution of intestinal enzyme activities along the intestinal tract of cultured Nile tilapia, Oreochromis niloticus L. Aqua., 182: 317-327.

Dat NT (2002). Investigation on parasite and bacteria diseases in giant freshwater prawns (Macrobrachium rosenbergii) cultured in pond and rice-prawn with low density. Msc. Thesis (in Vietnamese).

European Agency for the Evaluation of Medicinal Products (EAEMP) (2009). Studies to evaluate the metabolism and residue kinetics of veterinary drugs in food-producing animals: marker residue depletion studies to establish product withdrawal periods. EMEA, London, United   Kingdom. http://www.ema.europa.eu/pdfs/vet/vich/46319909en.pdf

Fairgrieve WT, Masada CL, McAuley WC, Peterson ME, Myers MS, Strom MS (2005). Accumulation and clearance of orally administered erythromycin and its derivative, azithromycin, in juvenile fall Chinook salmon Oncorhynchus tshawytscha. Dis. of Aqua. Org., 64: 99-106.

Isaac OK, Jos H, Eugene R, Hubert V, Ludo V, Maurice D, Claude VP, Pierre P, Georgette L (1985). Antibacterial activities of erythromycins A, B, C, and D and some of their derivatives. Anti. Ag. Chem., 28: 630-633.

Lalitha KV, Surendran PK (2006). Microbiological quality of farmed tropical freshwater prawn (Macrobrachium rosenbergii). J. Aqua. F. Pro. Tech., 15: 71-82.

Na L, Yunlong Z, Jian Y (2008). Effects of water-borne copper on digestive and metabolic enzymes of the giant freshwater prawn Macrobrachium rosenbergii. Arch. Environ. Contam. Tox., 55: 86-93.

Nagla FG, Safinaz GMI, Khalil RH, Soliman MK (2005). Studies on Edwardsiella infection in Oreochromis niloticus. Egypt. J. Aqua. Res., 31: 460-471.

Bowser PR, Kosoff RE, Chen CY, Wooster GA, Getchell RG (2009). Florfenicol residues in Nile tilapia after 10-d oral dosing in feed: effect of fish size. J. Aqua. AniM. I H., 21: 14-17.

Shih-Chu C, Yu-De L, Li-Ling L, Pei-Chi W (2001). Lactococcus garvieae infection in the giant freshwater prawn Macrobranchium rosenbergii confirmed by polymerase chain reaction and 16S rDNA sequencing. Dis. Aqua. Org., 45: 45-52.

Tran TTH, Dang THO, Nguyen TP (2002). Study on diseases in giant freshwater prawns (Macrobrachium rosenbergii): A review.

Weihai X, Xiaobin Z, Xinting W, Liping D, Gan Z (2006). Residues of enrofloxacin, furazolidone and their metabolites in Nile tilapia Oreochromis niloticus. Aquaculture, 254: 1-8.

William TF, Cyndy LM, Mark EP, McAuley WC, McDowell GC, Strom MS (2006). Concentrations of erythromycin and azithromycin in mature Chinook salmon Oncorhynchus tshawytscha after intraperitoneal injection, and in their progeny. Dis. Aquac. Org., 68: 227-234.

Winton C, Jiam-Chu C (1998). Isolation and characterization of an Enterococcus like bacterium causing muscle necrosis and mortality in Macrobrachium rosenbergii in Taiwan. Dis Aquac. Org., 34, 93-101.

11

Selected heavy metals and electrolyte levels in blood of workers and residents of industrial communities

Babalola O. O[1]* and Babajide S. O[2]

[1]Department of Biochemistry, Faculty of Science, Obafemi Awolowo University Ile Ife Nigeria.
[2]Department of Science Laboratory Technology, Moshood Abiola Polytechnic, P. M. B 2210, Abeokuta, Nigeria.

The study focused on the determination of the levels of lead, cadmium, sodium and potassium in the blood of selected industrial workers (n = 36), residents of the neighboring communities (n = 36) as well as the residents of the communities further away from any industrial setting (which controls) (n = 12). The subjects were recruited from the granite served as the, ceramic and cement industries at Ewekoro, Abeokuta North and South Local government areas of Ogun state, Nigeria. The blood lead (BPb) and cadmium (BCd) were determined by atomic absorption spectrophotometry, while blood sodium (BNa$^+$) and potassium (BK$^+$) were determined by flame photometry. The weight and height of all respondents were measured in addition to other information obtained through a structured questionnaire and their body mass indices computed. The mean BCd, BNa$^+$ and BK$^+$ for controls and the test subjects were not significantly different from each other. Significant difference was only observed in the level of BPb between the test and the control. This was also observed when comparing the mean of all the measurements in the blood of residents of the neighboring communities with that of the control subjects. These results revealed that workers and the residents of the neighboring communities are at the risk of lead poisoning to which they were exposed.

Key words: Industrial communities, heavy metals level, electrolyte levels.

INTRODUCTION

Industrialization is central to economic development and improved prospects for human well being. If proper abatement technology is not used, industry becomes a major source of air and water pollution, hazardous waste and noise. Industrial operations have the potential to affect the health of workforces and the general environment thereby impacting negatively on the health of nearby populations and sometimes the populations that are farther away (Sandoz, 1994).

The area of concern is the routine discharge of pollutants into the environment as well as accidents with serious environmental consequences. The variety of occupational health hazards makes quantification of their associated health risks and impacts at the global level very difficult (Mikheev, 1994). Some estimates have been based on the occupational injuries and diseases reported in official statistics (Leigh, 1996), but a large number of injuries and diseases caused by work place hazards are not reported, adjustment is therefore necessary. Making such adjustment, estimate that there might be as many as 125 million cases of occupational injuries and diseases each year, resulting in 220,000 fatalities. (ILO, 1997)

Working conditions, nature of work, vocational and professional status, and geographical location of industries and employment have a profound impact on the social status and social well-being of workers. In many countries, social policy and protection are closely linked with employment and unemployment.

As the mobility of workers increases leading to high numbers of migrant workers in some countries, their health, well being and social support require special attention which is the key issues for sustainable development (Silbergeld and Tonat, 1994).

This study therefore focused on the determination of the levels of lead, cadmium, sodium and potassium in the blood of selected industrial workers, residents of the neighboring communities as well as the residents of the distant communities away from the industrial settings and which served as the controls.

*Corresponding author. E-mail: doctorbablo@yahoo.com.

Table 1. Comparison of anthropometric indices and levels of lead, cadmium, sodium and potassium in blood samples of control, workers of granite industry and their neighbours.

	+Control (n=12)	+GI(n=12)	t_1	P_1sig. level	+ AGN (n=12)	t_2	P_2sig. level	t3	P_3 sig. Level
Age	34.67 ±11.87	34.33 ±5.31	0.103	NS	35.50±6.39	-0.223	NS	-.565	NS
Height (m)	1.65 ±0.08	1.62 ±0.07	1.162	NS	1.66±0.03	-0.339	NS	-.732	NS
Weight (kg)	62.42 ± 5.48	62.50 ±6.50	-0.036	NS	60.50±3.12	0.914	NS	-.042	NS
BMI (kg/m^2)	22.99 ±1.75	23.45 ±2.52	-1.155	NS	21.92±1.57	1.954	NS	2.419	NS
BPb (µg/dL)	31.0 ±0.10	43.0 ±0.12	-3.148	0.009*	46.0±0.19	-3.046	0.011*	-.483	NS
BCd (µg/dL)	22.0 ± 0.07	22.0 ± 0.05	-0.011	NS	19.0± 0.05	0.774	NS	8.611	.000 *
BK$^+$ (mM/L)	4.63 ± 0.50	4.66 ±0.30	-0.178	NS	4.53±0.29	0.702	NS	1.181	NS
Bna$^+$(mM/L	138.76 ±4.31	139.71 3.11	-0.532	NS	136.00±2.70	2.304	NS	2.792	NS

GI = Workers of granite industry, AGN = Neighbours of granite industry, BMI = Body mass index, BPb = Blood lead, BCd = Blood cadmium, BK+ = Blood potassium, BNa+ = Blood sodium. *= Mean ± SD * significant p<0.05, t_1 and P_1 = Comparison between control and workers, t_2 and P_2 = Comparison between control and neighbours, t_3 and P_3 = Comparison between workers and neighbours.

MATERIALS AND METHODS

Study population

A total of 84 male adult subjects (age range between 23 and 50) were voluntary to participate in the study. Informal consents were obtained from respondents after which they were properly educated about the benefits of the study. Respondents with history of cigarette smoking and drug addiction were excluded from the study. In addition respondents with medical history of hepatic and renal pathology were also excluded from the study.

Eighty four (84) male adult subjects (age range between 23 and 50) were selected to participated in the study. These comprised of twelve respondents each from three different factories namely cement factory, granite industry and porcelain industry, all located within South-Western Nigeria. In addition, twelve respondents each living within the rural communities, 1 to1.5 km radius around each of these factories were randomly selected as residents of industrial communities. Another twelve respondents were selected as the control, who were living and working in a relatively clean environment within the government reservation area.

Anthropometric indices

The ages, body weights and heights of the subject were noted. Heights (m) and body weights (kg) were used to calculate the body mass index (BMI) (kg/m^2).

Collection of blood

10 ml of venous blood was obtained from the antecubital vein using disposable pyrogen-free needle and syringes. The blood samples were dispensed into plain vacutainer tubes containing EDTA to prevent coagulation. Samples were kept frozen at -70°C until analysed.

Determination of lead and cadmium from blood samples

Blood digestion

The blood samples were retrieved and allowed to thaw. 1 ml of the blood sample was pipetted using micropipette into conical flask. 10 ml conc. HNO$_3$ was added to the blood sample in the conical flask and then heated on a hot plate.

When the fume became clear and the solution was almost colourless, the solution was removed and allowed to cool. After this, the solution was made up to 25 ml by adding deionised water and stirring.

Determination of lead and cadmium from digested blood samples

The samples digested were analyzed using Alpha 4 AAS (CHEMTECH 4200) at wavelength 217 and 228.8 nm for lead and cadmium respectively at the Obafemi Awolowo University Central Laboratory, Ile-Ife.

Determination of sodium and potassium in blood plasma

Blood sodium (BNa$^+$) and potassium (BK$^+$) were determined by flame photometry (CORNING Clinical Flame Photometer 410C) at wavelength 589 and 768 nm for sodium and potassium respectively at the Obafemi Awolowo University Central laboratory, Ile Ife.

Statistical analysis

Using SPSS 10.00, results were expressed as mean ± standard deviation (SD), student's t-test was used to determine significant difference between means. The 5% (p<0.05) level of significant difference using two tailed 't' table was used to compare the calculated and critical 't' value from the table and thus statistical difference. Analysis of variance was also calculated using SPSS 10.00.

RESULTS

Table 1 showed the comparison of mean values of anthropometric indices and levels of leads, cadmium, sodium and potassium in controls and granite industry workers. There was no significant difference (P>0.05) between the means of age, height, weight, BMI, BCd, BK$^+$ and BNa$^+$ of the controls and granite industry workers. The mean value of the BPb 43.0 µg/dL of the granite industry workers was significantly different (P = 0.009, t = -3.148) from that of control (31.0 µg/dL).

Similarly there was no significant difference (P>0.05) in

Table 2. Comparison of anthropometric indices and levels of lead, cadmium, sodium and potassium in blood samples of control, workers of ceramic industry and their neighbours.

	[+]Control(n=12)	[+]Ceramic(n=12)	t_1	P_1sig.level	[+]PN(n = 12)	t_2	P_2sig.level	t_3	P_3sig.level
Age	34.67 ±11.87	31.00 ±5.17	1.114	NS	32.83±6.19	0.411	NS	-1.379	NS
Height(m)	1.65 ±0.08	1.61 ±0.10	1.215	NS	1.64±0.05	0.699	NS	-.736	NS
Weight (g)	62.42 ±5.48	67.67 ±9.39	-1.961	NS	63.58±7.07	-.392	NS	1.872	NS
BMI (kg/m^2)	22.99 ±1.75	26.28±4.32	-2.195	NS	23.67±1.81	-.811	NS	2.153	NS
BPb (µg/dL)	31.0 ±0.10	44.0 ±0.12	-3.006	0.012 *	45.0± 0.06	-.314	0.00*	-.164	NS
BCd (µg/dL)	22.0±0.07	22.0 ±0.06	0.105	NS	23.0±0.04	-.294	NS	0.90	.NS
BK$^+$ (mM/L)	4.63± 0.50	4.87±0.41	-1.144	NS	4.72±0.27	-.473	NS	2.163	NS
Bna$^+$ (mM/L)	138.76 ±4.31	139.57 ±3.37	-0.585	NS	136.07±4.72	1.197	NS	1.126	NS

Ceramic = Workers of ceramic industry, PN = Neighbours of porcelein wares industry, BMI = Body mass index, BPb = Blood lead, BCd = Blood cadmium, BK+ = Blood potassium, BNa+ = Blood sodium. [+]= Mean ± SD* significant p<0.05, t_1 and P_1 = Comparison between control and workers; t_2 and P_2 = Comparison between control and neighbours; t_3 and P_3 = Comparison between workers and neighbours.

Table 3. Comparison of anthropometric indices and levels of lead, cadmium, sodium and potassium in blood samples of control, workers of cement factory and their neighbours.

	[+]Control(n=12)	[+]Cement(n=12)	t_1	P_1sig.level	[+]CN(n = 12)	t_2	P_2sig.level	t_3	P_3sig.level
Age	34.67 ±11.87	36.25±9.82	-0.660	NS	35.67±9.71	-0.233	NS	.816	NS
Height(m)	1.65 ±0.08	1.55±0.08	4.832	NS	1.64±0.07	0.422	NS	-2.484	NS
Weight(kg)	62.42 ±5.48	65.35±10.23	-1.047	NS	62.83±3.64	-0.277	NS	.184	NS
BMI(kg/m^2)	22.99 ±1.75	27.07±3.72	-3.126	0.010*	23.49±1.64	-0.763	NS	2.414	NS
BPb(µg/dL)	31.0 ±0.10	43.0±0.10	-4.461	0.001*	44.0±0.07	-3.979	0.002*	-.857	NS
BCd(µg/dL)	22 .0± 0.07	23.0±0.06	-0.458	NS	22.0±0.05	0.150	NS	.232	NS
BK$^+$(mM/L)	4.63 ± 0.50	4.37±0.54	1.017	NS	4.60±0.27	0.167	NS	-2.416	NS
Bna$^+$(mM/L)	138.76 ±4.31	139.53±2.50	-0.673	NS	137.43±4.00	0.679	NS	2.051	NS

Cement = Workers of cement company, CN = Neighbours of cement industry, BMI = Body mass index, BPb = Blood lead, BCd = Blood cadmium, BK+ = Blood potassium, BNa+ = Blood sodium. [+]= Mean ± SD *significant p<0.05, t_1 and P_1 = Comparison between control and workers, t_2 and P_2 = Comparison between control and neighbours; t_3 and P_3 = Comparison between workers and neighbours.

age, height and weight, BMI, BCd, BK$^+$ and BNa$^+$, of the control and neighbours of granite industry but there was significant difference (P = 0.011, t =-3.046) in BPd.

Comparison of the mean values of measured parameters of granite industry workers and their neighbours also indicated a significant difference (P = 0.00, t = 8.611) in their Cadmium (Table 1)

For the ceramic industry respondents, there was no significant difference (P>0.05) between the mean values of age, height, weight, BMI, BCd, BK$^+$ and BNa$^+$ of control and ceramic industry workers. There was significant difference (P = 0.012, t = -3.006) between the mean values of BPb of control and workers of the ceramic industry (Table 2).

There was not significant difference (P>0.005) in age, height, weight, BMI, blood Cd, blood K$^+$ and blood Na$^+$ of the control and neighbours of ceramic industry but there was significant difference (P = 0.0, t = -6.3.14) in their blood Pd (Table 2)

There was no significant difference (P>0.05) between the mean values of age, height, weight, BCd, BK$^+$ and BNa$^+$ of control and cement industry workers but there

were significant difference (P = 0.01, t = -3.126 and P = 0.001, t = -4.461 respectively) for BMI and BPd. (Table 3)

DISCUSSION

In this study, occupational exposure to selected toxic heavy metals was investigated and exposure to lead has once again been identified as a major occupational hazard. This may be as a result of the increasing abundance of lead in our environment, and the consequences of industrialisation.

Lead toxicity is of great concern to public health due to its persistence in the environment and has been reported to affect many human organs. It has special affinity for bone and brain tissues (Granjean, 1975). All known effects of lead on biological system are deleterious. Many physiological system including those of the renal, nervous, heamopoietic, immune, reproduction and endocrine are principal targets of this environmental toxicant (Anetor et al., 2002)

Granite is said to contain lead and cadmium and during the process of breaking the rocks into smaller units, dust are released into the atmosphere. The majority of lead in rocks is present in silicate structures, for elements like calcium; it is unavailable until weathering breaks down the primary mineral (O,Nail, 1985). However, when rocks are being blasted for the production of granite and granite dust, lead is released into the atmosphere through dust. This may place the workers at the risk of exposure to these heavy metals. The residents of the neighbouring communities are at greater risk because the workers might be using some protective devices, which are not in place for the villagers around these industries. Moreover, the workers of these industries spent few hours at work and within the environment while the residents of the neighbouring communities spend almost twenty-four hours of the day around the industries. Hence the period of exposure to dust might be the reason for high level blood lead. Additionally many people are less knowledgeable about significance of lead exposure and preventive measures. Polivka (1999) also observed that there are gaps in lead poisoning prevention knowledge among rural residents. This may also be responsible for the high level of blood lead of the residents of the neighbouring communities

The most significant source of lead exposure is dust. Although dusts are complex in nature and might be categorized consistently into few categories relating to human exposure, household dust, soil dust, street dust and occupational dust (Gots, 1992). Occupational dust is the reason for the test carried out on these industrial workers and the residents of the neighbouring communities

In each case, the lead in dust arises from a complex mixture of fine particle of soil, flaked paint and airborne particles of industrial or automotive origin. Dust is deposited in windowsills from outdoor sources. The particles characteristically accumulate on exposed surface and also trapped in the fibres of clothing and carpets (USEPA, 1986). Lead-glazed pottery and lead pigment in toys and pencils is other route of exposure (Fincleman 1996). Lead silicate is used for the manufacture of glass and ceramic frits, which are used in introducing lead into glass and ceramic finish. This might be the major source of exposure in ceramic industries. When lead is released into the environment, it has along resident time compared with other pollutants. Lead and its compounds tend to accumulate in soil and sediments. They will remain bioavailable far into the future due to their low solubility and relative freedom from microbial degradation (Kataba-Pendias and Pendias, 1992). This also may be another major reason for the significant difference in blood lead. Another reason may be that, most of the arable crops being consumed by the residents of the neighbouring communities might have taken up lead from the soil. Lead from dust and gases from various industrial sources such as these factories can contaminate soil and plants.

Conclusively, there should be an appropriate control technology to eliminate or reduce pollution arising from industrial process and operations, both in and out of workplace, in order to protect the general environment and the health of workers and surrounding communities. Full application of control technology comprises not only its design and implementation but also its operation, maintenance and management. Incorporating environmental considerations into design phase of products and processes can also do much to prevent and minimise pollution.

Workers in factories generally should be encouraged to observe workplace safety rules and follow procedures to protect themselves. When leaving the workplace, they should follow decontamination procedures to avoid wearing contaminated clothing that could potentially expose other persons. Measure should be taken to limit access to toxic products. Balance nutrition should be encouraged both in workers and neighbours that are exposed to pollution. Eating fresh fruits, vegetables, grains, lean meat and cold-water fish can do this.

Industrial workers should consider the use of supplemental antioxidants, herbs, minerals, amino acids, phytoextracts, detoxifying agents, protective agents and fibre as adjuncts to a healthy diet to enhance vital organ functioning and to aid in body's natural detoxifying actions. Garlic, for example has been valued for centuries for its medicinal properties. Research has shown that garlic can protect us from various pollutants and heavy metals (Cha, 1987).

REFERENCES

Anetor JI, Babalola OO, Adeniyi FAA, Akingbola TS (2002). Observations on the heamopoietic system in tropical lead poisoning. Nig. J. Physiol. Sci. 17(1-2): 9-15

Cha CW (1987). A study on the effect of garlic to heavy metals poisoning of rat. J. Korean Med. Sci. 2(4), 213-224.

Finkcleman J (1996). Phasing out Leaded Gasoline Will Not End Lead Poisoning in Developing countries. Environmental Health Perspectives. 104 (1), 1-2

Goyer RA (1996). Results of Lead Research: Prenatal Expoture and Neurological Consequences. Environmental Health Perspectives. 104 (10): 1050-1051.

Grandjean A (1975) Lead in Danes, Historical and Toxicological Studies. Env. Quality and Safety Suppl. 2: 6-9

International Labour Organisation (ILO) (1997). The Director-General's programme and budget proposals for 1998-1999. Geneva, ILO.

Kabata-Pendias A, Pendias H (1992). Trace Element in Soils and Plants 2nd Edition. CRS Press London pp.131-144

Leigh JP (1996). Costs of occupational injuries and illnesses. Final report to NIOSH (Cooperative Agreement U60/902886).

Mikheev M (1994). New Epidemics: The challenge for International health work. In: New epidemics in occupational health. Helsinki, Finnish Institute of Occupational Health.

Polivka BJ (1999). Rural residents' knowledge of Lead poisoning prevention. J. Community Health. 24(5): 393-408.

Silbergeld E, Tonat K (1994). Investin in Prevention: Opportunities to Prevent Disease and Reduce Health Care Costs by Identifying Environmental and Occupational causes of noncancer disease. Toxicology and Industrial Health. 10(6): 677-678.

United State Environmental Protection Agency (USEPA) (1996). National air quality and emissions trends report, 1995. Research Triangle Park, NC, USEPA. Office of air Quality Planning and Standards, Emissions Monitoring and Analysis Division, Air Quality trends Analysis Group (EPA 454/R96-005).

12

Metabolic engineering of an ethanol-tolerant *Escherichia coli* MG1655 for enhanced ethanol production from xylose and glucose

Ruiqiang Ma[1,2], Ying Zhang[2,3], Haozhou Hong[2], Wei Lu[2], Wei Zhang[2] ,Min Lin[2] and Ming Chen[2*]

[1]College of Biological Sciences, China Agricultural University, Beijing 100193, China.
[2]Biotechnology Research Institute, Chinese Academy of Agricultural Science, Beijing 100081, China.
[3]Zhengzhou Fruit Research Institute, Chinese Academy of Agriculture Sciences, Henan 450009, China.

Efficient ethanol production will require a recombinant to able to ferment a variety of sugars (pentoses, and hexoses), less formation of by-products, as well as to tolerate high ethanol stress. In this study, a mutant (MGE) that can grow in 60 g ethanol/l was selected from *Escherichia coli* MG1655 by enrichment method with increasing concentrations of ethanol. The ethanol-tolerant mutant was used as the host to develop the ethanologenic recombinant by knockout of pyruvate formate lyase (*pflB*) and lactate dehydrogenase (*ldhA*) genes, and expression of *Zymomonas mobilis* alcohol dehydrogenase and pyruvate decarboxylase genes in plasmid pZY507bc. The resultant recombinant (GMEPLbc) showed the genetic stability of *Z. mobilis* genes in glucose medium without antibiotics under anaerobic conditions, and generated little acetic acid (3.6 mM), no formic acid and lactic acid. The ethanol production by GMEPLbc were 41.6 and 35.8 g ethanol/l from 100 g/L glucose and 100 g/L xylose during fermentation in M9 mineral medium, 37.0 and 36.5% more than that of the ethanol-sensitive strain carrying pZY507bc alone, respectively. Our results indicated that enhancement of ethanol tolerance and inactivation of *pflB* and *ldhA* are advantageous in the production of ethanol.

Key words: *Escherichia coli*, ethanol production, ethanol-tolerance, Gene knockout, metabolic engineering.

INTRODUCTION

Hexose and pentose sugars, as major components of lignocellulosic biomass, are abundant in nature (Deanda et al., 2003), and represent an inexpensive and readily available source for ethanol production. Efficient conversion of biomass to ethanol requires the development of microorganisms capable of fermenting the components in lingocellulosic materials and that tolerating high concentrations of ethanol (Zaldivar et al., 2001). *E. coli* is widely recognized as the modern workhorse for industrial bio-technology. Strains of this organism can naturally metabolize all the sugars present in lignocellulosic hydrolysates, but normally produce only small amounts of ethanol.

Zymomonas mobilis efficiently ferment carbohydrates

using the Enter-Doudoroff pathway and pyruvate decarboxylase (PDC; *pdc*), alcohol dehydrogenase II (ADHII; adhB) are the key enzymes in the formation ethanol. PDC catalyses the decarboxylation of pyruvic acid to acetaldehyde, then ADHII catalyses the conversion between alcohols and aldehydes. Introduction of *Z. mobilis* alcohol dehydrogenase (*adhB*) and pyruvate decarboxylase (*pdc*) genes into *E. coli* has significantly improved ethanol productivity (Ingram et al., 1987). *E. coli* strains have been constructed, which feature traits that are advantageous in the production ethanol using lignocellulose sugars, however, they lack the high ethanol tolerance as *S. cerevisiae* (100 g/L) or *Z. mobilis* (110 g/L) (Ghareib et al., 1988; Rogers et al., 2007). Since the molecular basis for ethanol tolerance is not clearly understood, and the tolerance is a multigenic trait, underlying mechanisms are best explored using a variety of approaches (Alper and Stephanopoulos, 2007). Classic random mutation techniques have been proven helpful

*Corresponding author. E-mail: chenmingbio@hotmail.com.

random mutation techniques have been proven helpful in isolating 35 g/L ethanol-tolerant mutants (Yomano et al., 1998). Ethanol-resistant mutants of *E. coli* that can grow in 38 g ethanol/l have been isolated after nitrosoguanidine-mediated mutagenesis (VA and Novick, 1973).

The recombinant *E. coli* strains, containing *Z. mobilis* genes for the ethanol pathway, was developed for the fermentation of sugars from cellulose and hemicellulose into ethanol, but ethanologenic strains should be further improved by enhancing the ethanol-toloerance, reducing the by-product formation and increasing the phenotypic stability. In this study, we used an ethanol tolerant mutant (MGE), tolerant to 60 g ethanol/l, as the cloning host, which was selected by enrichment method with increasing concentrations of ethanol. To redirect the energy and carbon flux toward ethanol metabolism, two genes encoding lactate dehydrogenase and pyruvate formate lyase were knocked out by homologous recombination and genes encoding enzymes of the ethanol pathway from *Z. mobilis* were incorporated into the mutants. The recombinants were investigated by measuring the intracellular metabolite concentrations and ethanol production from hexose and pentose sugars.

MATERIALS AND METHODS

Strains, plasmids, and culture conditions

The strains and plasmids used in this study are listed in Table 1. The *E. coli* strains were grown at 37ºC in Luria-Bertani medium (LB). The pZY507bc plasmid containing *Z. mobilis* pyruvate decarboxylase (*pdc*) and alcohol dehydrogenase II (*adhB*) genes under the control of the *lac* promoter was constructed in our previously study (Wang et al., 2008). For the fermentation experiments, the strains anaerobic cultured at 37℃ in mineral medium M9 (Na_2HPO_4 64 g/l, KH_2PO_4 15 g/l, NH_4Cl 5 g/l, $MgSO_4$ 0.24 g/l, $CaCl_2$ 0.011 g/l) containing 100 g/L glucose or xylose in 100 ml volumetric flasks containing 80 ml broth and fitted with drilled rubber septa to allow gas to escape. Fermentation flasks were immersed in a temperature-controlled water bath and stirred by a magnetic stirrer at 100 rpm. When required, the medium was supplemented with 50 µg kanamycin (Kan)/ml, or 34 µg chloramphenicol (Cm)/ml. Cell growth was measured as optical density at 600 nm (OD600) after 24 - 96 h of incubation.

Selection and isolation of ethanol-tolerant mutants

A combination liquid and solid medium was used to successfully enrich the strains with mutations, which increase ethanol tolerance and maintain the genetically engineered trait of efficient ethanol production (Yomano et al., 1998). Strains were tested for their tolerance to ethanol (30 g/l) in M9 medium containing 100 g/L glucose. MG1655 was chosen for further improving the ethanol-tolerance. The strain was inoculated and transferred serially into LB medium containing increasing concentrations of ethanol (30 - 60 g/l). After every 3 - 5 liquid transfers, the cultures were diluted and spread on solid medium and the single colonies were re-selected in higher concentration of ethanol, and tested for their ability to grow in the presence of ethanol. One of the best clones (designated MGE) was selected to develop for enhancement of ethanol production.

Construction of strains in *E. coli*

A DNA fragment (1.0 kb) containing part of the *pflB* was amplified by polymerase chain reaction with PFLF (5'-GTCCGAGCTT AATGAAAAG-3') and PFLR (5'-AAGTCCACTG GATAGCTT-3') primers. A Kan[R] gene was inserted in the *Eco*RI site within *pflB*. The disrupted *pflB* gene fragment with its flanking sequences was electroporation transferred into MGE to generate the *pflB* mutant (MGEP).

The 5'-flanking region of *ldhA* was amplified with primers LdhaF1 (5'-ATTCATTAAA TCCGC CAGCT TATAAG-3') and LdhaR1 (5'-GAAGCAGCTC CAGCCTACAG AAAGTAGCCG CGTTTGTTGC-3')-3'); the 3'-flanking region of *ldhA* gene was amplified with primers LdhaF2 (5'-GGACCATGGC TAATTCCCAT GGCGCAACCT TCAACTGAA C-3') and LdhaR2 (5'-ATCCAGGTGT TAGGCAGCAT G-3'). The Cm[R] cassette fragment from PKD3 was amplified with oligonucleotide primers CmrF (5'-GTGTAGGCTG GAGCTGCTTC-3', complementary to LdhaR1) and CmrR (5'-ATGGGAATTA GCCATGGTCC-3', complementary to LdhaF2). The Cm[R]-marked *ldhA* deletion fragment connected by overlap extension PCR was transformed into MGEP. The Cm[R] gene and one of the FRT sequences from the deletion strains constructed by this method were removed as described by Datsenko and Wanner (2000) using the yeast FLP recombinase. The resulting double deletion mutant was designated MGEPL. The mutants (MGEP and MGEPL) were validated by nested PCR.

The plasmid pZY507bc carrying *Z. mobilis* *pdc* and *adhB* genes was transformed into MG1655, MGE, MGEP and MGEPL to yield MG1655bc, MGEbc, MGEPbc and MGEPLbc, respectively. Strains were serially transferred to LB medium containing 100 g/L glucose without antibiotics for more than 60 generations at 37℃. Appropriate dilutions of cultures were plated containing 2% glucose (without antibiotics) after 48 and 120 h. Colonies were screened on 2% glucose-LB plates containing 34 µg Cm/ml and on aldehyde indicator plates.

Analysis of carbon substrate utilization

Strains were tested for their ability to utilize carbon substrates using Biolog GN2 plates (BIOLOG Inc., Hayward, CA, USA). GM1655 and GME strains were grown overnight in LB medium plate at 37℃. Cells were scraped from the surface of the plate using a cotton swab, and suspended in GN/GP-IF inoculating fluid (BIOLOG Inc., Hayward, CA, USA) at a cell density equivalent to 61% transmittance on a Biolog turbidimeter. Suspensions with an OD_{600} of 0.2 to 0.3 were transferred to the GN2-Plate (150 µl per well) at 37℃ for 24 h. The amount of tetrazolium violet dye in each well was determined spectrophotometrically at 590 nm with subtraction of a 750 nm reference reading (A590-750) using a Biolog plate reader.

Fermentation

Cells were first grown in 100 ml LB medium at 37℃ and shaking until the OD_{600} reached 1.0. After centrifuging and washing twice with M9 medium, seed cultures were resuspended in 100 ml M9 medium containing 100 g/L glucose or xylose. Ethanol was assayed using gas chromatography (GC) with a glass column (0.26 × 200 cm) filled with Porapak Type QS (80 - 100 mesh, Waters, Milford, MA) at 180℃; nitrogen was the carrier gas (40 ml/min), and an FID detector was used (Lapaiboon et al., 2007).

Concentrations of glucose, xylose, and organic acids were determined by high- pressure liquid chromatography (HPLC) (Agilent Technologies Palo, Alto, CA, USA) (Hespell et al., 1996) equipped with a refractive index detector, using an Aminex 87H column (Biorad) maintained at 65℃. The mobile phase used was 0.5 mM $H2SO4$ at a flow rate of 0.6 ml/min. The standard components used

Table 1. *E. coli* strains and plasmids used in this study.

Strains	Relevant characteristics	References or resources
E. coli		
JM109	recA1 supE44 endA1 hsdR17 gyrA96 relA1 thiΔ(lac-proAB)F' [traD36 proAB+ lacIq lacZΔM15]	TransGen Biotech
MG1655	Wild type	Novagen
MGEP	Δ(pflB)- Kan^R	This work
MGEPL	Δ(pflB)- Kan^R, ΔldhA	This work
Plasmid		
pKD46	Phage λ red recombinase expression vector, Kan^R, help recombination	Gift from Yale University
pKD3	Template plasmid for gene disruption, Cm^R gene is flanked by FRT sites	Gift from Yale University
pCP20	Ap^R, Cm^R, FLP recombinase expression	Gift from Yale University
pZY507bc	Cloning vector, Cm^R, carrying pdc and adhB of Z. mobilis	Wang Zc et al., 2008
pMD18-T simple	Cloning vector, Amp^R	Takara
pGEM-T	Cloning vector, Amp^R	Promega

nents used for calibration were glucose, xylose, formic acid, acetic acid and lactic acid. These were purchased from Sigma (St. Louis, MO). The chromatograms obtained had the results in terms of micro-refractive index units. Triplicate analyses were performed on each sample to guarantee the reliability of the analysis.

RESULTS

Selection and isolation of ethanol-tolerant mutants of E. coli

To choose an ethanol tolerant *E. coli* strain as the starting strain for ethanol production, we compared 12 *E. coli* strains for their ability to tolerate ethanol. The strain MG1655 was superior to others *E. coli* strains examined. Enrichment method with increasing ethanol concentrations was performed to further improve the tolerance of MG1655. After a further series of transfers, an ethanol-tolerant mutant, MGE, was isolated. MGE showed the normal growth in presence of 50 g ethanol/l (Figure 1), as compared with MG1655 or MGE in M9 medium lacking ethanol (control). With 60 g ethanol/l, MGE maintained growth to reach an OD600 value of 2.47 ± 0.13 (about 80% of the OD600 value of control (3.05 ±0.07)). While MG1655 displayed only 50% relative growth rate in 30 g ethanol /l, and failed to grow in 50 g ethanol /L (Figure 1).

Plasmid stability

The strains of *pflB* mutant (MGEP), *pflB/ldhA* double mutant (MGEPL) grew as well as their parent strain MGE in M9 medium containing glucose under aerobic conditions. Under anaerobic conditions, however, the growth rate of MGEP was only approximately 50% of that of MGE, while the MGEPL strain was incapable of anaerobic

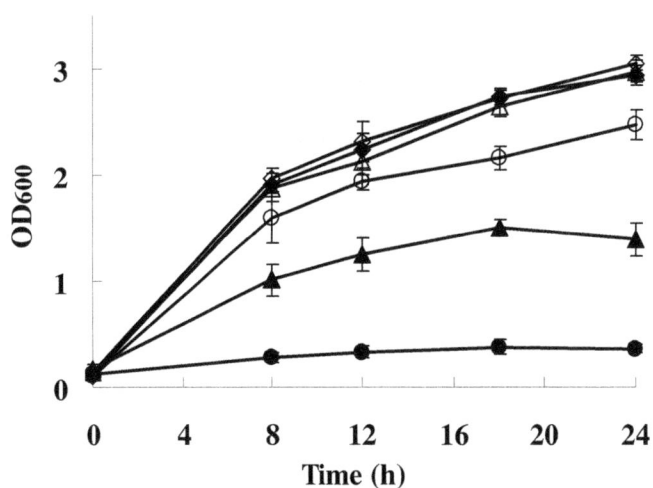

Figure 1. Effect of ethanol on growth. Standard deviations for three experiments are represented by error bars. ◆, ▲, and ● represent the growth of MG1655 in absence of ethanol, 3% and 5% of ethanol respectively; ◇, △, and ○ show the growth of GME in 0%, 5%, and 6% ethanol.

growth (Figure 2). The results showed that both *pflB* and *ldhA* genes are essential for anaerobic growth of *E. coli*. Transforming the mutants with plasmid pZY507bc containing *Z. mobilis pdc* and *adhB* genes totally restored the fermentative growth under anaerobic conditions (Figure 2). The expression of these ethanologenic enzymes caused dramatic increase in final cell density (about 2-fold increase compared with the parent strain MGE).

During anaerobic growth, the mutants were fully complemented by pZY507bc. We therefore, examined plasmid stability after growth in glucose medium without antibiotics for 60 generations under anaerobic conditions.

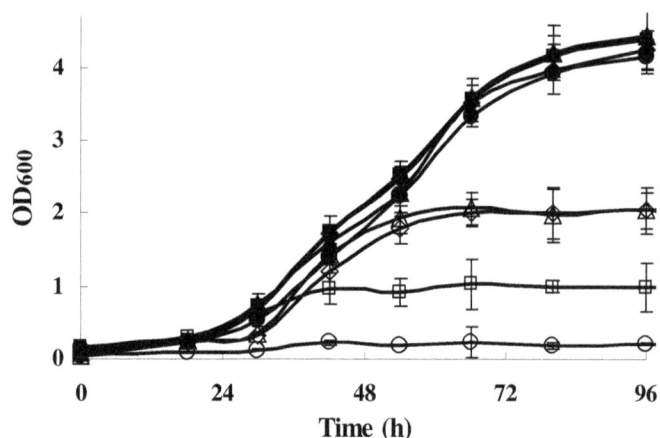

Figure 2. The growth of *E. coli* strains in M9 medium with 100 g/L glucose under anaerobic conditions. Standard deviations for three experiments are represented by error bars. Symbols: ◇, MG1655 ; △, MGE ; □, MGEP ; ○, MGEPL ; ◆, MG1655bc ; ▲, MGEbc ; ■, MGEPbc ; ●, MGEPLbc.

Table 2. Organic acids formed during fermentation of 100 g/L glucose by *E. coli* strains after 96 h.

Strains	Organic acids (mM)		
	Formic acid	Acetic acid	Lactic acid
MG1655	14.2 ± 5.2	60 ± 8.1	460 ± 37.4
MGE	12.4 ± 1.5	24.6 ± 2.4	82.5 ± 7.4
MGEP	ND	5.4 ± 0.6	102.4 ± 10.1
MGEPbc	ND	2.7 ± 0.2	40.3 ± 6.3
MGEPLbc	ND	3.6 ± 0.5	ND

ND = not detected. The values shown represent the mean ± standard deviations for 3 independent measurements

The plasmid pZY507bc was well maintained and showed negligible loss in the MEPLbc strain. In MGEPbc, the plasmid pZY507bc was relatively stable, with 98% of the population retaining both the antibiotic resistance gene and the genes from *Z. mobilis*.

Production of organic acids

The *E. coli* strain ferments a wide range of sugars, including pentoses and hexoses, and in doing so yields a variety of organic acids. After a 96 h fermentation of 100 g/L glucose, MGE generated less organic acids than the ethanol sensitive strain MG1655 (Table 2). Inactivation of *pflB* clearly resulted in a decrease in the production of acetic acid (5.4 ± 0.6 mM), and loss of formic acid production, but the lactic acid concentration (102.4 ± 10.1 mM) in MGEP was higher than that of MGE (82.5 ± 7.4 mM).

The expression of *Z. mobilis* *pdc* and *adh* genes in *E. coli* resulted in the production of ethanol as the primary fermentation product during anaerobic growth. MGEPbc produced less acetic acid (2.7 ± 0.2 mM) and lactic acid (40.3 ± 6.3 mM) than the MGE or MGEP strains. Inactivation of *ldhA* further reduced the formation of organic acids. MGEPLbc generated relatively small amounts of acetic acid (3.6 ± 0.5 mM), and no formic acid or lactic acid were detected (Table 2). The data indicate that metabolic engineering of *E. coli* resulted in a redirection of the energy and carbon flux to ethanol biosynthesis.

Ethanol production

The efficiency of fermentation in M9 medium with glucose

or 100 g/L xylose at pH 7.0 and 37°C is presented in Figure 3. In spite of the fact that the growth rates, glucose-consumption rates and xylose-consumption rates were similar, clear differences in ethanol production in all the ethanolgenic *E. coil* strains were observed. The final concentrations of ethanol in MGEbc cultures grown with 100 g/L glucose or xylose were 33.8 and 28.9 g/L, a 100 g/L increase compared with the ethanol-sensitive strain MG1655bc (30.3 and 26.2 g ethanol/l). Inactivation of the *pflB* and *ldhA* genes blocks the main pyruvate assimilation pathway under anaerobic conditions (Clark, 1989; Mat-Jan et al., 1989). The growth defects of strain MGEPL were compensated for by the introduction of pZY507bc (Figure 3), implying the redirection of energy and carbon flux to pyrvate-to-ethanol pathways. The maximum ethanol levels produced by MEPbc and MGEPLbc were 35.4 and 41.6 g/L from 100 g/L glucose, 30.5 and 35.8 g/L from 100 g/L xylose in M9 medium. Disrupting *pflB* and *ldhA* genes in the pyruvate assimilation pathway resulted in a significant decrease in the formation of formic acid, acetic acid and lactic acid, and a concomitant increase in ethanol levels (16 and 37% more than that of MG1655bc from glucose and xylose, respectively).

DISCUSSION

It is highly desirable to increase the ethanol-tolerance of *E. coli* strains undergoing ethanologenic cell growth activity and ethanol production. Several approaches such as mutagenesis and genetic modifications have been used to develop ethanol-tolerant strains (Alper and Stephanopoulos, 2007; Yomano et al., 2001). Perhaps the best example of increased ethanol tolerance of E. coli strain was described by Yomano (1998). The ethanol tolerant mutants were successful generated by enrichment method. In this study, we used a similar approach to generate the ethanol-tolerant mutants. However, in considering selection of the best candidates with tolerance to harsh conditions (e.g. high ethanol concentration) for enhancement of ethanol production by

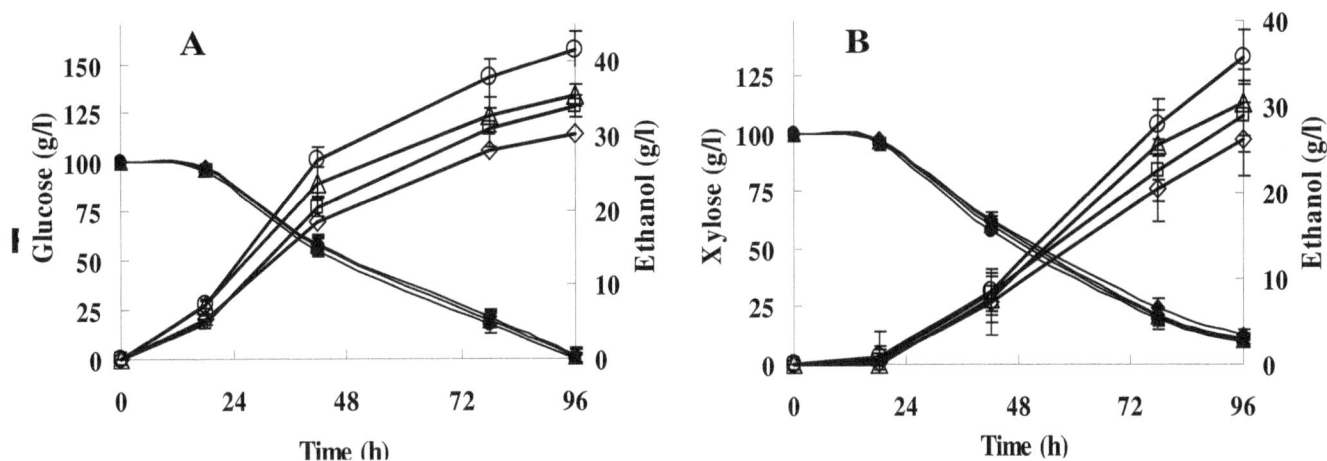

Figure 3. Fermentation of *E. coli* strains in M9 medium with 100 g/L glucose (A) and xylose (B). Vertical bars represent standard deviations for three experiments. Solid symbols represent the sugar concentrations in the cultures, and open symbols represent ethanol levels in the cultures. Diamond symbols (◆ and ◇), MG1655bc, square symbols (■ and □), MGEbc; triangle symbols (▲ and △), MGEPbc; cycle symbols (● and ○), MGEPLbc.

metabolic engineering, we tested the ethanol tolerance of different *E. coli* strains, and selected MG1655 to develop ethanol-tolerant mutants by enrichment method.

Ethanol tolerance is a multigenic trait involving a number of biochemical reactions and physiological processes. Adaptive responses of bacteria to ethanol stress correlates with increases in fatty acid chain length (Ingram et al., 1987), the levels of *trans*-fatty acids (Heipieper and Bont, 1994; Pinkart and White, 1997), and changes in phospholipid composition (Clark and Beardl, 1979). In this study, the ethanol-tolerant strain *E. coli* MGE showed 100 and 80% relative growth rate at 50 and 60 g ethanol/l by serial transfers into increasing concentrations of ethanol, respectively. Compared with the MG1655bc ethanol-sensitive strain, the engineered ethanol-tolerant mutant (MGEPLbc), had a higher rate of ethanol production and higher ADHB and PDC activities, especially in high concentrations of ethanol. Our results show that directed evolution of a bacterial train is an effective means of eliciting a desired trait, the result is in agreement with previous observations (Ingram, 1986; Yomano, *et al.,* 2001).

To further improve ethanol production and to block undesirable metabolic pathways, the *pflB* and *pflB/ldhA* double mutants of *E. coli* were created. These mutants were incapable of anaerobic growth because of deficiencies in the fermentative lactate dehydrogenase and pyruvate formate lyase (Clark, 1989; Hasona et al., 2004). The plasmid containing *Z. mobilis adhB* and *pdc* genes can complement the mutants for anaerobic growth (Hespell et al., 1996).

Therefore, the plasmid in the mutants was quite stable and should not require antibiotics in the medium to maintain genetic stability and high ethanol production traits (Ohta et al., 1991; Yomano et al., 2001). Elimination of

antibiotics requirement for recombinants to carry the *Z. mobilis* genes, now provides additional opportunities for the commercial application of ethanologenic *E. coli*. Our results showed that the plasmid pZY507bc was quite stable in MGEPbc and MGEPLbc and diverted the carbon flow of pyruvate from organic acids to ethanol production (Jarboe et al., 2007). Inactivation of *pflB* and *ldhA* genes resulted in a significant decrease of the formation of organic acids, and a concomitant increase of the ethanol levels.

ACKNOWLEDGMENTS

The authors wish to thank Dr. Russell A. Nicholson for critical reading of the manuscript. This work was supported by the National Basic Research Program of China (Grant No. 2007CB707805); National High-Tech Research and Development Plan of China (Grant No. 2006AA02Z229) and National High-Tech Research and Development Plan of China (Grant No. 2006AA020101).

REFERENCES

Alper H, Stephanopoulos G (2007). Global transcription machinery engineering: a new approach for improving cellular phenotype. Metab. Eng. 9: 258-267.

Clark DP, Beard JP (1979). Altered phospholipid composition in mutants of *Escherichia coli* sensitive or resistant to organic solvents. J. Gen. Microbiol., 113: 267-274.

Clark DP (1989). The fermentation pathways of *Escherichia coli*. FEMS. Microbiol. Rev., 63: 223-234.

Deanda K, Zhang M, Eddy C, Picataggio S (1996). Development of an arabinose-fermenting *Zymomonas mobilis* strain by metabolic pathway engineering. Appl. Environ. Microbiol. 62(12): 4465-4470.

Ghareib M, Youssef KA, Khalil AA (1988). Ethanol tolerance of Saccharomyces cerevisiae and its relationship to lipid content and composition. Folia. Microbiol., 33: 447-452.

Hasona A, Kim Y, Healy FG, Ingram LO, Shanmugam KT (2004). Pyruvate formate lyase and acetate kinase are essential for anaerobic growth of Escherichia coli on xylose. J. Bacteriol., 186: 7593-7600.

Heipieper HJ, JAM de Bont (1994). Adaptation of Pseudomonas putida S12 to ethanol and toluene at the level of fatty acid composition of membranes. Appl. Environ. Microbiol., 60: 4440-4444.

Hespell RB, Wyckoff H, Dien BS, Bothast RJ (1996). Stabilization of pet operon plasmids and ethanol production in Escherichia coli strains lacking lactate dehydrogenase and pyruvate formate-lyase activities. Appl. Environ. Microbiol., 62: 4594-4597.

Ingram LO (1986). Microbial tolerance to alcohols: role of the cell membrane. Trends. Biotechnol., 4: 40-44.

Ingram LO, Conway T, Clark DP, Sewell GW, Preston JF (1987). Genetic engineering of ethanol production in Escherichia coli. Appl. Environ. Microbiol., 53: 2420-2425.

Jarboe LR, Grabar TB, Yomano LP, Shanmugan KT, Ingram LO (2007). Development of ethanologenic bacteria. Adv. Biochem. Eng. Biotechnol., 108: 237-61.

Lapaiboon LP, Thanonkeo P, Jaisil, Laopaiboon P (2007). Ethanol production from sweet sorghum juice in batch and fed-batch fermentation by Saccharomyces cerevisiae. World. J. Micriobiol. Biotechnol., 23: 1497-1501.

Mat-Jan F, Alam KY, Clark DP (1989). Mutants of Escherichia coli deficient in the fermentative lactate dehydrogenase. J. Bacteriol., 171: 342-348.

Ohta K, Beall DS, Mejia JL, Shanmugam KT, Ingram LO (1991). Genetic improvement of Escherichia coli for ethanol production: chromosomal integration of Zymomonas mobilis genes encoding pyruvate decarboxylase and alcohol dehydrogenase II. Appl. Environ. Microbiol., 57: 893-900.

Pinkart HC, White DC (1997). Phospholipid biosynthesis and solvent tolerance in Pseudomonas putida strains. J. Bacteriol. 179: 4219-4226.

Rogers PL, Jeon YJ, Lee KJ, Lawford HG (2007) Zymomonas mobilis for fuel ethanol and higher value products. Adv. Biochem. Eng. Biotechnol. 108: 263-288.

Va F, Novick A (1973). Organic solvents as probes for the structure and function of the bacterial membrane: effects of ethanol on the wild type and an ethanol-resistant mutant of Escherichia coli K-12. J. Bacteriol. 114: 239-248.

Wang ZC, Chen M, Xu YQ, Li SY, Lu W, Ping SZ, Zhang W, Lin M (2008). An ethanol-tolerant recombinant Escherichia coli expressing Zymomonas mobilis pdc and adhB genes for enhanced ethanol production from xylose. Biotechnol. Lett., 30(4): 657-63.

Yomano LP, York SY, Ingram LO (1998). Isolation and characterization of ethanol-tolerant mutants of Escherichia coli KO11 for fuel ethanol production. J. Ind. Microbiol. Biotechnol., 20: 132-138.

Changes in the tissue antioxidant enzyme activities of palm weevil (*Rynchophorous phoenicis*) larva by the action of 2, 2-dichlorovinyl dimethyl phosphate

Olufemi Bamidele[1]*, Joshua Ajele[1], Ayodele Kolawole[1] and Akinkuolere Oluwafemi[2]

[1]Department of Biochemistry, The Federal University of Technology, P.M.B. 704, Akure, Nigeria.
[2]Department of Biology, The Federal University of Technology, P.M.B. 704, Akure, Nigeria.

The aim of this study was to determine the changes caused by the action of 2,2-dichlorovinyl dimethylphosphate (DDVP) on the three tissues, fat body (FB), gut (GT) and head (H), and the antioxidant enzymes of African black palm weevil, *Rynchophorous phoenicis* (Coleoptera: Curculionidae) larva. Larvae of *R. phoenics* were injected with varied concentrations (0 to 0.66 µg g^{-1}) DDVP solution and antioxidant enzymes, ascorbate peroxidase (APX), catalase (CAT), glutathione peroxidase (GPX) and superoxide dismutase (SOD) activities of the recovering DDVP intoxicated larvae 48 h after treatment were measured. DDVP lethal dose (LD_{50}) was 0.38 µg g^{-1}. APX, CAT and SOD activities were altered with increased DDVP concentrations. Residual GPX activities in the FB, GT and H tissues were 19.65, 15.12, and 9.83 µmolmin^{-1}mg^{-1} protein. Removal of cellular-damaging DDVP-related reactive oxygen species (ROS) in which CAT and SOD play important roles supported the possible role of gut of larva as a detoxifier of toxic substances.

Key words: 2, 2-dichlorovinyl dimethyl phosphate (DDVP); Ascorbate peroxidase (APX); Catalase (CAT); Glutathione peroxidase (GPX); Superoxide dismutase (SOD); palm weevil larvae.

INTRODUCTION

The African black palm weevil, *Rynchophorous phoenicis* (Coleoptera: Curculionidae) is a major insect pest of palm trees in Nigeria. This insect is also a secondary pest of coconut and raffia palm as well as the sago palm. The life cycle of *R. phoenicis* takes place within the host tree (oil palm) and other similar hosts (Thomas et al., 2004). During these developmental stages, the larvae of *R. phoenicis* can excavate cavities of more than a metre in length which may lead to the death of the tree after three to four months of infestation (Faleiro et al., 1999).

However, several control methods are currently in use to control palm weevil. Biological control (Peter, 1989), chemical control (Barranco et al., 1998), cultural and sanitary (Azam and Razvi, 2001), pheromones and other behavioral chemical (Faleiro et al., 1999) methods have been used in the Western and some African countries. In Nigeria, there is no clear-cut pest control method that could be awarded 100% successful but chemical control method seems to be most often than not used for pest control. The search light is on the detection of effective insecticide or pesticide with potent ingredients for the control of insect pest vis-à-vis insect resistance.

Action of insecticides starts from its ingestion either by direct or indirect exposure. This may eventually lead to generation of reactive oxygen species (ROS), which are detrimental to insects. Toxicity may also occur directly as a result of insecticides being converted to free radicals or via the formation of superoxide as by-product of their metabolism. These ROS and/or their reaction products alter the normal metabolic processes in cells by causing oxidative stress through triggering mainly the peroxidation of membrane lipids and consequent severe damages to biological molecules including nucleic acids, lipids and proteins and/or cell death (Apel and Hirt, 2004; Miller et al., 2008; Gill and Tuteja, 2010; Anjum et al., 2010, 2011). In response to ROS generated by the active insecticide component, insects have evolved defense mechanisms, especially towards ROS.

R. phoenicis, though a dangerous pest, is used in some local communities in Nigeria as food. Deliberate action of some youths and rural farmers in rearing this insect has paved way for indiscriminate cutting down of raffia and oil palm trees (Fasoranti and Ajiboye, 1993). Not only does this action destroys palm plantation but also gives rise to the emergence of R. phoenicis adult from abandoned culture trees. Since several hundreds of this insect could be released to the environment under suitable conditions, there is the possibility of outbreak if rate of R. phoenicis rearing is not minimized. Therefore, there is the need to put in place measures to control existing insect pests and possible R. phoenicis outbreak. In putting up control measures, information on the metabolic adaption of R. phoenicis larva to its host environment will serve as a guide to formulating suitable insecticide against the insect. 2, 2-dichlorovinyl dimethyl phosphate (DDVP) is an organophosphate pesticide that is widely used to treat domestic animals and livestock for internal and external parasites, control insects commercially and in the home, and to protect crops from insect infestation. It occurs as an oily colorless to amber liquid, with an aromatic chemical odor (ATSDR, 1997; USEPA, 2006). A characteristic vapor-producing property of DDVP is considered in the choice of this insecticide for the research.

Hence, one purpose in this research was to determine the metabolic alterations caused by the action of DDVP on both the antioxidant and non-antioxidant enzymes in the major tissues of R. phoenicis larva. In this study, the connection between the generation of ROS by DDVP and status of antioxidant defense system in larvae of R. phoenicis was investigated.

MATERIALS AND METHODS

Chemicals

2, 4-dinitrophenylhydrazine-thio-urea-CUSO$_4$ (DTC), 5, 5'-dithiobis-2-nitrobenzoic acid (DTNB), metaphosphoric acid, reduced nicotinamide adenine dinucleotide (NADPH), nitroblue tetrazolium salt (NBT), trichloroacetic acid (TCA), folin ciocalteau reagent, sodium-potassium tartarate were purchased from Sigma Chemical Company, (St. Louis, Mo, USA). All other chemicals used were of analytical grade.

The biochemical assays were carried out using Biochrome 4060 UV/Visible spectrophotometer (Pharmacia, Sweden).

Insecticide

The tested compound - "Sniper 1000 EC" containing 2, 2-dichlorovinyl dimethyl phosphate (DDVP, 1000 g/L) was manufactured by Hubei Sanonda Co. Ltd, Shashi Hubei, China and supplied by Saro Agrosciences Limited, Apapa, Lagos, Nigeria.

Insect larvae collection

Palm weevil (R. phoenicis) larvae (of about 9.89 g mean body weight) were collected from Igbokoda palm plantation, Ondo State in February, 2012. The larvae were transported to the laboratory in an insect box constructed with iron-wire mesh (25 × 40 × 45 cm). The insect larvae were acclimatized in an air-conditioned laboratory room at 27°C, relative humidity (75%) and exposed to 12 h L: D for one week prior to the experiment. The larvae were reared on degraded palm fibre collected from infested palm tree. Identification of larva was carried out in the Entomology Research Laboratory, The Federal University of Technology, Akure. The last larval instar was used for the experiment.

Larvae grouping

Individual larva weight was taken and recorded prior to the experiment. Random selection method was used (based on recorded weights) to place active and stress-free larvae in groups. The larvae were divided into six groups consisting of 25 larvae per group. Group one was used as the control. The larvae were of an average weight of 9.75 g and the body length range of 5 to 7 mm in each group.

Injection of larvae with DDVP solution

Injection of larvae with DDVP solution was carried out according to Kostaropoulos et al. (2001). Stock DDVP solution was prepared by dissolving "Sniper" solution in acetone and was diluted with normal saline (0.9% NaCl) to the desired concentrations of DDVP before injection. R. phoenicis larvae were injected 2 μL with different concentrations of DDVP (0, 0.22, 0.33, 0.44, 0.55 and 0.66 μg g^{-1} body weight) at two to three abdominal segments using a microsyringe. Care was taken to avoid puncturing the alimentary tract. Control larvae received 2 μL of normal saline containing 40% acetone (saline/acetone). Each DDVP concentration was repeated three times and was injected into 25 individuals. Knocking down effect was recorded 48 h after treatment. The insect larvae were separated into two groups: one consisted of live larvae and was used for the analysis of APX, CAT, GPX and SOD and those in the other group were dead. After the insects had been separated, they were immediately stored at -4°C. The number of dead insects was determined and used for the calculation of LD$_{50}$ with probit analysis (Finney, 1970). All tests were conducted at room temperature.

Dissection and preparation of larvae cytosolic fraction

Surviving larvae (n=5) from each group 48 h after exposure were demobilized by freezing, quickly dissected and separated each into three fractions; fat body (FB), gut (GT) and head (H) tissues using dissecting kit. Fractions were stored below -20°C until use. The tissues were thereafter homogenized 1:3 (w/v) in ice-cold buffer: (25 mM potassium phosphate buffer, pH 7.2 containing 1 mM EDTA, and 1 mM 2-mercaptoethanol). Crude cytosolic antioxidant enzyme was prepared subsequently by differential centrifugation. Homogenate was centrifuged at 10,000 g for 30 min at 4°C; the floating lipid was carefully removed from supernatant through a funnel plugged with glass wool. The supernatant obtained after filtration was stored in aliquots at below -4°C and subsequently used as crude enzyme. Protein concentration was determined by the method of Lowry et al. (1951) using bovine serum albumin (BSA) as the standard.

Enzyme assays

Estimation of ascorbate peroxidase (APX) activity

APX (EC 1.11.1.11) activity was measured according to Nakano and Asada (1981). The assay depends on the decrease in absorbance at 290 nm as ascorbate was oxidized over a 3 min period at 25°C. The APX activity was expressed as micromoles of H_2O_2 reduced per minute per milligram of protein, using an extinction coefficient of 2.8 mM^{-1} cm^{-1}.

Measurement of catalase (CAT) activity

Catalase (CAT, EC 1.11.1.6) activity was measured according to Aebi (1984) assaying the hydrolysis of H_2O_2 and decreasing absorbance at 240 nm over a 3 min period at 25°C. The CAT activity was expressed as micromoles of H_2O_2 reduced per minute per milligram of protein, using an extinction coefficient of 0.0394 mM^{-1} cm^{-1}.

Measurement of glutathione peroxidase (GPX) activity

The GPX (EC 1.11.1.9) activity was measured with H_2O_2 as substrate according to Paglia and Valentine (1987). This reaction was monitored indirectly as the oxidation rate of NADPH at 340 nm for 3 min. Enzyme activity was expressed as micromoles of NADPH consumed per minute per milligram of protein, using an extinction coefficient of 6.220 $M^{-1}cm^{-1}$. A blank without homogenate was used as a control for the non-enzymatic oxidation of NADPH upon addition of hydrogen peroxide in 0.1 M Tris buffer, pH 8.0.

Measurement of superoxide dismutase (SOD) activity

Superoxide dismutase (SOD) was assayed according to Beauchamp and Fridovich (1971). The reaction mixture contained 1.17 µM riboflavin, 0.1 M methione, 0.2 µM potassium cyanide (KCN) and 0.56 µM nitroblue tetrazolium salt (NBT) dissolved in 3 ml of 50 mM sodium phosphate buffer (pH 7.8). 3 ml of the reaction medium was added to 1 ml of enzyme extract. The mixtures were illumineted in glass test tubes by two sets of Phillips 40 W fluorescent tubes in a single row. Illumination was started to initiate the reaction at 30°C for 1 h. Identical solutions that were kept under dark served as blanks. The absorbance was read at 560 nm in the spectrophotometer against the blank. SOD activity was expressed in units (U mg^{-1} protein). One unit is defined as the amount of change in the absorbance by 0.1 h^{-1} mg^{-1} protein.

Estimation of ascorbic acid (AA) content in larva

AA content was assayed as described by Omaye et al. (1979); the extract was prepared by pulverizing 1 g of fresh larva tissue in 5 ml of 10% TCA, centrifuged at 3500 g for 20 min, re-extracted twice and the supernatant made up to 10 ml and used for the assay. To 0.5 ml of extract, 1 ml of 2,4-dinitrophenyl hydrazine-thio-urea-CuSO$_4$ (DTC) reagent was added, incubated at 37°C for 3 h and 0.75 ml of ice-cold 65% H_2SO_4 was added, allowed to stand at 30°C for 30 min, resulting colour was read at 520 nm in spectrophotometer. The AA content was determined using standard curve prepared with AA and the results were expressed in µg g^{-1} fresh weight (FW).

Estimation of glutathione (GSH) content in larva

The GSH content was assayed as described by Griffith and Meister, (1979). Two hundred milligram fresh larva tissue was pulverized in 2 ml of 2% metaphosphoric acid and centrifuged at 10,000 g for 10 min. 0.6 ml of 10% sodium citrate added to neutralize the supernatant. One milliliter of assay mixture was prepared by adding 100 µl extract, 100 µl distilled water, 100 µl 5, 5-dithio-bis-(2-nitrobenzoic acid) and 700 µl NADPH. The mixture was stabilized at 25°C for 3 to 4 min. Then 10 µl of glutathione reductase was added, and the absorbance was read at 412 nm in spectrophotometer and the GSH contents were expressed in µg g^{-1} fresh weight (FW).

Statistical analysis

The LD_{50} value of DDVP and the 95% confidence intervals were analyzed by probit analysis (Finney, 1970). Data of the experiment were analyzed statistically, and the results were expressed as mean ± SEM of five independent measurements of enzyme activity. All analyses were carried out with IBM-SPSS (20).

RESULTS

Insecticide toxicity bioassay

Bioassay was conducted on 25 R. phoenicis larvae. The relationship between DDVP concentrations and the mortality rate (calculated as a percentage after 48 h of DDVP treatment) was linear (Figure 1). The mortality of R. phoenicis larvae increased significantly ($P < 0.05$) with increased concentration of DDVP. The LD_{50} value (0.38 µg g^{-1}) of DDVP and the 95% confidence intervals (0.365 to 0.401 µg g^{-1}) were analyzed by probit analysis (Table 1).

APX activity

The effect of DDVP on APX activities in the FB, GT and H tissues of R. phoenicis is shown in Figure 2. APX activity increased significantly ($P < 0.05$) in the FB tissue of R. phoenicis with increased DDVP concentration. A concentration of 0.22 µg g^{-1} DDVP caused 33.26 µmol min^{-1} mg^{-1} protein of APX activities in the FB tissue when compared with the control. In the GT tissues of R. phoenicis, APX activity (113.91 µmol min^{-1} mg^{-1} protein) was observed at concentration of 0.22 µg g^{-1} DDVP. At higher DDVP concentrations, the enzyme activities decreased significantly ($P < 0.05$) in the GT tissue of R. phoenicis larva when compared with the control. APX activity in the H tissue of larva increased and later decreased with increased DDVP concentrations. APX

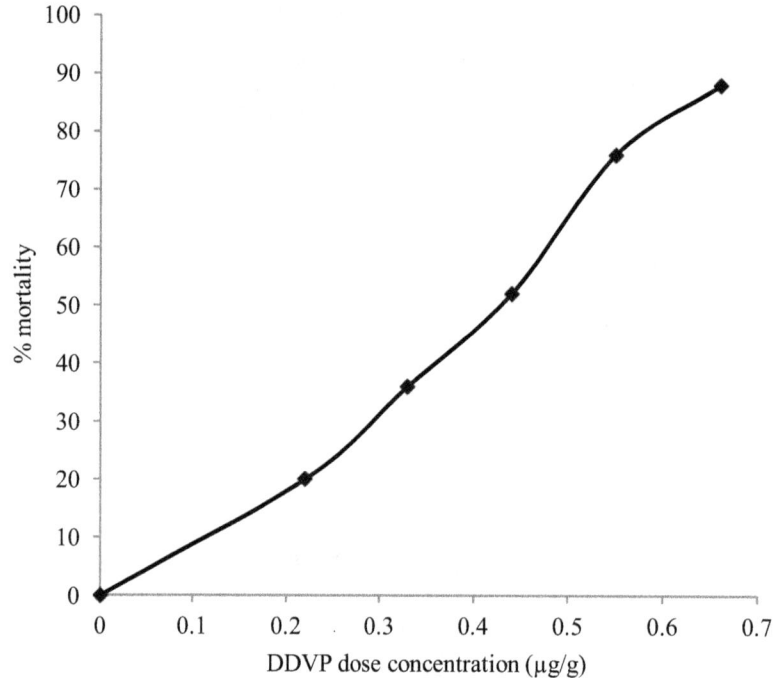

Figure 1. The relationship between DDVP concentration and the mortality rate of *Rynchophorous phoenicis* larvae after 48 h after treatment. DDVP toxicity was tested on larvae divided into six groups consisting of 25 larvae per group.

Table 1. Toxicity results (expressed as LD$_{50}$) of DDVP on *Rynchophorous phoenicis* larvae 48 h after exposure.

Lethal dose	Concentration$^{\Phi}$	95% confidence limit$^{\Phi}$
*LD 10	0.064	0.049 - 0.079
LD 50	0.383	0.365 - 0.401
LD 95	1.351	1.179 - 1.606

Each experimental group consisted of 25 larvae. *LD, Lethal dose, Unit is (µg g-1).

activity in the GT tissue of larva at varied concentrations of DDVP was well expressed than the APX activities of the FB and H tissues.

CAT activity

Figure 3 shows the effect of DDVP on the catalase activities in the FB, GT and H tissues of *R. phoenicis* larva. CAT activity in the FB tissue of larva increased with initial DDVP concentrations but later declined significantly ($P < 0.05$) with increased DDVP concentrations when compared with the larvae in the control group. CAT activity (25.36 µmol min^{-1} mg^{-1} protein) was recorded at 0.22 µg g^{-1} DDVP concentration in the FB tissue. In the GT tissue of *R. phoenicis* larva, increased DDVP concentrations caused progressive increase in CAT activity when compared with CAT activity in the GT tissue of *R.*

phoenicis larva of the control group. CAT activity in the H tissue of DDVP-treated *R. phoenicis* larva initially increased at concentration of 0.22 µg g^{-1} and decreased significantly ($P < 0.05$) with increasing DDVP concentrations. Highest CAT activity was observed in the GT tissue of *R. phoenicis* larva.

GPX activity

Effect of DDVP on glutathione peroxidase activities in the FB, GT and H tissues of *R. phoenicis* larva is shown in Figure 4. In the FB tissue of DDVP-treated *R. phoenicis* larvae, GPX activities increased significantly ($P < 0.05$) with increased DDVP concentrations when compared with the control. Concentration of 0.33 µg g^{-1} DDVP caused 24.94 µmolmin^{-1}mg^{-1} protein of GPX activity in the FB tissue of larva. In the GT tissue of larva, initial DDVP

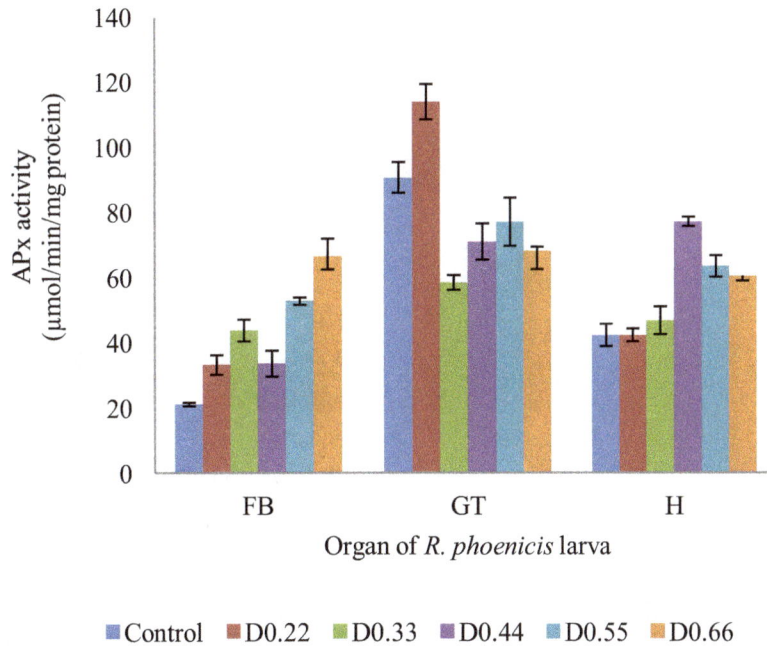

Figure 2. Effect of DDVP on ascorbate peroxidase activities in the fat body (FB), gut (GT) and head (H) tissue of *Rynchophorous phoenicis* larva. Each value is the mean of five repetitions ± SEM. Values are significantly different at (P<0.05). Control, larva treated with acetone in normal saline; D0.22, larva treated with 0.22 µg g^{-1} DDVP concentration; D0.33, larva treated with 0.33 µg g^{-1} DDVP concentration; D0.44, larva treated with 0.44 µg g^{-1} DDVP concentration; D0.55, larva treated with 0.55 µg g^{-1} DDVP concentration; D0.66, larva treated with 0.66 µg g^{-1} DDVP concentration.

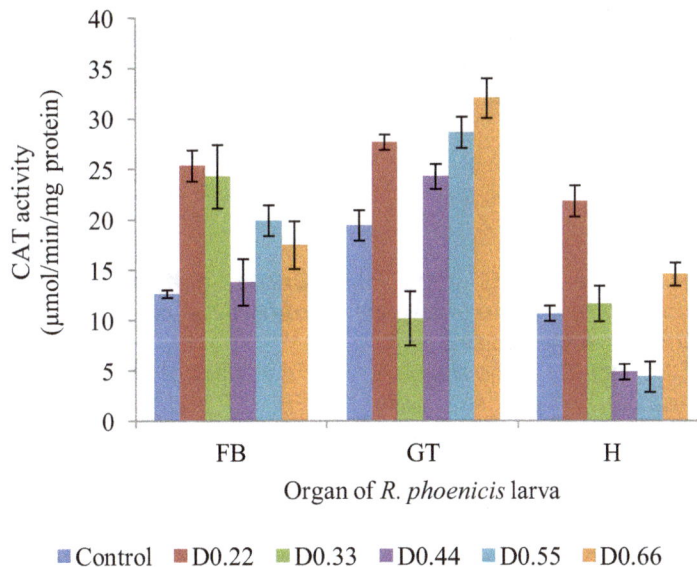

Figure 3. Effect of DDVP on catalase activities in the fat body (FB), gut (GT) and head (H) tissue of *Rynchophorous phoenicis* larva. Each value is the mean of five repetitions ± SEM. Values are significantly different at (P<0.05). Control, larva treated with acetone in normal saline; D0.22, larva treated with 0.22 µg g^{-1} DDVP concentration; D0.33, larva treated with 0.33 µg g^{-1} DDVP concentration; D0.44, larva treated with 0.44 µg g^{-1} DDVP concentration; D0.55, larva treated with 0.55 µg g^{-1} DDVP concentration; D0.66, larva treated with 0.66 µg g^{-1} DDVP concentration.

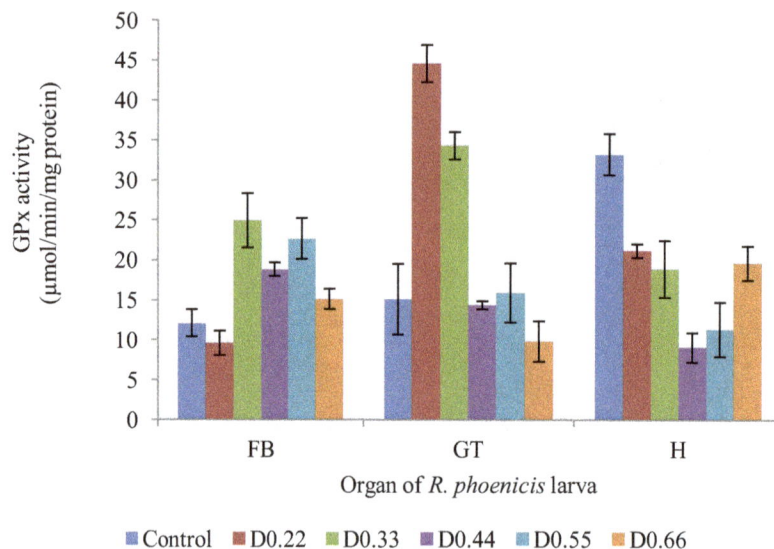

Figure 4. Effect of DDVP on glutathione peroxidase activities in the fat body (FB), gut (GT) and head (H) tissues of *Rynchophorous phoenicis* larva. Each value is the mean of five repetitions ± SEM. Values are significantly different at (P<0.05). Control, larva treated with acetone in normal saline; D0.22, larva treated with 0.22 µg g^{-1} DDVP concentration; D0.33, larva treated with 0.33 µg g^{-1} DDVP concentration; D0.44, larva treated with 0.44 µg g^{-1} DDVP concen-tration; D0.55, larva treated with 0.55 µg g^{-1} DDVP concentration; D0.66, larva treated with 0.66 µg g^{-1} DDVP concentration.

concentrations (0.22 and 0.33 µg g^{-1}) caused increased GPX activity while higher concentrations caused a decline in GPX activity when compared with the control. GPX activity in the H tissue of larva decreased significantly ($P < 0.05$) with increased DDVP concentrations when compared with the control. Highest GPX activity was recorded in the GT tissue of larva.

SOD activity

Superoxide dismutase activity in the FB, GT and H tissues of DDVP-treated *R. phoenicis* larvae are shown in Figure 5. The effect of DDVP generally caused significantly (P < 0.05) increased SOD activity in the major *R. phoenicis* tissues with increased DDVP concentrations. Highest SOD activity (50.44 U/mg protein) was observed in the GT tissue of larva treated with 0.55 µg g^{-1} DDVP solution when compared with the control. SOD activity in the larval GT tissue was about five folds more than the observable SOD activities in the FB and H tissues.

Crude ascorbic acid content in the tissues *R. phoenicis* larva

The effect of DDVP on crude ascorbic acid in the FB, GT and H tissues of *R. phoenicis* larva is shown in Figure 6.

The crude ascorbic content in the FB tissue of *R. phoenicis* larva decreased significantly ($P < 0.05$) with increased DDVP concentrations when compared with the control. Similarly, crude ascorbic contents in the GT and H tissues of larva decreased significantly ($P < 0.05$) with increased DDVP concentrations. The range of observed crude ascorbic acid contents in the FB, GT and H tissues of DDVP-treated larva were 29.47 to 42.10, 24.21 to 35.78 and 24.73 to 47.10 µg g^{-1}, respectively.

Crude glutathione content in the tissue of *R. phoenicis* larva

The effect of DDVP on crude glutathione contents in the FB, GT and H tissue of *R. phoenicis* larva is shown in Figure 7. Crude GSH content in the FB tissue of *R. phoenicis* larva decreased significantly ($P < 0.05$) with increased DDVP concentration when compared with the control. In the GT tissue, the amount of GSH was almost three times higher than those found in the FB and H tissues. Highest GSH content (55.45 µg g^{-1}) in the GT tissue of larva was observed at 0.44 µg g^{-1} DDVP when compared with the control. GSH content in the H tissue of *R. phoenicis* larva at 0.22, 0.55 and 0.66 µg g^{-1} DDVP concentrations showed similar result except for higher GSH content (29.09 µg g^{-1}) at 0.44 µg g^{-1} DDVP concen-trations when compared with the control.

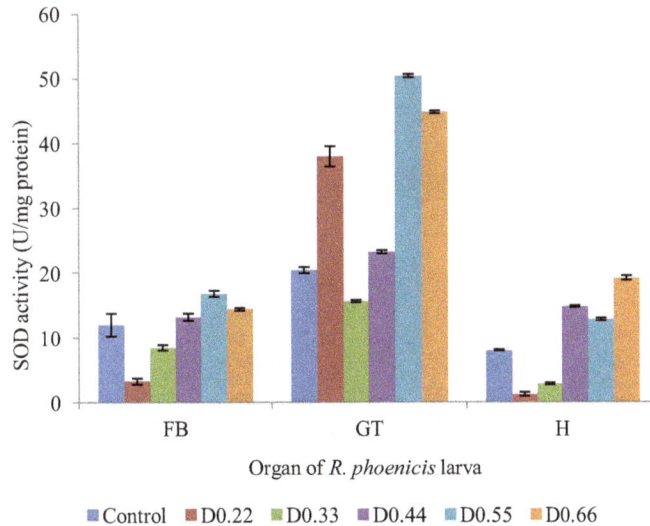

Figure 5. Effect of DDVP on suproxide dismutase activities in the fat body (FB), gut (GT) and head (H) tissue of *Rynchophorous phoenicis* larva. Each value is the mean of five repetitions ± SEM. Values are significantly different at (P<0.05). Control, larva treated with acetone in normal saline; D0.22, larva treated with 0.22 µg g^{-1} DDVP concentration; D0.33, larva treated with 0.33 µg g^{-1} DDVP concentration; D0.44, larva treated with 0.44 µg g^{-1} DDVP concentration; D0.55, larva treated with 0.55 µg g^{-1} DDVP concentration; D0.66, larva treated with 0.66 µg g^{-1} DDVP concentration.

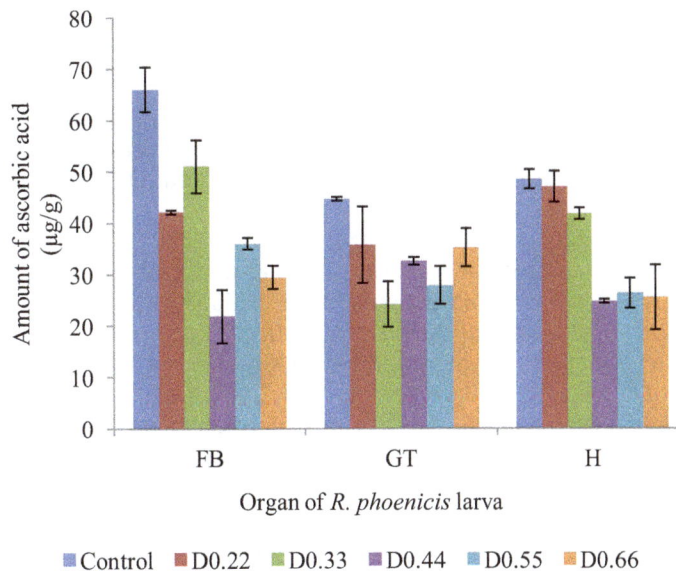

Figure 6. Effect of DDVP on crude ascorbic acid content in the fat body (FB), gut (GT) and head (H) tissues of *Rynchophorous phoenicis* larva. Each value is the mean of five repetitions ± SEM. Values are significantly different at (P<0.05). Control, larva treated with acetone in normal saline; D0.22, larva treated with 0.22 µg g^{-1} DDVP concentration; D0.33, larva treated with 0.33 µg g^{-1} DDVP concentration; D0.44, larva treated with 0.44 µg g^{-1} DDVP concentration; D0.55, larva treated with 0.55 µg g^{-1} DDVP concentration; D0.66, larva treated with 0.66 µg g^{-1} DDVP concentration.

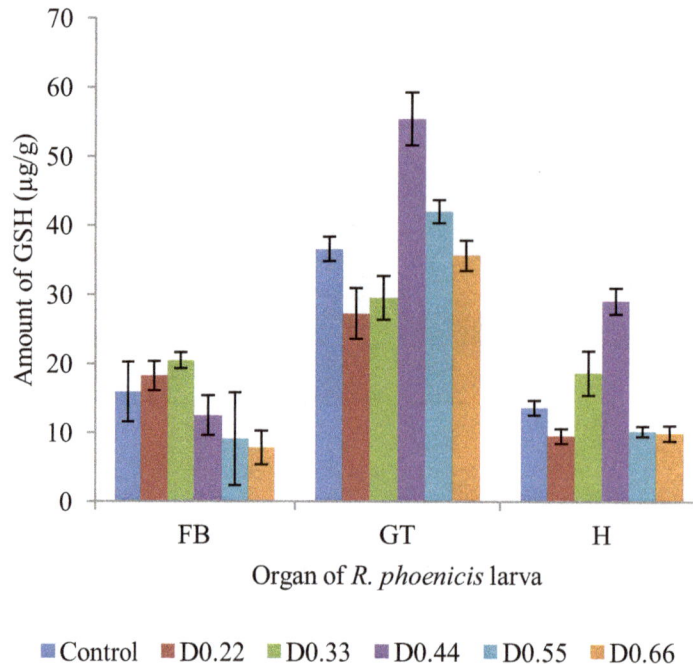

Figure 7. Effect of DDVP on crude glutathione content in the fat body (FB), gut (GT) and head (H) tissues of *Rynchophorous phoenicis* larva. Each value is the mean of five repetitions ± SEM. Values are significantly different at (P<0.05). Control, larva treated with acetone in normal saline; D0.22, larva treated with 0.22 µg g^{-1} DDVP concentration; D0.33, larva treated with 0.33 µg g^{-1} DDVP concentration; D0.44, larva treated with 0.44 µg g^{-1} DDVP concentration; D0.55, larva treated with 0.55 µg g^{-1} DDVP concentration; D0.66, larva treated with 0.66 µg g^{-1} DDVP concentration.

DISCUSSION

In this research work, the use of DDVP is aimed at providing information on the H_2O_2-scavanging enzymes in the major tissues of *R. phoenicis* at this developmental stage of its host plant amidst other protective features exhibited by the insect larva. DDVP vapour is as actively toxic to insects and other higher animals as its solution (Miles et al., 1962). Though injection method was employed in the exposure of *R. phoenicis* to DDVP, the vapour characteristic of the insecticide was also considered. The exposure of *R. phoenicis* larvae to DDVP action resulted in observable tissue physiochemical changes. The action of DDVP caused 10 and 25% body weight loss which could be traced to huge loss of body fluids. Additionally, lipid layer of fat body turned black, a deviation from the usual yellow coloration of the fat body. Leakage of fluid similar to that of the gut contents was observed during tissue damage in some cases. Tissue damages eventually resulted in the death of larvae. These observable changes indicated severe cellular damaging interactions of DDVP metabolites with various tissue organic macromolecules of *R. phoenicis* larvae. It was established by some authors (Bagchi et al., 1995; Gultekin et al., 2000)

that oxidative stress could be an important component of the mechanism of toxicity of organophosphate (OP) insecticides. OP insecticides may induce oxidative stress leading to a generation of free radicals and alterations in antioxidants or reactive oxygen species (ROS)-scavenging enzymes *in vivo* and *in vitro*. Furthermore, Lucić et al., (2002) established that DDVP caused changes in butylcholine esterase activity and lipid and lipoprotein metabolism in rats.

The present study was also aimed at getting more information on the possible interactions between DDVP and tissue macromolecules of *R. phoenicis* larva with their associated antioxidant enzymes. The significance of these *in vivo* interactions was to evaluate the possible response strategy of *R. phoenicis* larvae against DDVP and other related insecticides. Initially several concentrations of DDVP solution were used during the exposure of larvae to the insecticide. These attempts eventually resulted in the use of concentration range of 0.220 to 0.660 µg g^{-1} DDVP solution. Lethal dose (LD$_{50}$) was observed at DDVP concentration of 0.38 µg g^{-1}. Though mortality was linear with concentrations, gradual recovery of larvae from DDVP intoxication was observed even at the highest concentration. Similar result was reported by Kostaropoulos et al. (2001). Slama and Miller (1987) observed recovery of insect from insecticide poisoning weeks after exposure.

Effect of DDVP action was evaluated from the major organs of the surviving *R. phoenicis* larvae 48 h after insecticide exposure. From the observable changes in the activities of antioxidant and non-antioxidant enzymes, the test chemical compound was highly toxic to the larva. The degree of DDVP toxicity varied from one larva to another within the same group and this was cumulatively evaluated to give possible overall effect of DDVP on larvae tissue macromolecules particularly in the major organs of *R. phoenicis* larva. Interestingly, activities of APX, CAT, and GPX were probably induced or inhibited by the action of DDVP metabolites in the organs evaluated. In these organs, results of tissue damages- both visual analysis and chemical changes suggested oxidative stress. This might be due to the action of DDVP metabolites generated during the insecticide metabolism in the larvae. Oxidative stress was observed by Ridnour et al. (2005) as the overall cellular-damaging effect of free radicals (also called reactive oxygen species) generated by toxic chemical compounds in the biological system having deleterious interactions with cellular macromolecules such as protein, membrane carbohydrates and lipids and DNA molecules. Reactive oxygen species (ROS) are products of normal cellular metabolism. ROS are well recognized for playing deleterious species, since they can be either harmful or living systems (Valko et al., 2006).

The effect of DDVP on both the antioxidant and non antioxidant enzyme activities in the FB tissues of *R. phoenicis* larva indicated possible generation of reactive oxygen species from the insecticide. SOD activity increased with increasing DDVP concentration, suggesting that SOD was probably stimulated by scavenging superoxide radicals ($°O_2^-$). Tissue APX, GPX and CAT activities suggested increased level of hydrogen peroxide (H_2O_2) generation in the FB tissue larva exposed to the insecticide. Hydrogen peroxide is most efficiently scavenged by the enzyme GPX and APX which require GSH and ascorbic acid (ASA) respectively as the electron donors (Noctor and Foyer, 1998). Together with the antioxidants ascorbic acid and glutathione, these enzymes provide cells with highly efficient machinery for detoxifying H_2O_2. Reduced GSH and AA contents in the FB tissue of *R. phoenicis* clearly indicated utilization of the two antioxidants by GPX, APX and possibly other GSH - / AA - utilizing enzymes. The balance between superoxide-converting enzymes (SODs) and the different H_2O_2-scavenging enzymes in cells is considered to be crucial in determining the steady-state level of O^{2-} and H_2O_2 (Bowler, 1991). The cellular pools of the ASA and GSH are maintained in their reduced state by a set of enzymes capable of using NAD(P)H to regenerate oxidized glutathione or ascorbic acid e.g. monodehydroascorbate reductase, dehydroascorbate reductase and glutathione

reductase. Though increased APX and GPX activities were observed in the FB tissue of *R. phoenicis* larva with increased DDVP concentrations, increased cellular concentration of H_2O_2 probably had inhibitory effects on the antioxidant enzymes thereby reducing the enzyme activity in some cases. This inhibitory effect was observed more on CAT activity. Consequently, the recovery process of *R. phoenicis* larva from DDVP intoxication could not be traced to the antioxidant activities in the FB tissue of *R. phoenicis* larva.

Expression of APX, GPX and CAT activities in the GT tissues of *R. phoenicis* larvae were unexpectedly on the high side when compared with activities of theses enzymes in the two other tissues (Figures 2, 3 and 4). Though suspected cellular-damaging DDVP-related ROSs were initiated in the FB tissues of the larva, effect of DDVP could be mentioned here as not being seriously felt in the GT compared with the two other major larval tissues. This indication is probably on the account of the role of GT tissue in the removal of toxic substance in contact with the insecticide directly or indirectly. The GT tissue of *R. phoenicis* larva revealed significant increase in APX activity (Figure 1) with increased DDVP concentrations and this might be explained from the point of view of the H_2O_2-scavaging activities of the APX. The result of APX activity then shifted our attention to the activities of CAT and GPX in GT tissue of *R. phoenicis* larva. CAT activity (Figure 3) in the GT tissue of larva seems to be mainly involved in the elimination of H_2O_2 as suggestively indicated by the increased activity of this antioxidant enzyme with increased DDVP concentration. On the contrary, GPX activity (Figure 4) in the GT tissue was probably inhibited by DDVP metabolite possibly accumulated in the GT tissue for elimination. Since GSH is required by GPX as the first substrate for its activity, GSH depletion (Figure 6) could be a possible reason amongst others for the observed reduced GPX activity in the GT tissue of larva. Furthermore, the antioxidants AA and GSH are capable of non enzymatic interactions with generated deleterious ROSs in the system- a form through which the depletion of antioxidants occur though not a major route.

A completely different picture of the tissue-DDVP damaging effect was revealed in the H tissue of *R. phoenicis* larva. The H tissue of larva was speculated to be composed of antioxidant and non antioxidant enzymes. This speculation was based on the fact that the head is the control center of larva biological activities (such as feeding, movement and so on) and constantly involved in creation of tunnel in the host plant. The results on the effect of DDVP on the activities of APX, CAT and GPX revealed that these major H_2O_2-scavaging enzymes in the H tissue of *R. phoenicis* larva were inhibited. APX showed a promising H_2O_2-scavaging ability in the H tissue than CAT and GPX. However the expression level of CAT and GPX activities indicatively participated in H_2O_2-scanvaging in the H tissue prior to their inhibition. In addition, the antioxidant AA and GSH were equally depleted probably as a result of increased H_2O_2 and other related ROSs.

The results indicate cellular macromolecular-damaging effects of DDVP on the major tissue antioxidant and non-antioxidant enzymes of the African black palm weevil (*R. phoenicis*) larva. DDVP was observed to be a ROS generating compound which caused oxidative stress consequently leading to the death of treated insects. However, the recovering larvae demonstrated diverse expression of H_2O_2-scanvaging enzyme (APX, CAT, and GPX) activities. Removal of DDVP metabolite in which CAT and SOD play important roles supported larval gut as a detoxifier of toxic substances. Though antioxidant AA and GSH were depleted, they also play important role in the major tissues of *R. Phoenicis* larva. Though alteration of antioxidant enzyme activities occurred in the larva, there is the likelihood of resistance in this developmental stage of the insect. Other factors are being considered to further explain the connection between the effect of DDVP on the cellular macromolecules and the adaptability of *R. phoenicis* to its environment - host plant in relation to insecticide control. These will help put in place a better control method and improve oil palm product yield.

REFERENCES

Aebi H (1984). Catalase, *in vitro*. Methods Enzymol. 105: 121-126.

Anjum NA, Umar S, Ahmad A (2011). Oxidative stress in plants: Causes, consequences and tolerance. IK International Publishing House Pvt. Ltd., New Delhi. pp. 50-59.

Anjum NA, Umar S, Chan MT (2010). Ascorbate - glutathione pathway and stress tolerance in plants. Springer, Dordrecht, The Netherlands.

Apel K, Hirt H (2004). Reactive oxygen species: metabolism, oxidative stress, and signal transduction. Ann. Rev. Plant Biol. 55: 373- 399.

Azam KM, Razvi SA (2001). Infestation of Rhynchophorus ferrugineus in relation to off-shoots on date palm trunk and its management. In Proceedings of the Meeting on *Rhynchophorus ferrugineus*.

Bagchi D, Bagchi M, Hassoun EA, Stohs SJ (1995). In vitro and in vivo generation of reactive oxygen species: DNA damage and lactate dehydrogenase leakage by selected pesticides. Toxicology 104:129-140.

Barranco P, De la Pena J, Martin MM, Cabello T, De la Pena J (1998). Efficiency of chemical control of the new palm pest Rhynchophorus ferrugineus. Bolet. Sanid. Veget. Plagas. 24:301-306.

Beauchamp C, Fridovich I (1971). Improve assays and an assay applicable to acrylamide gels. Anal. Biochem. 44: 276-287

Bowler C (1991). Manganese superoxide dismutase can reduce cellular damage mediated by oxygen radicals in transgenic plants. Embo J. 10: 1723 – 1732.

Faleiro JR, Al-Shuaibi MA, Abraham VA, Kumar TP (1999). A technique to assess longevity of the pheromone (Ferrolure) used in trapping Rhynchophorous ferrugineus. J. Sci. Res. Agric. Sci. 4: 5-9.

Fasoranti JO, Ajiboye DO (1993). Some Edible Insects of Kwara State, Nigeria. Am. Entomol. 39: 113-116.

Finney DJ (1970). Probit Analysis, 3rd ed. Cambridge University Press, London.

Gill SS, Tuteja N (2010). Reactive oxygen species and antioxidant machinery in abiotic stress tolerance in crop plants. Plant Physiol. Biochem. 48: 909- 930.

Griffith OW, Meister A (1979). Potent and specific inhibition of glutathione synthesis by buthionine sulfoximine(s-n-butylhomocystine sulfoximine). J. Biol. Chem. 254:7558-7560.

Gultekin F, Ozturk M, Akdogan M (2000). The effects of organophosphate insecticide chlorpyrifos-ethyl on lipid peroxidation and antioxidant enzymes (*in vitro*). Arch. Toxicol. 74: 533-538.

Kostaropoulos I, Papadopoulos AI, Metaxakis A, Boukouvala E, Papadopoulou-Mourkidou E (2001). Glutathione S- transferase in the defense against pyrethroids in insects. Insect Biochem. Mol. Biol. 31: 313- 319.

Lowry OH, Rosebrough NJ, Farr AL, Randall RJ (1951). Protein measurement with the Folin phenol reagent. J. Biol. Chem. 193: 265-275.

Lucić A, Bradamante V, Radić B, Peraica M, Domijan A, Fuchs R, Stavljenić- Rukavina A (2002). The effect of dichlorvos treatment on butyrylcholinesterase activity and lipid metabolism in rats. Arch. Toxicol. 53: 275- 282.

Miles JW, Pearce GW, Woehst JE (1962). Stable formulations for sustained release of DDVP. Agric. Sci. Food Chem. 10: 240-244

Miller G, Shulaev V, Mittler R (2008). Reactive oxygen signaling and abiotic stress. Plant Physiol. 133: 481- 489.

Nakano Y, Asada K (1981). Hydrogen peroxide is scavenged by ascorbate-specific peroxidase in spinach chloroplasts. Plant Cell Physiol. 22: 867- 880.

Noctor G, Foyer CH (1998). Ascorbate and glutathione: keeping active oxygen under control. Ann. Rev. Plant Physiol Plant Mol Biol. 49: 249- 279

Omaye ST, Turnbull JD, Sauberilich HE (1979). Selected methods for the determination of ascorbic acid in animal cells tissues and fluids. Methods in Enzymology, Academic press, New York. pp. 3-11.

Paglia DE, Valentine WN (1967). Studies on the quantitative and qualitative characterization of erythrocytes glutathione peroxides. J. Lab. Clinic. Med. 70(1): 158-169.

Peter C (1989). A note on the mites associated with Rhynchophorus ferrugineus in Tamil Nadu. J. Insect Sci. 2: 160-161.

Ridnour LA, Isenberg JS, Espey MG, Thomas DD, Roberts DD, Wink DA (2005). Nitric oxide regulates angiogenesis through a functional switch involving thrombospondin-1. Proc. Natl. Acad. Sci. USA 102: 13147- 13152.

Slama K, Miller TA (1987). Insecticide poisoning: disruption of a possible autonomic function in pupae of Tenebrio molitor. Pest. Biochem. Physiol. 29: 25- 34.

Thomas CN, Ogbalu OK, Okwakpam BA (2004). Oviposition of Rhynchophorus phoenicis (f.) (Coleoptera: curcuuonidae) in Palms of the Niger delta. Indian J. Agric. Res. 38: 126 - 130.

U.S. Centers for Disease Control Agency for Toxic Substances and Disease Registry (ATSDR 1997).Toxicological Profile for Dichlorvos. http://www.atsdr.cdc.gov/toxprofiles/tp88.pdf.

U.S. Environmental Protection Agency (USEPA 2006). Interim Reregistration Eligibility Decision for Dichlorvos (DDVP). http://www.epa.gov/oppsrrd1/REDs/ddvp_ired.pdf

Valko M, Rhodes CJ, Moncol J, Izakovic M, Mazur M (2006). Free radicals, metals and antioxidants in oxidative stress-induced cancer. Chem. Biol. Interactions. 160: 1- 40.

Effects of ethanolic and aqueous leaf extracts of *Landolphia owariensis* on the serum lipid profile of rats

Nwangwu Spencer C.[1*], Ike Francisca[1], Olley Misan[2], Oke James M.[3], Uhunmwangho Esosa[1], Amegor, O. F.[4] Ubaoji Kingsley[5] and Nwangwu Udoka[5]

[1]Department of Biochemistry, Faculty of Basic Medical Sciences, Igbinedion University, P. M. B. 0006, Okada, Nigeria.
[2]Pathology Department, Igbinedion University Teaching Hospital, P. M. B. 0006, Okada, Nigeria.
[3]College of Pharmacy, Department of Pharmaceutical Chemistry, Igbinedion University, P. M. B. 0006, Okada, Nigeria.
[4]Department of Medical Laboratory Sciences, Igbinedion University, P. M. B. 0006, Okada, Nigeria.
[5]Department of Biochemistry, Faculty of Natural Sciences, Nnamdi Azikiwe University Awka, Nigeria.

The lipid profile of normal adult male rats administered both ethanolic and aqueous leaf extracts of *Landolphia owariensis* (P. Beauv) were determined. The animals were distributed into two sets of four groups with five animals in each group. Each set had one group, which served as control while the other three groups in the two sets were administered different concentrations of the ethanolic and aqueous leaf extracts. The control groups were administered normal saline and the other groups' 100, 200 and 300 mgkg^{-1} of the ethanolic and aqueous extracts respectively, twice daily for two weeks. The Total cholesterol (TC), Triacylglyceride (TAG), High Density Lipoprotein-Cholesterol (HDL-C), Low Density Lipoprotein- Cholesterol (LDL-C), and Very Low Density Lipoprotein (VLDL) levels were determined in both sets by colorimetric methods. The ethanolic extract showed a marked reduction of 87.45% in LDL-C level with the 100 mgkg^{-1} dose, though effect of all the three concentrations were significant but depreciated with increase in concentration. Animals administered both the extracts at all three concentrations increased in their HDL-C levels, but effect was pronounced in 100 and 200 mgkg^{-1} with 15 and 150% increases respectively, in the aqueous extract group. There were dose-dependent reductions of TC levels, with the 100, 200 and 300 mgkg^{-1} with reduction of 40.78, 37.59 and 34.56% respectively, in ethanolic extracts. There were 50.55 and 55.33% reduction in 100 and 200 mgkg^{-1} of the aqueous extracts on TAG level. The results are indicative of the hypocholesterolaemic potentials of *L. owariensis* leaf extracts.

Key words: Aqueous extract, ethanolic extract, Hypocholesterolaemia, *Landolphia owariensis*.

INTRODUCTION

Most primitive tribes possess expert knowledge of medicinal plants, which number at times in hundreds (Singerist, 1951). Although, modern medicine may be available in developing countries, but the use of herbs for treatment and management of diseases has often maintained popularity for historical and cultural reason. The practice has gained more grounds, as traditional medicine has become a topic of global importance (Zhang, 1999). This age long practice (Sofowora, 1982) encouraged research into pharmacologic activities of plant secondary metabolites and has improved modern

pharmacotherapeutics around the world (Nwaogu et al., 2007).

Landolphia owariensis commonly called vine rubber is widely used for treatment of many ailments (Owoyele et al., 2001). It is widely used in the sub-Sahara of Africa. The decoction of its leaves is used as a purgative and treatment of malaria (Gill, 1992). In some tribes, extracts of roots soaked in local gin is used to cure gonorrhoea (Gill, 1992). The leaf extracts has been proven to have anti-inflammatory and analgesic activities (Owoyele et al., 2001).

Also, Lewis and Lewis (1977) demonstrated the use of the stem bark as vermifuge. People of the French Equatorial Africa use the latex as enema for intestinal worms (Irvine, 1961). The latex is also used as natural preserva-

*Corresponding author. E-mail: scon333@yahoo.com.

Table 1. Effect of ethanolic extracts of *L. owariensis* on the serum TC, TAG, HDL-C, LDL-C and VLDL levels (mg/dl).

	Control	100 mg/kg	200 mg/kg	300 mg/kg
TC	92.09 ± 13.2	54.54 ± 4.8*	57.47 ± 9.2*	60.26 ± 10.3*
TAG	93.70 ± 7.1	131.13 ± 12.5	68.13 ± 11.0	88.15 ± 12.7
HDL-C	11.03 ± 8.2	18.85 ± 10.4*	28.47 ± 18.3*	19.07 ± 12.1*
LDL-C	75.31 ± 3.1	09.45 ± 4.7*	15.33 ± 8.2*	23.06 ± 10.4*
VLDL	17.70 ± 2.9	26.23 ± 7.2*	13.63 ± 23.7	17.63 ± 3.6

Values * given as mean ± standard deviation had significant differences, when compared with the control ($p \leq 0.05$) as determined by t-test.

preservative (Anthony, 1995). The leaf extracts has been shown to have antimicrobial activity (Ebi and Ofoefule, 1997). This was corroborated by the works of Nwaogu et al. (2007).

This study was aimed at the investigation of the effect of different concentrations of aqueous and ethanolic leaf extracts of *L. owariensis* on the serum lipid profile of rats. This is with the view of exercising restraint in the use and exploitation of other possible potentials of the plant that is beginning to gain wider acceptance in Nigeria.

MATERIALS AND METHODS

Experimental animal

The experimental animals (*Rattus norvegicus*), all male, which weighed 120 - 160 g used for the research work was obtained from the animal house of the College of Health Sciences, Igbinedion University, Okada. They were acclimatized and housed in plastic cages. The rats were fed and given water *ad libitum*. The temperature of the room was maintained at $25 \pm 2°C$ throughout the whole experimental period.

Experimental design

The animals were randomly selected and grouped. There were a total of eight groups with five animals per group. The animals were distributed into two sets of four groups. Each set had one group which served as control while the other three groups in the two sets were administered different concentrations of the ethanolic and aqueous leaf extracts. The control groups were administered normal saline while the other groups were administered 100, 200 and 300 mgkg[-1] of the ethanolic and aqueous extracts respectively twice daily for two weeks. The ethanolic and aqueous extracts were administered orally in normal saline. The animals were sacrificed after two weeks by cervical dislocation and blood samples were collected by cardiac puncture.

Lipid profile assay

The determination of the Total cholesterol (TC) in the serum was by the method of Searcy and Berquist (1960), which utilized the cholesterol oxidase. The LDL-cholesterol and HDL-cholesterol were determined according to the methods described by Friedwald et al. (1972). Serum triacylglyceride (TAG) was determined using the method of Tiez (1990) while VLDL-cholesterol was calculated using the formula of Friedwald et al. (1972).

Statistical analysis

The results obtained in the research work were expressed as mean ± standard deviation. The difference between mean values was assessed for significance by student T-test (SPC-XL, 2000) at $P \leq 0.05$ level of significance.

RESULTS

The results of the effect of ethanolic extracts of *L. owariensis* on lipid profile of rats as shown in Table 1 shows that the ethanolic extract of *L. owariensis* reduced the serum Total cholesterol (TC) of rats at all three concentration. This result suggests that, the reducing ability of the extract increased as the concentration of extract reduced making the list concentration administered (100 mgkg[-1]) the most potent. The same pattern of result was seen with Low Density Lipoprotein- Cholesterol (LDL-C). Though the ethanolic extract of *L. owariensis* increased the serum High Density Lipoprotein-Cholesterol (HDL-C), the result did not show any dependence on concentration. These results show fluctuations in the serum Triacylglyceride (TAG), which was not significant at any concentration when compared with the control. The Very Low Density Lipoprotein (VLDL) serum level increased significantly only at the 100 mgkg[-1] and fluctuated non-significantly with other concentrations.

The results of the effect of aqueous extracts of *L. owariensis* on lipid profile of rats as shown in table 2 shows that the serum Total cholesterol (TC) reduced significantly at 100 and 300 mgkg[-1] concentrations. The reduction was concentration dependent, though there was no reduction with 200 mgkg[-1] concentration. The Low Density Lipoprotein-Cholesterol (LDL-C) serum levels reduced significantly at all three concentrations administered.

The 100 mgkg[-1] extract reduction was drastic as shown in Table 2. The administration of aqueous extracts of *L. owariensis* showed an increase in the serum High Density Lipoprotein-Cholesterol (HDL-C). The increase was concentration dependent. The Triacylglyceride (TAG) and Very Low Density Lipoprotein (VLDL) serum levels as shown in Table 2 reduced significantly with the 100 and 200 mgkg[-1]. The reduction seems to be concentration dependent with the two concentrations.

Table 2. Effect of aqueous extracts of *L. owariensis* on the serum TC, TAG, HDL-C, LDL-C and VLDL levels (mg/dl).

	Control	100 mg/kg	200 mg/kg	300 mg/kg
TC	93.29 ± 11.4	61.17 ± 4.7*	92.85 ± 19.9	48.42 ± 10.3*
TAG	95.50 ± 14.3	48.18 ± 13.2*	43.66 ± 12.6*	92.68 ± 12.7
HDL-C	13.02 ± 9.2	29.7 ± 19.8*	29.06 ± 18.3*	25.85 ± 10.1*
LDL-C	77.21 ± 3.1	24.9 ± 8.6*	58.11 ± 11.1*	52.07 ± 10.4*
VLDL	19.81 ± 2.9	11.38 ± 13.2*	10.48 ± 2.6*	20.28 ± 7.5

Values * given as mean ± standard deviation had significant differences, when compared with the control ($p \leq 0.05$) as determined by t-test.

DISCUSSION

The results obtained from this work suggests that though the two extracts affected the TC, LDL-C, HDL-C, TAG and VLDL serum levels in the rats, the degree varied from one extract to the other. The aqueous extracts seem to increase the HDL-C and reduce TAG and VLDL more effectively when compared with the ethanolic extracts. The ethanolic extracts were found to reduce the TC and LDL-C better than the aqueous extracts. To that effect, it will not be out of place to suggest that this degree of variation is dependent on the solvent used for extraction.
It is well known that, Cardiovascular Diseases (CVD) is the leading cause of death of men and women in developed countries (Kummerow, 1982). Hypercholesterolemia has been identified as a primary risk factor in the development of CVD. Therefore, preventing or reducing the increase in serum cholesterol is associated with reducing the risk of CVD (Onyeneke et al., 2008). In this work the ethanolic and aqueous extracts showed reduction TC and LDL-C, while there was an increase in the HDL-C serum levels. This may be attributed to changes in serum lipids, which are controlled by the enzymes responsible for lipid metabolism (Onyeneke et al., 2008).

There was increase in HDL-C serum levels of rats administered both ethanolic and aqueous extracts at all three concentration, which was not dose dependent. The aqueous extracts seem to be more effective in increasing the HDL-C levels from the result. This may be as result of the influence of the extracts on the activities of Lecithin Cholesterol acyl transferase (LCAT), a serum enzyme that esterifies free cholesterol, primarily at the surface of the HDL particle after which the cholesteryl ester molecules migrate to the inner core of this lipoprotein. Through this action, LCAT plays a key role in the maturation of HDL particles (Glomset, 1968).

This may also be attributed activities secondary plant metabolites, such as saponins, alkaloids, phenols, flavonoids and tannins detected in *L. owariensis* leaf extract, which exhibit varied biochemical and pharmacological actions in animals and even micro organisms when ingested (Trease and Evans, 1983). The ethanolic extracts of *L. owariensis* at all doses were able to reduce the serum TC more effectively than the aqueous extract. The activity may have resulted from lowering effect that can be attributed to the gut intra-lumenal interactive effect of saponins (Igwe et al., 2008).

LDL-C serum levels of rats administered both ethanolic and aqueous extracts showed decrease at all three concentration, which was dose dependent, increasing with dose. Unlike in the HDL-C, the aqueous extracts were more effective in the reduction of serum LDL-C. This agrees with Igwe et al. (2008) who surmise the extract even at the lowest dose used significantly reduced LDL-C concentration. This may affect the Low-density lipoprotein transport of cholesterol to the arteries where they can be retained by artierial proteoglycans, starting the formation of plaques. LDL-C poses a risk for cardiovascular disease when it invades the endothelium and become oxidized, since the oxidized form is more easily retained by the proteoglycans (Cromwell and Otvos, 2004).

Hypertriglyceridemia, a major risk factor for CVD which has been demonstrated to be associated with increased LPL and TGL activity (Richards et al., 1985; Kanter et al., 1985). The effect may be alleviated by effect of aqueous extracts of *L. owariensis*, which reduce TAG levels in rat serum in the experiment. The ethanolic extracts had no significant effect on the TAG levels indicating the polarity of the active ingredients.
The different extracts, showed varied effects on the lipid profile indicating variation in polarity of active ingredients. Therefore, improved extraction methods will improve efficacy of extracts. In consideration of the availability of plants and the popularity gained by phytomedicine, it is necessary to trade the path of caution in the use of these extracts whose full potentials has not been harnessed.

ACKNOWLEDGEMENTS

Special thanks to the Forestry Research Institute of Nigeria, Ibadan and the Igbinedion University Teaching Hospital for their support.

REFERENCES

Anthony CD (1995). Natural preservatives from *Landolphia owariensis*. Afr. Dev. J. 2(1): 21-22.

Cromwell WC, Otvos,JD (2004). Low density lipoprotein particles number and risk for cardiovascular disease. Curr. Atheroscler. Rep. 6: 381-387.

Doerge RF, Wilson CO, Grivol O (1971). Textbook of Organic Medicine and Pharmaceutical Chemistry. 6th edition, Lippinncot Company, Philadelphia. pp. 34-37.

Ebi GC, Ofoefule, SI (1997). Investigation into the Folkloric Antimicrobial Activities of *Landolphia owariensis*. Phytother. Res. 11(2): 149-151.

Friedwald WT, Levy RT, Fredickson DS (1972). Estimation of the concentration of low density lipoprotein cholesterol in plasma, without use of preparative ultracentrifuge. Clin. Chem. 18:499-502.

Gill LS (1992). Ethnomedical uses of Plants in Nigeria. Uniben Press. pp. 145-146.

Glomset JA (1968). The Plasma Lecithin: Cholesterol acyl transferase reaction. J. Lipid. Res. 9:155-167.

Igwe CU, Ojiako OA, Nwaogu LA, Onyeze GOC (2008). Lipid Lowering Effect Of Aqueous Leaf Extract Of Spondias Mombin Linn . Internet J. Pharmacol.(6) 1.

Irvine FR (1961). The Woody Plants of Ghana. Oxford University Press, London. pp. 26-28.

Kanter MA, Bianchini A, Bernier D, Sady SP, Thompson PD (1985). Androgen reduce HDL2-cholesterol and increase hepatic triglyceride lipase activity. Med. – Sci. Sports Exerc. 17:462-465.

Kummerow FA (1982). The possible involvement of dietary fats in atherosclerosis. Prog. Lipid. Res, 21: 743-746.

Lewis WH, Lewis MPE (1977). Medical Botany. John Wiley and Sons Publishers, New York. pp. 21-22.

Nwaogu LA, Alisi CS, Ibegbulem CO, Igwe CU (2007). Phytochemical and antimicrobial activity of ethanolic extract of *Landolphia owariensis* leaf. African J. Biotechnol. 6(7): 890-893.

Onyeneke EC, Oluba OM, Adeyemi O, Aboluwoye C, Eriyamremu CE, Ojeaburu SI, Adebisi K E, Adeyemi O (2008). Effect of Soy Protein On Serum Lipid Profile And Some Lipid-Metabolizing Enzymes In Cholesterol Fed Rats. Internet J. Alternative Med.(5) 2.

Owoyele BV, Olaleye SB, Oke JM, Elegbe RA (2001). Anti-inflammatory and Analgesic Activities of Leaf Extracts of *Landolphia owariensis*. African. J. Biomed. Res. 4: 131-133.

Richards EG, Grundy SM, Cooper K (1989). Influence of plasma triglycerides on lipoprotein patterns in normal subjects and in patients with coronary artery disease. Am. J. Cardiol; 63: 1214-1220.

Searcy RL, Berquist LM (1960). A new colour reaction for the quantification of serum cholesterol. Clin. Chim. Acta, 5, 192-199.

Singerist HE (1951). A history of medicine. Oxford University Press, New York. (1) 180

Sofowara, EA (1982). Medicinal Plants and Traditional Medicines in Africa. John Wiley and Sons Ltd, Nigeria. pp 64-79.

Tiez NW (1990). Clinical guide to laboratory tests. 2nd edn. W.B. Saundrs company, Philadelphia, U.S.A. 554.

Trease GE, Evans WC (1983). Textbook of Pharmacognosy. 12 th ed. Balliese Tindall and Company Publishers, London. pp. 343-383.

Zhang T(1999). WHO monograph on selected medicinal plants. 1: 4-1.

15

Oxidative stress biomarkers in young male rats fed with stevioside

Hala A. Awney

Department of Environmental Studies, Institute of Graduate Studies and Research, Alexandria University, Alexandria, Egypt.

Stevioside is a natural non-caloric sweetener refined from *Stevia rebaudiana Bertoni* leaves. The introduction of stevioside as a sugar substitute in the diets of diabetics and others on carbohydrate-controlled diets has been suggested, but safety issues have prevented implementation. The aim of this study was to examine antioxidant status changes in the sera, livers and kidneys of young male rats fed with low doses of stevioside (SL) or high doses of stevioside (SH) for 12 weeks. We investigated oxidative stress biomarkers such as the levels of reduced glutathione (GSH), thiobarbituric acid reactive substances (TBARS), the activities of superoxide dismutase (SOD), catalase (CAT) and glutathione reductase (GR). Our results show that SH treatment causes significant induction of TBARS in liver and kidney, accompanied by a significant reduction in SOD, CAT, GR and GSH levels in the same organs. SL treatment causes insignificant changes in SOD and CAT, but significant reduction was observed in GR, GSH and TBARS in serum and all tested organs when compared with the control. In conclusion, features of oxidative stress were detected in the liver and kidney of young male rats treated with SH for 12 weeks, whereas no significant changes in SOD and CAT were detected after SL treatment.

Key words: Stevioside, safety of natural sweeteners, food safety, oxidative markers.

INTRODUCTION

Stevioside, a white, odorless, crystalline powder, is a major diterpenoid glycoside from the leaves of *Stevia rebaudiana Bertoni* that has gained worldwide attention due to its non-caloric potent sweetness (250 to 300 times sweeter than sucrose). It has also been used in Japan and several South American countries as both a medicinal herb and a non-caloric sweetener for a variety of foods and beverages (Kinghorn et al., 1984). The US FDA has denied several attempts to market stevia as a

food additive, but steviol glycosides have been allowed as a dietary supplement since 1995 (FDA, 1995).

Furthermore, there were no objection letters from the FDA regarding a specific steviol glycoside (rebaudioside A, purity higher than 97%) after two independent self-conducted generally recognized as safe (GRAS) determinations (FDA, 2008). The Joint FAO/WHO expert committee on food additives (JECFA, 1999, 2000, 2005, 2006, 2007, 2008 and 2009) reviewed the safety of steviol glycosides and established an acceptable daily intake (ADI) of 4 mg/kg body weight (bw)/day (expressed as steviol). Steviol glycosides are chemically defined as mixtures that comprise not less than 95% stevioside and/or rebaudioside A with smaller amounts of rebaudiosides B, C, D, E and F, steviolbioside, rubusoside and dulcoside A. In April 2010, after considering all data related to stability, degradation products, metabolism and toxicology, the European food

*Corresponding author. E-mail: hawney@alex-igsr.edu.eg.

Abbreviations: SL, Low doses of stevioside; **SH,** high doses of stevioside; **GSH,** reduced glutathione; **GR,** glutathione reductase; **SOD,** superoxide dismutase; **CAT,** catalase, **TBARS,** thiobarbituric acid reactive substances.

safety authority (EFSA, 2010) concluded that steviol glycosides are not carcinogenic, genotoxic or associated with any reproductive or developmental toxicity. Complying with JECFA, the EFSA established the same ADI for steviol glycosides. This ADI is based on the application of a 100-fold uncertainty factor to the No observed adverse effect level (NOAEL) determined in a 2-year carcinogenicity study of rats fed with 2.5% stevioside, which is equal to 967 mg stevioside/kg bw/day (corresponding to approximately 388 mg steviol equivalents/kg bw/day). Although EFSA stated that steviol and some of its oxidative derivates show clear evidence of genotoxicity *in vitro*, particularly in the presence of a metabolic activation system, they reported that any concern raised by the *in vitro* genotoxicity profile of steviol is fully addressed by the fact that the genotoxic potential of steviol is not expressed *in vivo* and by the negative genotoxicity findings for steviol glycosides *in vitro* and *in vivo*.

In the past several years, studies of steviol glycoside metabolism in animals and humans have demonstrated that these compounds are poorly absorbed after oral exposure until they are hydrolyzed to steviol by the microflora in the colon (Wingard et al., 1980; Hutapea et al., 1997; Koyama et al., 2003). A large amount of steviol is absorbed; the rest is excreted in the feces so that little or no stevioside is absorbed into the blood. In the liver, steviol undergoes conjugation with glucuronic acid to form steviol glucuronide. The only interspecies difference is that the glucuronide is excreted primarily via the urine in humans and *via* the bile in rats. No accumulation of steviol glycoside derivatives occurs in the body. Aside from steviol glucuronide, no other derivatives could be detected in the urine of humans exposed orally to steviol glycosides (Gardana et al., 2003; Geuns, 2003; Geuns et al., 2007).

Oxidative stress represents an imbalance between the production of reactive oxygen species (ROS) and the biological system's ability to readily detoxify the reactive intermediates or to repair the resulting damage. An imbalance in the normal redox state of tissues can cause toxic effects through the production of ROS, such as the superoxide ion (O_2^-) and the hydroxyl ion (OH^-). These chemically unstable compounds carry free electrons that react with and destabilize other molecules, thereby inducing chain reactions. In particular, ROS damage DNA, essential cellular proteins and lipid membranes. This damage may lead to mutagenesis, carcinogenesis and cell death (Kasai, 1997). Although there is evidence linking oxidative stress with chronic diseases such as cardiovascular disease and cancer (Hoeschen, 1997; Klaunig et al., 1997; Dhalla et al., 2000), oxidative stress has not been utilized as a tool for either toxicity tests or no observed adverse effect level (NOAEL) assessment.

Stevioside is a non-caloric natural sweetener that does not induce a glycemic response since the purported

mechanism of action for steviol glycosides involves enhanced secretion of insulin from the pancreas when there is impaired response to glucose stimulation. It is an attractive sweetener for diabetics and others, such as obese people, on carbohydrate-controlled diets. Increasing evidence from both experimental and clinical studies suggests that oxidative stress plays a major role in the pathogenesis of obesity and both types of diabetes mellitus (Baynes, 1991; Ihara et al., 1999; Furukawa et al., 2004). A low caloric diet that induces minimal oxidative stress could reduce the incidence of complications from obesity and diabetes and is therefore of great interest. The introduction of stevioside as a non-nutritive and non-caloric sweetener in diets has been investigated concerning its safety, stability during different processing and storage conditions and interaction with other food ingredients or food additives (Chang and Cook, 1983; Kroyer, 1999, 2010; Clos et al., 2008). However, the *in vivo* effects of stevioside on oxidative stress have not received any research attention.

In the present study, we wanted to gain a better understanding of the antioxidant status changes in the serum, liver and kidney of young male rats exposed to low doses of stevioside (SL) or high doses of stevioside (SH) in drinking water for 12 weeks. We therefore investigated oxidative stress biomarkers such as the levels of reduced glutathione (GSH) and thiobarbituric acid reactive substances (TBARS) and the activity of the antioxidant enzymes superoxide dismutase (SOD; EC 1.15.1.1), catalase (CAT; EC 1.11.1.6) and glutathione reductase (GR; EC 1.8.1.7).

MATERIALS AND METHODS

Purity of stevioside

Stevioside (imported from Stevia Pac Food Innovation in Singapore) and stevioside standard 98% (from Sigma-Aldrich, USA) were donated to our laboratory by Professor Tarek Al-nemr of the Faculty of Agriculture, Alexandria University, Egypt. The degree of purity of the stevioside was determined by HPLC (high performance liquid chromatography) as described in Vanek et al. (2001). Analytical instrumentation consisted of the Shimadzu Solvent Delivery Module (System LC 10 AD model) and a UV–visible detector (Chromatopac CR 6A model) with a photodiode array. Analyses were performed on a reverse phase Shimpac column (250 × 4.6 mm) packed with CL–C8 (5 µm, VP-ODS, Shimadzu Corporation) at 25°C. The solvent system used was water: acetonitrile with a linear gradient from 75:15 to 50:50 over 30 min. The flow rate was 1.0 ml/min, and the amount of stevioside applied was 20 µl. All solvents and samples were filtered through a 0.45 µm filter prior to use in HPLC. All chromatograms were collected at 205 nm. Stevioside in the samples was identified by its characteristic retention time and UV spectra, and the identification was confirmed by the addition of a standard to the injected samples. Our results show that the stevioside sample was 97.8% pure. The low dose of stevioside (SL) used in this experiment was 15 mg/kg body weight/day, and the high dose (SH) was 1500 mg/kg body weight/day (equivalent to 100 times the low dose).

Animal studies

Immature (age 21 days) male Sprague–Dawley rats (n = 24) were obtained from the Institute of Graduate Studies and Research animal house in Alexandria University, Egypt. The local committee approved the design of the experiment, and the protocol conforms to the guidelines of the national institutes of health (NIH). Four animals per cage were housed in a room kept at 21°C with a 12 h light/dark cycle. All animals were given *ad libitum* access to distilled water and a standard diet that meets the nutrient requirements for growing rats. After 7 days acclimation, animals were randomly assigned (n = 8 rats/group) to three groups. The animals in these groups drank distilled water (control), low dose stevioside solution (SL) or high dose stevioside solution (SH) for 12 weeks as the sole source of liquid. The SH dose is equivalent to 1/10 of the acute oral LD_{50} for stevioside. Fluid intake was recorded daily, and the intake of the substance being tested (mg/kg body weight/day) was calculated from the mean amount of fluid consumed (ml/kg body weight/day) and the concentration of the tested substance in the solution. Solution concentrations were adjusted weekly based on the average weight of the animals and their current fluid consumption.

Blood and tissue preparation

At the end of the experiment, the rats were ether-anesthetized and sacrificed, and blood samples were collected. Serum samples were obtained by centrifugation at 860 g for 20 min and stored at -20°C until being assayed. The liver and kidney were immediately removed and weighed, then washed using chilled saline solution. Each tissue was minced and homogenized to yield a final concentration of 10% w/v in an ice-cold solution of 1.15% KCl and 0.01 M sodium in potassium phosphate buffer (pH 7.4) in a Potter–Elvehjem type homogenizer. The homogenate was centrifuged at 10,000 g for 20 min at 4°C, and the resultant supernatant was stored at -70°C until being used for enzyme assays.

Oxidative stress measurements

Plasma, liver and kidney glutathione reductase (GR; EC 1.6.4.2) activity was determined according to the method of Mannervik and Carlberg (1985). Reduced glutathione (GSH) was measured based on the reduction of 5,5′-dithiobis(2-nitrobenzoic acid) by the addition of GSH to produce a yellow compound whose concentration is directly proportional to that of GSH and whose absorbance can be measured at 405 nm (Beutler et al., 1964). Thiobarbituric acid reactive substances (TBARS) were measured in plasma, liver and kidney at 532 nm using 2-thiobarbituric acid (2, 6-dihydroxypyrimidine-2-thiol; TBA). An extinction coefficient for TBA of 156,000 M^{-1} cm^{-1} was used and the calculations were performed as described by Tappel and Zalkin (1959). Superoxide dismutase (SOD; EC 1.15.1.1) activity was measured according to the method of Nishikimi et al. (1972). Catalase (CAT; EC 1.11.1.6) activity was determined using the decomposition of hydrogen peroxide (Aebi, 1984).

Statistical analysis

Data collected was recorded, analyzed and expressed as the means ± SE (standard error). The significance of differences among experimental groups was tested by analysis of variance (ANOVA) or paired and unpaired Student's *t* test as appropriate. A *p* value of 0.05 was considered statistically significant.

RESULTS

Superoxide dismutase (SOD) activity

Animals treated with SL showed insignificant changes in SOD activity levels in the serum, liver and kidney when compared with the control. Animals treated with SH showed significant decreases in SOD activity levels in the liver (0.38 ± 0.03 U/mg protein) and kidney (0.20 ± 0.011 U/mg protein). No difference was detected in serum (0.22 ± 0.02 U/mg protein) when compared with the corresponding control (Figure 1).

Catalase (CAT) activity

As shown in Figure 2, there was no significant difference in CAT activity between the SL treatment group and the control group in serum and all other organs. SH treatment causes significant decreases in CAT activity in the liver and kidney when compared with the control. The change in serum CAT activity was insignificant.

Glutathione reductase (GR)

GR activity is significantly reduced after both SL and SH treatments in the serum, liver and kidney (Figure 3). Acute reductions were detected in the liver following both SL treatment (89.61 ± 5.14 U/mg protein) and SH treatment (82.56 ± 3.03 U/mg protein) when compared with the control (139.34 ± 9.17 U/mg protein) and in the kidney following both SL treatment (9.90 ± 2.68 U/mg protein) and SH treatment (3.01 ± 0.71 U/mg protein) compared with the control (42.17 ± 6.18 U/mg protein). Moderate reductions in GR activity were detected in the serum of rats treated with either SL or SH compared with the control. SL and SH groups had serum GR activity of 59.54 ± 8.6 U/mg protein and 49.32 ± 6.3 U/mg protein, respectively, whereas the control group had a serum GR activity of 67.98 ± 2.9 U/mg protein.

Reduced glutathione (GSH)

Reduced glutathione (GSH) occupies a prominent position as the main water-soluble non-enzymatic antioxidant in the cell membrane. GSH levels are affected by SL and SH treatments in the serum, liver and kidney (Figure 4). After SH treatment, a severe reduction in GSH levels was detected in the serum (5.94 ± 0.78 mmol/g tissue), liver (6.66 ± 0.44 mmol/g tissue) and kidney (0.42 ± 0.15 mmol/g tissue) when compared with the corresponding controls. A smaller reduction in GSH levels was detected in the SL group in the serum (10.87 ± 2.44 mmol/g tissue), liver (19.97 ± 1.00 mmol/g tissue)

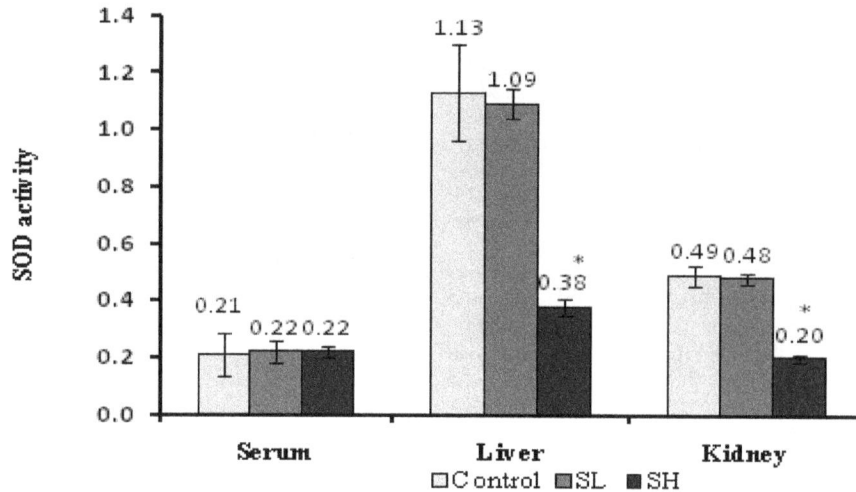

Figure 1. Superoxide dismutase (SOD) activities (U/ mg protein) in serum, liver and kidney of young male rats treated with SL (15 mg/kg body weight) or SH (1500 mg/kg body weight) comparing with control after 12 weeks. SOD values were determined as described in Materials and methods section and were expressed as the mean ± SE of eight independent determinations. *: Significantly different from the corresponding control value ($p < 0.05$).

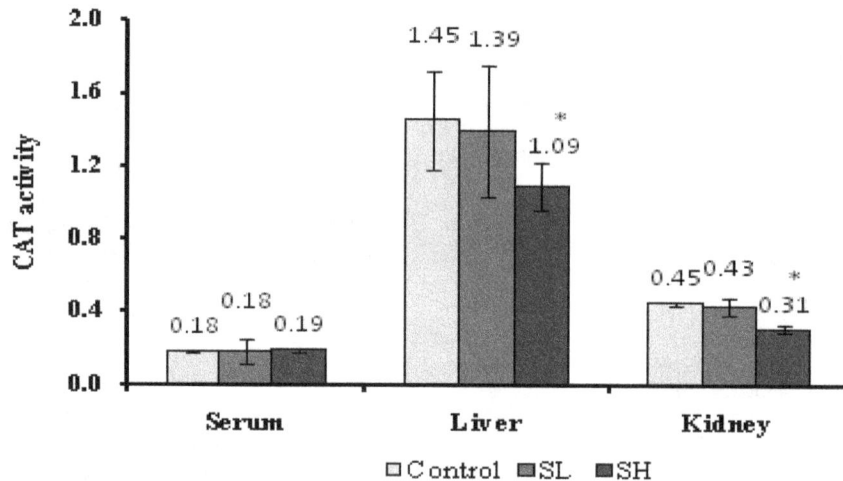

Figure 2. Catalase activities (CAT) (U/ mg protein) in serum, liver, kidney and brain of young male rats treated with SL (15 mg/kg body weight) or SH (1500 mg/kg body weight) comparing with control after 12 weeks. CAT values were determined as described in Materials and methods section and were expressed as the mean ± SE of eight independent determinations. *: Significantly different from the corresponding control value ($p < 0.05$).

and kidney (0.56 ± 0.13 mmol/g tissue) when compared with control values.

Thiobarbituric acid reactive substances (TBARS)

As shown in Figure 5, there were insignificant changes in the levels of serum TBARS in SL- and SH-treated rats (4.66 ± 0.07 and 6.01 ± 0.40 nmol/mg protein, respectively) when compared with control rats (5.25 ± 0.20 nmol/mg protein). In rat liver homogenates, the level of TBARS decreased following SL treatment (36.72 ± 2.99 nmol/g tissue) but increased after SH treatment

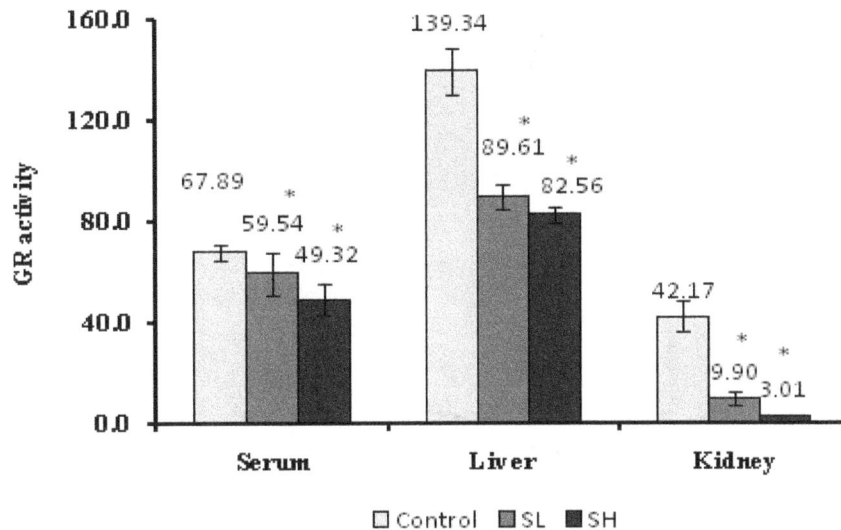

Figure 3. Glutathione reductase (GR) activities (U/ mg protein) in serum, liver, kidney and brain of young male rats treated with SL (15 mg/kg body weight) or SH (1500 mg/kg body weight) comparing with control after 12 weeks. GR values were determined as described in Materials and methods section and were expressed as the mean ± SE of eight independent determinations. *: Significantly different from the corresponding control value ($p < 0.05$).

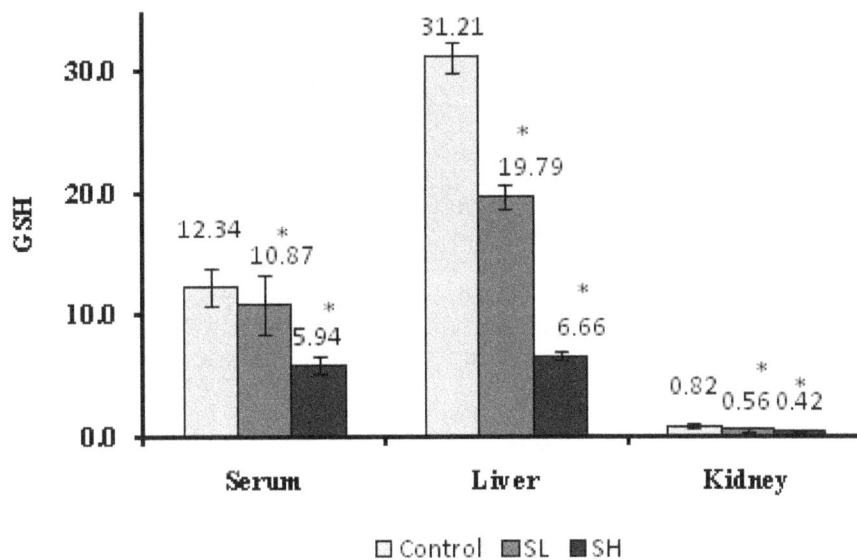

Figure 4. Reduced glutathione (GSH) in serum (mmol/g protein), liver (mmol/g tissue), kidney (mmol/g tissue) and brain (mmol/g tissue) of young male rats treated with SL (15 mg/kg body weight) or SH (1500 mg/kg body weight) comparing with control after 12 weeks. GSH values were determined as described in Materials and methods section and were expressed as the mean ± SE of eight independent determinations. *: Significantly different from the corresponding control value (p < 0.05).

(52.37 ± 2.84 nmol/g tissue) when compared with the control (45.93 ± 3.08 nmol/g tissue). The same pattern of results was obtained in kidney homogenates: the level of TBARS was significantly decreased following SL treatment (37.04 ± 0.41 nmol/g tissue) and increased after SH treatment (68.99 ± 9.27 nmol/g tissue) when

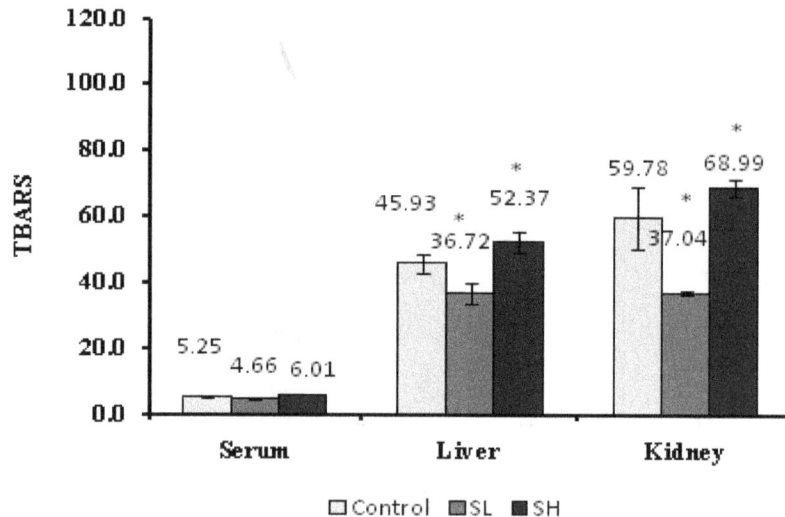

Figure 5. Thiobarbituric acid-reactive substances (TBARS) in serum (nmol/mg protein), liver (nmol/g tissue), kidney (nmol/g tissue) and brain (nmol/g tissue) of young male rats SL (15 mg/kg body weight) or SH (1500 mg/kg body weight) comparing with control after 12 weeks. TBARS values were determined as described in Materials and methods section and were expressed as the mean ± SE of eight independent determinations. *: Significantly different from the corresponding control value ($p < 0.05$).

compared with the corresponding control (59.78 ± 2.34 nmol/g tissue).

DISCUSSION

Several toxicological assessments of stevioside have suggested that it is a relatively safe compound, and this assessment has been continually reviewed by different food safety authorities (FDA, 1995; FSANZ, 2004; JECFA, 2008; EFSA, 2010). Stevioside has a very low acute oral toxicity with an LD_{50} value > 15 g/kg body weight in rodent species (JECFA, 1999) and an ADI for steviol glycosides, expressed as steviol equivalents, of 4 mg/kg bw/day. However, steviol and some of its oxidative derivates show clear evidence of genotoxicity in vitro but that the genotoxic potential of steviol is not expressed in vivo (EFSA, 2010). In this study, we examined oxidative stress biomarkers in young male rats fed high doses (1500 mg/kg body weight/day) or low doses (15 mg/kg body weight/day) of stevioside for 12 weeks during their early stage of life.

Our data shows that serum SOD and CAT activities are not significantly different when either the SH or SL group is compared with the control group (Figures 1 and 2). Furthermore, SL treatment did not affect SOD or CAT activity levels in the liver and kidney. However, SH treatment lowered SOD and CAT activity levels in liver and kidney tissues when compared with the corresponding controls. These results suggest that SH

treatment affects the activities of SOD and CAT in the liver and kidney of young male rats. SOD and CAT are metalloenzymes involved in the cellular defense against oxygen cytotoxicity. It seems reasonable to assume that these two enzymes act in a concerted fashion because SOD catalyzes O_2^- dismutation, producing H_2O_2, while CAT removes H_2O_2 (Mavelli et al., 1982).

We also observed remarkable decreases in GR activity and GSH levels in the serum, liver and kidney of rats treated with SH, but only moderate reductions in the same tissues in SL-treated rats compared with corresponding controls (Figures 3 and 4). GR is the enzyme that reduces glutathione disulfide (GSSG) to the sulfhydryl form (GSH). GSH plays an important protective antioxidant role against free radicals and is a main water-soluble non-enzymatic antioxidant in the cell membrane. As its concentration tends to decrease during oxidative insults, GSH is a relevant biomarker of oxidative stress conditions, and determination of its concentration is very informative (Swiderska-Kołacz et al., 2007). The reduction of GR activity and therefore the subsequent reduction in GSH in serum might be due to the lack of adequate amounts of NADPH in cells. NADPH is required for GR to reduce the GSSG to GSH that the cells need as an antioxidant. The original source of NADPH is from the breakdown of glucose in the pentose phosphate pathway to generate glucose-6-phosphate dehydrogenase. Lack of cellular glucose could be expected in animals fed diet containing SH dose (1500 mg/kg equivalent to 592.6 mg/kg steviol). The effects of stevioside (MW 804.9) and steviol (MW 318) on glucose absorption have been investigated by Toskulkao et al. (1995). Steviol at a concentration of 1 mM inhibits glucose absorption by about 40% which indirectly lead to an inadequate amount of NADPH and reduce the GR activity in cells.

High levels of TBARS were detected in the liver and kidney of the SH group when compared with the control group. These data support the hypothesis that SH treatment may induce lipid peroxidation as a pro-oxidant while at SL treatment has an anti-oxidant effect in the liver and kidney of young male rats. The detection and measurement of lipid peroxidation is the evidence most frequently cited to support the involvement of free-radical reactions in toxicology and tissue damage (Gutteridge, 1995). Histopathological changes in the livers of rats treated with stevioside were reported by Aze et al. (1991), but the EFSA panel considered these effects to be nonspecific because of the lack of a dose–response relationship. In addition, Nunes et al. (2007) used a comet assay to show that Wistar rats treated with stevioside (4 mg/kg) through oral administration (*ad libitum*) for 45 days had chromosomal lesions in peripheral total blood as well as in liver, brain and spleen cells. However the EFSA panel noted a number of factors that limit the interpretability and utility of this study in assessing the safety of stevioside.

In a previous study, Awney et al. (2010) recognized that there was a significant decrease in body weight gain and feed intake in the SH group (1500 mg/kg bw/day) when compared with the control group, which could lead to malnutrition in SH-treated young male rats. In addition, the liver weight to body weight ratio was significantly lower in the SH group than in the control group, while significant increases in testes, epididymis, kidney and brain were observed in the SH treatment group when compared with the control group.

The data from this study when combined with our previous data indicate that oxidative stress is associated with the reduction in body weight gain and organ weight in young male rats fed high doses of stevioside (1500 mg/kg body weight/day) for 12 weeks. Further studies are needed to clarify the role of steviol glycosides in affecting the oxidative stress biomarkers of living organisms.

ACKNOWLEDGMENT

The author is grateful to Dr. Mona Massoud, Sugar Crops Research Institute, Agriculture Research Center, Ministry of Agriculture, Egypt, for her distinguished assistance during the preparation of this work.

REFERENCES

Aebi H (1984). Catalase *in vitro*. Methods Enzymol., 105: 121-126.
Awney H, Massoud M, El-Maghrabi S (2010). Long-term feeding effects of stevioside sweetener on some toxicological parameters of growing male rats. J. Appl. Toxicol., 31: 431-438.
Aze Y, Toyoda K, Imaida K, Hayashi S, Imazawa T, Hayashi Y, Takahashi M (1991). Subchronic oral toxicity study of stevioside in F344 rats. Eisei Shikenjo Hokoku.109: 48-54.
Baynes JW (1991). Role of oxidative stress in development of complications in diabetes.Diabetes, 40(4): 405-412.
Beutler E, Duron O, Kelly MB (1964). Improved method for the determination of blood glutathione. J. Lab. Clin. Med., 61: 882-888.
Chang SS, Cook JM (1983). Stability studies of stevioside and rebaudioside A in carbonated beverages. J. Agric. Food Chem., 31: 409-412.
Clos JF, DuBois GE, Prakash I (2008). Photostability of rebaudioside A and stevioside in beverages. J. Agric. Food Chem., 56: 8507-8513.
Dhalla N, Temsah RM, Netticadan T (2000). Role of oxidative stress in cardiovascular diseases. J. Hypertens., 18(6): 655-673.
FDA (Food and Drug Administration) (2008). Center for Food Safety and Applied Nutrition (CFSAN)/Office of Food Additive Safety, December 17, 2008. Agency Response Letter GRAS Notice No. GRN 000253.
FDA (Food and Drug Administration) (1995). Center for Food Safety and Applied Nutrition (CFSAN) /Office of Food Additive Safety. Attachment revised 12/19/95. IA#45-06, REVISED 2/2/96.
EFSA (2010). Scientific Opinion on safety of steviol glycosides for the proposed uses as a food additive. EFSA J., 8(4): 1537-1622.
FSANZ (Food Standards Australia New Zealand) (2004). Consumption of intense sweeteners in Australia and New Zealand: Benchmark Survey 2003. Evaluation Report Series No. 8. ISBN: 0 642 34598 8.
Furukawa S, Fujita T, Shimabukuro M, Iwaki M, Yamada Y, Nakajima Y, Nakayama O, Makishima M, Matsuda M, Shimomura I (2004). Increased oxidative stress in obesity and its impact on metabolic syndrome. J. Clin. Invest., 114(12): 1752-1761.
Gardana C, Simonetti P, Canzi E, Zanchi R and Pieta P (2003). Metabolism of stevioside and rebaudioside A from Stevia rebaudiana extracts by human microflora. J. Agric. Food. Chem., 51: 6618-6622.
Geuns MC (2003). Molecules of Interest, Stevioside. Phytochemistry, 64: 913-921.
Geuns MC, Buyse J, Vankeirsbilck A and Temme EM (2007). Metabolism of stevioside by healthy subjects. Exp. Biol. Med., 232: 164-173.
Gutteridge CJ (1995). Lipid Peroxidation and Antioxidants as Biomarkers of Tissue Damage. Clin. Chem., 41(12): 1819-1 828.
Hoeschen RJ (1997). Oxidative stress and cardiovascular disease. Can. J. Cardiol., 13(11): 1021-1025.
Hutapea AM, Toskulkao C, Buddhasukh D, Wilairat P, Glinsukon T (1997). Digestion of stevioside, a natural sweetener, by various digestive enzymes. J. Clin. Biochem. Nutr., 23: 177-186.
Ihara Y, Toyokuni S, Uchida K, Odaka H, Tanaka T, Ikeda H, Hiai H, Seino Y, Yamada Y (1999). Hyperglycemia causes oxidative stress in pancreatic beta-cells of GK rats, a model of type 2 diabetes. Diabetes, 48 (4): 927-932.
JECFA (1999). Sweetening agent: stevioside. In: 51st Meeting Joint FAO/WHO Expert Committee on Food Additives (JECFA). World Health Organization,. WHO Food Additive Series. 42: 119-143.
JECFA (2000). Evaluation of Certain Food Additives. Fifty-first Report of the Joint FAO/WHO Expert Committee on Food Additives. Geneva, Switzerland. WHO Technical Report Series, 891: 35-37.
JECFA (2005). *Evaluation of Certain Food Additives*. Sixty-third Report of the Joint FAO/WHO Expert Committee on Food Additives, Geneva, Switz. WHO Technical Report Series, 928: 34-39-138.
JECFA (2006). Safety evaluation of certain food additives. Prepared by the 63rd meeting of the Joint FAO/WHO Expert Committee on Food Additives. WHO Food Additives Series, 54: 117-144.
JECFA (2007). Evaluation of Certain Food Additives and Contaminants. Sixty-eighth Report of the Joint FAO/WHO Expert Committee on Food Additives. World Health Organization (WHO), Geneva, Switzerland. WHO technical report series 947: 50-54.
JECFA (2008). Steviol glycosides. in: Compendium of Food Additive Specifications Joint FAO/WHO Expert Committee on Food Additives - 69th meeting of FAO JECFA, Rome, Italy, Monographs 5: 75-78.
JECFA (2009). Safety evaluation of certain food additives. Prepared by

the 69th meeting of the Joint FAO/WHO Expert Committee on Food Additives. WHO Food Additives Series, 66: 183-220.

Kasai H (1997). Analysis of a form of oxidative DNA damage, 8-hydroxy-2'-deoxyguanosine, as a marker of cellular oxidative stress during carcinogenesis. Mutat. Res., 387(3): 147-163.

Kinghorn AD, Soejarto DD, Nanayakkara NP, Compadre CM, Makapugay HC, Hovanec-Brown JM, Medon PJ, Kamath SK (1984). A phytochemical screening procedure for sweet ent-kaurene glycosides in the genus Stevia. J. Nat. Prod., 47(3): 439-444.

Klaunig JE, Xu Y, Isenberg JS, Bachowski S, Kolaja KL, Jiang J, Stevenson DE, Walborg EF (1997). The role of oxidative stress in chemical carcinogenesis. Environ. Health Perspect, 106(1): 289-295.

Koyama E, Sakai N, Ohori Y, Kitazawa K, Izawa O, Kakegawa K (2003). Absorption and metabolism of glycosidic sweeteners of stevia mixture and their aglycone, steviol, in rats and humans. Food Chem. Toxicol., 41(6): 875-883.

Kroyer G (1999). The Low Calorie Sweetener Stevioside: Stability and Interaction with Food Ingredients. LebensmWiss Technol., 32(8): 509-512.

Kroyer G (2010). Stevioside and Stevia-sweetener in food: application, stability and interaction with food ingredients. J. Consumer Protection and Food Safety, 5(2): 225-229.

Mannervik BI, Carlberg C (1985). Glutathione Reductase. Method Enzymol., 113: 484-490.

Mavelli I, Rigo A, Federico R, Cirilolo M, Rotilio G (1982). Superoxide dismutase, glutathione peroxidase and catalase in developing rat brain. Biochem. J., 204: 535-540.

Nishikimi M, Appaji N, Yagi K (1972). The occurrence of superoxide anion in the reaction of reduced phenazine methosulfate and molecular oxygen. Biochem. Biophys. Res. Commun., 31(2): 849-854.

Nunes AP, Ferreira-Machado SC, Nunes RM, Dantas FJ, De Mattos JC, Caldeira-de-Arau A (2007). Analysis of genotoxic potentiality of stevioside by comet assay. Food Chem. Toxicol., 45: 662-666.

Swiderska-Kołacz G, Klusek J, Kołątaj A (2007). The effect of exogenous GSH, GSSG and GST-E on glutathione concentration and activity of selected glutathione enzymes in the liver, kidney and muscle of mice. Anim. Sci. Pap. Rep., 25(2): 111-117.

Tappel AL, Zalkin H (1959). Inhibition of lipid peroxidation in mitochondria by vitamin E. Arch. Biochem. Biophys., 80: 333-336.

Toskulkao C, Sutheerawattananon M, Piyachaturawat P (1995). Inhibitory effect of steviol, a metabolite of stevioside, on glucose absorption in everted hamster intestine in vitro. Toxicol. Lett., 80: 153-159.

Vanek T, Nepovím A, Valícek P (2001). Determination of Stevioside in Plant Material and Fruit Teas. J. Food Compos. Anal., 14: 383-388.

Wingard RE, Brown JP, Enderlin FE, Dale JA, Hale RL, Seitz CT (1980). Intestinal degradation and absorption of the glycosidic sweeteners stevioside and rebaudioside A. Experientia., 36(5): 519-520.

16

Synergistic effects of glucan and resveratrol

Vaclav Vetvicka[1]* and Zuzana Vancikova[2]

[1]Department of Pathology, University of Louisville, Louisville, KY 40202, USA.
[2]1st Medical Faculty, Department of Pediatrics, Thomayer University Hospital, Prague, Czech Republic.

Recent data showing that glucan stimulates defense reactions in plants through synthesis of resveratrol, led us to study the possible synergetic effects of a glucan-resveratrol complex on immune reactions in mice. We measured phagocytic activity, expression of CD4 marker on spleen cells, IL-2 secretion and antibody response. In all cases we confirmed the stimulatory effects of glucan. Resveratrol alone had either limit or has no effect. However, a combined preparation showed very strong synergetic effects. Our data support further studies of these two natural immunomodulators.

Key words: Glucan, resveratrol, phagocytosis, IL-2, immune reactions, macrophage, antibody.

INTRODUCTION

Glucans belong to a group of physiologically active compounds called biological response modifiers and represent highly conserved structural components of cell walls in yeast, fungi and seaweed. Glucan's role as a biologically active immunomodulator has been well documented for over 50 years. Initial interest in the immunomodulatory properties of polysaccharides was raised after experiments showing that a crude yeast cell preparation stimulated macrophages through activation of the complement system (Benacerraf and Sebestyen, 1957).

The best known effects of glucans consist of the augmentation of phagocytosis of professional phagocytes granulocytes, monocytes, macrophages and dendritic cells which direct stimulation of natural killer cells. In this regard, macrophages (Chihara et al., 1982; Vetvicka et al., 1996), considered to be the basic effector cells in host defense against bacteria, viruses, parasites and tumor cells, which play the most important role. There is evidence that glucan makes a considerable contribution toward the increased production of nitric oxide, one of the most effective reactive nitrogen species, by inducible nitric oxide synthase (iNOS) in macrophages (Ohno et al., 1996). Additional biological effects of glucans include stimulation of infectious immunity, activation of bone marrow cell production, anti-cancer effects and lowering of blood cholesterol (Kimura et al., 1994; Kogan, 2000; Vetvicka et al., 2009) for review see Novak and Vetvicka 2008).

Glucan is clearly not the only known immunomodulator in the entire world. Despite the fact that glucan, with over 10,000 published scientific papers, is the best studied and best documented natural modulator, other biologically active molecules exist. More and more manufacturers and retailers are experimenting with the preparation of various cocktails or mixtures of potentially bioactive powders. It is now very common to find glucan in combination with five or more ingredients, including Echinacea, Aloe vera, Astragalus and Goldenseal.

There are recent studies showing that some bioactive molecules have synergistic effects when combined with glucan. Numerous scientific studies have confirmed some beneficial effects when glucan was given in combination with vitamin C. The main reason why vitamin C shows synergistic effects is the fact that this vitamin has been proven to stimulate the exact same immune responses as glucan, that is, macrophage activities, natural killer cell activity and specific antibody formation. A mouse study revealed significant healing abilities of a glucan-vitamin C combination in the treatment of infection by *Mesocestoides corti*. The treatment resulted in positive modulation of liver fibrosis and pathophysiological changes (Ditteova et al., 2003).

Resveratrol (trans-3,4',5-trihydroxystilbene) is a non-flavonoid polyphenol found in various fruits and vegetables and is abundant in the skin of grapes. In addition to various biochemical, biological and

*Corresponding author. E-mail: vaclav.vetvicka@ louisville.edu.

pharmacological activities, resveratrol has been found to exhibit numerous immunomodulatory activities such as suppression of lymphocyte proliferation, changes in cell-mediated cytotoxicity, cytokine production (Gao et al., 2001) and induction of apoptosis (Losa, 2003).

Our study was based on a recent observation showing that seaweed-derived glucan elicited defense responses in grapevine and induced protection against *Botrytis cinerea* and *Plasmopara viticola* through the induction of production of two phytoalexins including resveratrol (Aziz et al., 2003). This led us to evaluate the possible synergetic effects of glucan and resveratrol on immune reactions.

MATERIALS AND METHODS

Animals

Female, 8 week old BALB/c mice were purchased from the Jackson Laboratory (Bar Harbor, ME). All animal work was done according to the University of Louisville IACUC protocol. Animals were sacrificed by CO_2 asphyxiation.

Materials

Yeast-derived insoluble glucan #300 were purchased from Transfer Point (Columbia, SC). Resveratrol was purchased from Suan Farma, Paramus, NJ. Based on the HPLC analysis, it is 98.2% pure transresveratrol isolated from *Polygonum cuspidatum*. Anti-mouse CD4 antibodies conjugated with FITC were purchased from Biosource (Camarillo, CA).

Phagocytosis

The technique employing phagocytosis of synthetic polymeric micro spheres was described earlier (Vetvicka et al., 1982; 1988). Briefly: peripheral blood cells or isolated peritoneal cells were incubated *in vitro* with 0.05 ml of 2-hydroxyethyl methacrylate particles (HEMA; 5×10^8/ml).

The test tubes were incubated at 37°C for 60 min with intermittent shaking. Smears were stained with Wright stain. The cells with three or more HEMA particles were considered positive. Mice were injected with glucan, resveratrol or PBS (control). All experiments were performed in triplicates. At least 200 cells in 60 high power fields were examined in each experiment.

Flow cytometry

Cells were stained with monoclonal antibodies in 12 × 75 mm glass tubes using standard techniques for 30 min on ice. After washing with cold PBS, the cells were resuspended in PBS containing 1% BSA and 10 mM sodium azide. Flow cytometry was performed with a FACScan (Becton Dickinson, San Jose, CA) flow cytometer and the datas from over 10,000 cell sample were analyzed.

IL-2 production

Purified spleen cells (2×10^6/ml in RPMI 1640 medium with 5% FCS) from mice injected with glucan or resveratrorol were added into wells of a 24-well tissue culture plate. After addition of 1 µg of

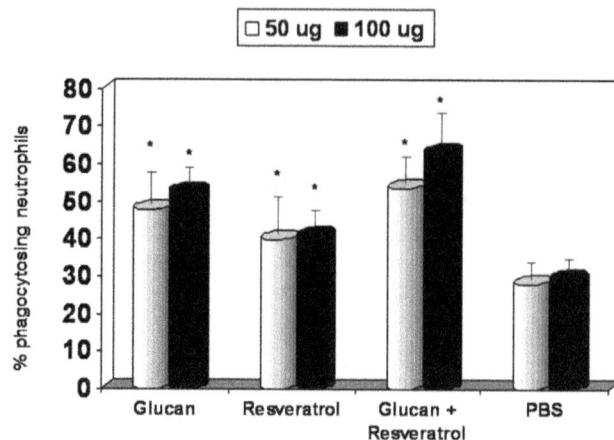

Figure 1. Effect of an ip. administration of glucan or resveratrol samples on phagocytosis by peripheral blood granulocytes. Each value represents the mean ± SD. *Represents significant differences between control (PBS) and tested samples at $P \leq 0.05$ level. All experiments were performed in triplicates.

Concanavalin A (Con A), cells were incubated for 48 h in a humidified incubator (37°C, 5% CO_2). At the endpoint of incubation, supernatants were collected, filtered through 0.45 µm filters and tested for the presence of IL-2 using a Quantikine mouse IL-2 kit (R&D Systems, Minneapolis, MN).

Antibody formation

Formation of antibodies was evaluated using ovalbumin as an antigen. Mice were injected twice (two weeks apart) with 100 µg of albumin and the serum was collected 7 days after last injection. Experimental groups were getting daily ip. injections of either glucan or resveratrol. Level of specific antibodies against ovalbumin was detected by ELISA. As positive control, combination of ovalbumin and Freund's adjuvant was used.

Statistics

Student's t-test was used to statistically analyse the data.

RESULTS

The effects of various glucans on macrophages are well established. However, in order to demonstrate that a new combination of immunomodulators really exhibits synergistic immunomodulatory characteristic, an evaluation of phagocytosis is necessary. We measured the effects of glucan or resveratrol on blood neutrophil phagocytosis of synthetic microparticles based on HEMA (Figure 1). Both glucan and resveratrol significantly stimulated the phagocytosis of synthetic particles. However, the combined preparation exhibited strong synergetic effect both on monocytes and neutrophils. Similar results were obtained when we measured the phagocytic ability of peritoneal macrophages *in vitro*. The effects of resveratrol alone

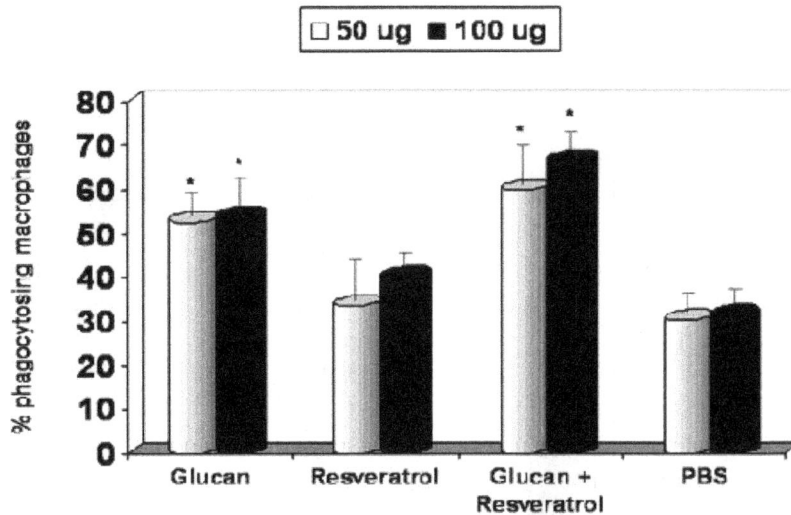

Figure 2. Effect of an ip. administration of glucan or resveratrol samples on phagocytosis by peritoneal macrophages. Each value represents the mean ± SD. *Represents significant differences between control (PBS) and tested samples at P ≤ 0.05 level. All experiments were performed in triplicates.

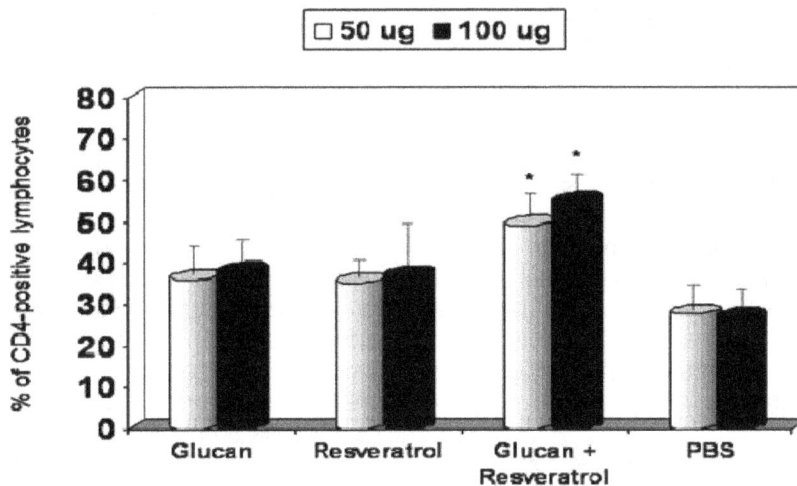

Figure 3. Effect of ip. injection of glucan or resveratrol on the expression of CD4 marker by spleen cells. The cells from three donors at each time interval were examined and the results given represent the means ± SD. *Represents significant differences between control (PBS) and samples at P ≤ 0.05 level. All experiments were performed in triplicates.

were less pronounced and not statistically significant, but again, there was strong statistically significant synergictic effect (Figure 2). Our preliminary experiments showed that these effects last up to 3 days after treatment (data not shown). These data were in agreement with previously published data using different types of glucan and resveratrol (Vetvicka and Yvin 2004; Vetvicka et al., 2007).

Next we evaluated the effects of the tested compounds on expression of CD4 marker on mouse splenocytes. Our

data summarized in Figure 3 showed that, whereas both glucan and resveratrol increased the expression of this marker, their simultaneous effects were much stronger and statistically significant. No changes between lower (50 µg) and higher (100 µg) doses were observed. The cellular of peritoneal cavity was not influenced by either glucan or resveratrol (data not shown), so the observed changes were not caused by the influx of cells.

Evidence of the immunomodulating activity was also demonstrated through effects on the production of IL-2 by

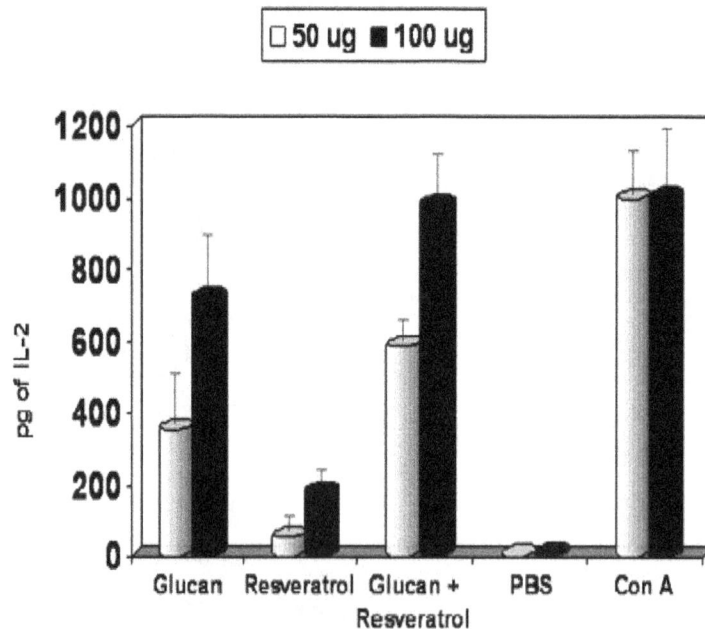

Figure 4. Effects of glucan or resveratrol on Con A-stimulated secretion of IL-2 by spleen cells. As the control (PBS) production of IL-2 is zero, all collumns represents significant differences between control (PBS) and samples at P ≤ 0.05 level. All experiments were performed in triplicates.

spleen cells (Figure 4). The production of IL-2 was measured after 48 h *in vitro* incubation of spleen cells isolated from control and treated mice. Whereas glucan showed strong and dose-dependent stimulation of IL-2 secretion, resveratrol alone was only slightly effective and only in the high dose. However, when used together, they stimulated IL-2 secretion as much as Con A, serving as a positive control. As there was no IL-2 production with PBS alone, every increase shown in Figure 4 was statistically significant.

We then focused on the use of glucan and resveratrol as an adjuvant. As an experimental model, we used immunization with ovalbumin. Glucan and resveratrol combination were applied simultaneously with two intraperitoneal doses of antigen, a commonly used Freund's adjuvant was used as additional positive control. The results (Figure 5) showed that resveratrol had no effects. Glucan alone significantly elevated the antibody response to 200%, whereas a combination of glucan and resveratrol increased the antibody production four times. It must be noted, however, that none of the tested substances potentiated the humoral immunity to the level of Freund's adjuvant.

DISCUSSION

The immune system is a system of cells, organs and soluble molecules working in unison to defend the body

against foreign pathogens. This system consists of numerous components, constantly on alert to find invading pathogens finding means to destroy them and eliminating them from our body. Individual cells interact with one another using physical contact or the secretion of various bioactive molecules. The innate system is considered to be the first line of defense and represents a significant part of the entire immune system. It includes mechanical barriers, cells such as macrophages and neutrophils and soluble factors such as the complement system and antimicrobial peptides. In many cases of infection, the innate immune mechanisms are sufficient to prevent full-blown infection.

The acquired immune system identifies the characterristic proteins of invading microorganisms and their toxins. This part of immunity consists of cells such as T and B lymphocytes and antigen-presenting cells (macrophages and dendritic cells). The lymphocytes recognize the invading pathogens by specific antibodies (B lymphocytes) or specific receptors (T lymphocytes). All types of immune cells work together symphoniously. They constantly interact with each other and function on the basis of information transferred using a complicated network of humoral factors such as enzymes and cytokines.

Immunomodulators usually offer systemic effects and details of the mechanisms of their effects are often unknown. This paper focused on hypothesis that glucan and resveratrol might together offer higher stimulation of

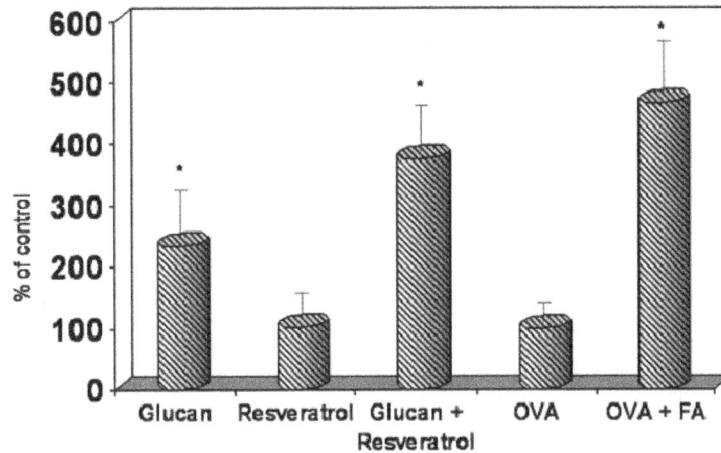

Figure 5. Effects of two ip. injections of glucan or resveratrol on formation of antibodies against ovalbumin. Mice were injected twice (two weeks apart) and the serum was collected 7 days after last injection. Level of specific antibodies against ovalbumin was detected by ELISA. As positive control, Freund's adjuvant was used. *Represents significant differences between control (ovalbumin alone) and samples at $P \leq 0.05$ level. All experiments were performed in triplicates.

immunity than individual molecules. Therefore we decided to monitor their effects on the most important reactions covering both branches of the immune reactions, that is, both cellular and humoral immunity.

Various types of immunomodulators, glucans in particular, are well established to stimulate phagocytosis (Abel et al., 1989). Therefore, the evaluation of this basic type of defense reaction is important for determining the effectiveness of any biologically active immunomodulator. We tested the peripheral blood leukocytes and peritoneal macrophages for changes in phagocytosis. Using synthetic microspheres based on 2-hydroxyethyl methacrylate, we found that both tested substances caused significant increase in phagocytosis, but the combined preparation showed significant synergetic effect on both macrophages and neutrophils. The data shown reflects the effects of a single injection of either glucan or resveratrol.

Observations of the effects on expression of cell surface CD4 marker present on splenocytes demon-strated that the numbers of CD4 + lymphocytes were significantly affected. Again, the combined preparation of glucan and resveratrol showed synergetic effects. Two days after the application, the numbers returned to normal. A similar increase in the number of CD4-positive cells after glucan application has been described for lentinan (Arinaga et al., 1992) and Phycarine (Vetvicka and Yvin, 2004).

In addition to the direct effect on various cells of the immune system, the immunostimulating action of β-glucans is caused by potentiation of a synthesis and release of several cytokines such as TNFα, IFNγ, IL-1,

and IL-2. This cytokine stimulating activity was found to be dependent on the triple helix conformation (Falch et al., 2000). It is hypothesized that glucans enhance leukocyte functions through increased cytokine secretion, particularly during early stages of infection. The potential effect of resveratrol on individual cytokines is much less clear. Some studies suggest that resveratrol can inhibit some IL-2 or TNF-α mediated functions (Lee and Moon, 2005; Kolgazi et al., 2006; Conover et al., 2006) represent only indirect proof of the resveratrol-cytokine interaction. Therefore, the evaluation of the potential systemic effect on IL-2 was particularly interesting. Using the previously established dose and time interval (Vetvicka and Yvin, 2004), we observed that resveratrol-induced only very low IL-2secretion. However, when used together with glucan, the synergetic effects were profound.

Glucans are usually considered modulators of the cellular branch of immune reaction and very little attention has been directed to their potential effects on antibody response. Regarding resveratrol, its effects on antibody response were never tested. We decided to take advantage of the recently published method of evaluating the use of glucan as adjuvant (Vetvicka and Vetvickova, 2007). Our results confirmed that glucan elevated the antibody response and that resveratrol alone has no activity. Surprisingly, simultaneous application of both agents showed very strong synergistic effects. Data presented in this paper represent further proof that combined preparations of glucan and resveratrol strongly stimulate both branches of immune reactions. A study attempting to reveal the exact mechanisms of these effects is currently under progress.

REFERENCES

Abel G, Szollosi J, Chihara G, Fachet J (1989). Effect of lentinan and mannan on phagocytosis of fluorescent latex microbeads by mouse peritoneal macrophages: a flow cytometric study. Int. J. Immunopharmacol. 11: 615-621.

Arinaga S, Karimine N, Takamuku K, Nanbara S, Nagamatsu M, Ueo H, Akiyoshi T (1992). Enhanced production of interleukin 1 and tumor necrosis factor by peripheral monocytes after lentinan administration in patients with gastric carcinoma. Int. J. Immunopharm. 14: 43-47.

Aziz A, Poinssot B, Daire X, Adrian M, Bezier A, Lambert B, Joubert J-M, Pugin A (2003). Laminarin elicits defense responses in grapewine and induces protection against Botrytis cinerea and Plasmopara viticola. MPMI 16: 1118-1128.

Benacerraf B, Sebestyen MM (1957). Effect of bacterial endotoxins on the reticuloendothelial system. Fed. Proc. 16: 860-867.

Chihara G, Maeda YY, Hamuro J (1982). Current status and perspectives of immunomodulators of microbial origin. Int. J. Tis. React. 4: 207-225.

Conover CA, Bale LK, Harrington SC, Resch ZT, Overgaard MT, Oxvig C (2006). Cytokine stimulation of pregnancy-associated plasma protein A expression in human coronary artery smooth muscle cells: inhibition by resveratrol. Am. J. Physiol. Cell. Physiol. 290: C183-C188.

Ditteova G, Velebny S, Hrckova G (2003). Modulation of liver fibrosis and pathological changes in mice infected with Mesocestoides corti (M. vogae) after administration of glucan and liposomized glucan in combination with vitamin C. J. Helmintol. 77: 219-226.

Falch BH, Espevik T, Ryan L, Stokke BT (2000). The cytokine stimulating activity of (1-3)-β-D-glucans is dependent on the triple helix conformation. Carbohydrate Res. 329: 587-596.

Gao X, Xu XY, Janakiraman N, Chapman RA, Gautam SC (2001). Immunomodulatory activity of resceratrol: suppression of lymphocyte proliferation, development of cell-mediated cytotoxicity, and cytokine production. Biochem. Pharmacol. 62:1299-1308.

Kimura Y, Tojima H, Fukase S, Takeda K (1994). Clinical evaluation of sizofilan as assistant immunotherapy in treatment of head and neck cancer. Acta Otolaryngol. 511 (Suppl.): 192-195.

Kogan G (2000). (1-3,1-6)-β-D-glucans of yeast and fungi and their biological activity. In: Atta-ur-Rahman C (ed). Studies in Natural Products Chemistry, Elsevier, Amsterdam, 23: 107-127.

Kolgazi M, Sener G, Cetinel S, Gedik N, Alican I (2006). Resveratrol reduces renal and lung injury caused by sepsis in rats. J. Surg. Res. 134: 215-321.

Lee B, Moon SK (2005). Resveratrol inhibits TNF-alpha-induced proliferation and matrix metalloproteinase expression in human vascular smooth muscle cells. J. Nutr. 135: 2767-2773.

Losa GA (2003). Resveratrol modulates apoptosis and oxidation in human blood mononuclear cells. Eur. J. Clin. Invest. 33: 818-823.

Novak M, Vetvicka V (2008). Beta-glucans, history, and the present: immunomodulatory aspects and mechanisms of action. J. Immunotoxicol. 5: 47-57.

Ohno N, Egawa Y, Hashimoto T, Adachi Y, Yadomae T (1996). Effect of beta-glucans on the nitric oxide synthesis by peritoneal macrophage in mice. Biol. Pharm. Bull. 19: 608-612.

Vetvicka V, Fornusek L, Kopecek J, Kaminkova J, Kasparek L, Vranova M (1982). Phagocytosis of human blood leukocytes: A simple micromethod. Immunol. Lett. 5: 97-100.

Vetvicka V, Holub M, Kovaru H, Siman P, Kovaru F (1988). Alpha-fetoprotein and phagocytosis in athymic nude mice. Immunol. Lett. 19: 95-98.

Vetvicka V, Thornton BP, Ross GD (1996). Soluble β-glucan polysaccharide binding to the lectin site of neutrophil or natural killer cell complement receptor type 3 (CD11b/CD18) generates a primed state of the receptor capable of mediating cytotoxicity of iC3b-opsonized target cells. J. Clin. Invest. 98: 50-61.

Vetvicka V, Volny T, Saraswat-Ohri S, Vashishta A, Vetvickova J (2007). Glucan and resveratrol complex - possible synergistic effects on immune system. Biomed. Pap. 151: 41-46.

Vetvicka V, Yvin JC (2004). Effects of marine β-glucan on immune reaction. Int. Immunopharmacol. 4: 721-730.

Vetvicka V, Vetvickova J (2007). An evaluation of the immunological activities of commercially available β1,3-glucans. JANA 10: 25-31.

Vetvicka V, Vetvickova J (2009). Effects of yeast-derived beta-glucans on blood cholesterol and macrophage functionality. J. Immunotoxicol. 6: 30-35.

17

Effect of *Hibiscus sabdariffa* anthocyanins on 2, 4-dinitrophenylhydrazine-induced hematotoxicity in rabbits

A. Ologundudu[1]*, A. O.Ologundudu[1], I. A. Ololade[2] and F. O. Obi[3]

[1]Department of Biochemistry, Adekunle Ajasin University, P.M.B. 001, Akungba Akoko, Ondo State, Nigeria.
[2]Department of Chemistry and Industrial Chemistry, Adekunle Ajasin University, P.M.B.001, Akungba Akoko, Ondo State, Nigeria.
[3]Department of Biochemistry, Faculty of Life Sciences, University of Benin, Benin City, Edo State, Nigeria.

In this study, the 2, 4-dinitrophenylhydrazine-induced biochemical and hematological changes in rabbits were examined under the administrative protocol of anthocyanin extract from *Hibiscus sabdariffa* calyces. Blood levels of reduced glutathione (GSH) and malondialdehyde (MDA) as well as red blood cell counts (RBC), white blood cell counts (WBC), packed cell volume (PCV) and hemoglobin (Hb) concentration were determined as indices of alteration and protection. Relative to control, 2, 4-dinitrophenylhydrazine (2, 4-DNPH) treatment significantly decreased ($P < 0.05$) blood level of GSH and significantly increased blood MDA level .It also significantly ($P < 0.05$) decreased RBC counts, PCV and Hb but increased WBC counts. On the other hand treatment of rabbits with *Hibiscus* anthocyanin extract led to significant ($P < 0.05$) increase in blood GSH, RBC counts, PCV and Hb and a decrease in MDA and WBC counts. These findings indicate that anthocyanin extract from dried calyces of *H. sabdariffa* protects the blood against 2, 4-DNPH lipoperoxidative and hemolytic effects.

Key words: Rabbit, *Hibiscus sabdariffa*, anthocyanin extract, 2,4 dinitrophenylhydrazine, reduced glutathione, malondialdehyde, complete blood count.

INTRODUCTION

A paradox in metabolism is that while the vast majority of complex life requires oxygen for its existence, oxygen is a highly reactive molecule that damage living organisms by producing reactive oxygen species (ROS) (Davies, 1995). Consequently, organisms contain a complex network of antioxidant metabolites such as vitamins C and E and enzymes such as catalase and superoxide dismutase (SOD) that work together to prevent oxidative damage to cellular components such as DNA, proteins and lipids (Sies, 1997; Vertuani et al., 2004).

Antioxidants can cancel out the cell-damaging effects of free radicals (Sies, 1997), and people who eat fruits and vegetables rich in polyphenols and anthocyanins have a lower risk of cancer, heart disease and some neu-

rological diseases(Stanner et al., 2004).Antioxidants work in 2 ways first is by chain-breaking which involves breaking the preventive side. Antioxidant enzymes like superoxide dismutase and catalase prevent oxidation by reducing the rate of chain initiation. The antioxidants scavenge initiating radicals and destroy them before oxidation is set in motion (Sies, 1997).

Hibiscus sabdariffa is a plant that finds various uses in traditional medicine. It is used as an antiseptic, diuretic, emollient and purgative agent (Truswell, 1992). It is a remedy for cancer, cough, heart ailments, hypertension and neurosis (Duke, 1985). The dried calyces of *H. sabdariffa* contain the flavonoids gossypetine, hibiscetin, anthocyanins and sabdaretine (Pietta, 2000). Certain amounts of delphinidin-3-monoglucoside and cyanidin-3-monoglucoside which constitute the anthocyanins are also present (Langenhoven et al., 2001). Flavonoids are phenolic compounds, they act as antioxidants in plants

---
*Corresponding author. E-mail: oluologundudu@yahoo.com.

(Robinson, 1975). There are indications that extracts from the red calyces of *H. sabdariffa Linn* contain antioxidant principles (Tseng et al., 1997; Wang et al., 2000; Ologundudu and Obi, 2005; Ologundudu et al., 2006a, b). It is therefore conceivable that the consumption of the extract may provide natural agents against oxidative tissue damage and other free radical-induced disease conditions (Harman, 1984; Wolff et al., 1986).

Phenylhydrazine and its derivative, 2, 4-dinitrophenylhydrazine are toxic agents. Their toxic action has been attributed to their ability to undergo auto oxidation. This increased oxidant potential enables them to oxidize enzymes, membrane proteins and hemoglobin. Phenylhydrazine initiates lipid peroxidation in membrane phosphorlipids (Jain and Hochstein, 1979) while 2,4-dinitrophenylhydrazine induces lipid peroxidation and other oxidative damage in rabbits(Ologundudu and Obi, 2005; Ologundudu et al., 2006a,b) and rats (Maduka et al,2003). The ability of 2,4-DNPH to induce lipid peroxidation and other free radical damage makes it an appropriate model toxicant for testing the claim that the anthocyanin extract of *H. sabdariffa Linn* calyces could probably protect tissues from oxidative stress-induced changes and other attendant biochemical consequences.

This research was therefore carried out to evaluate the hematoprotective properties of *H. sabdariffa* anthocyanin extract using the model of 2, 4-dinitrophenylhydrazine-induced oxidative stress in rabbits.

MATERIALS AND METHODS

Experimental animals and materials

Male rabbits (weight range 800 – 1000 g and four months old) used for this study were purchased from a local breeder in Benin City, Nigeria. 2,4-dinitrophenylhydrazine,trichloroacetic acid, NaCl and diethyl ether were purchased from BDH Chemical Company (Poole, England), 2,-thiobarbituric acid from Koch-Light Laboratories (England). HCl and absolute ethanol were obtained from WN Laboratories (US) and Chow (Growers mash) was obtained from Bendel Feed and Flour Mills, Ewu, Edo State, Nigeria.

Preparation of anthocyanin extract

Anthocyanin extract from *H. sabdariffa calyces* was prepared according to the method described by Hong and Wrolstad (1990a). 1 kg of *H. sabdariffa* calyces was pulverized and extracted with 10 l of 0.1% trifluoroacetic acid (TFA) solution for 12 h at 40°C. The extract was filtered through filter paper (Advantech filter paper no. 5C). The filtrate was applied to sepabeads SP-207 resin column (Mitsubishi Chemicals, Japan). The resin was washed with 3 l of water and then eluted with 50% ethanol solution containing 0.1 % TFA. The eluate was dried under vacuum at 40°C. The concentrated eluate was then subjected to high-speed liquid chromatography (HPLC) in order to identify its active principles.

HPLC analysis

The HPLC system consisted of a horizontal flow-through planar centrifuge with a multilayer coil (Pharma-Tech Research Co., Model CCC-1000, MD,USA), a pump (JASCO, 880-PU), a microflow pH sensor (Broadley-James, Model 14, CA, USA), a manual injection valve with a 20 ml loop, and a fraction collector (JASCO, SF-212N). The upper phase, consisting of a mixture of tert-butylmethylether: 1-butanol: MeCN: water (2:2:1:5 v/v) containing 0.2% of TFA, was used as the stationary phase, while the lower phase was as the mobile phase. A total of 300 mg of crude anthocyanin extract was dissolved in 20 ml of a mixture of the stationary phase: mobile phase (3:1 v/v) and introduced through the injection port. The mobile phase was pumped at 2.5 ml/min, while centrifugation was carried out 1000 rpm. 4 ml of each fraction was collected. A multi-wavelength detector (Waters, 490E) monitored the absorbance of the effluent at 515 nm.

Treatment of animals and collection of blood samples

Experimental rabbits were divided into 4 groups, 5 rabbits each and housed in standard cages. Rabbits were given free access to feed and water throughout the experiment period that lasted for 28 days. Group 1 and 2 were given a daily twice doses of 2.5 ml H_2O/kg body weight by gavage for 4 weeks. Similar treatment with anthocyanin extract at dose of 100 mg/kg body weight was given to rabbits in groups 3 and 4. After the 21st day of the experiment, rabbits in groups 2 and 4 were intraperitoneally administered with a dose of 28 mg/kg body weight of 2, 4-dinitrophenylhydrazine for 5 consecutive days. By the end of the experimental period, the animals were anaesthetized in a diethyl ether saturated chamber, and then dissected to expose the heart. Blood was obtained via cardiac puncture by means of a 5 ml hypodermic syringe and needle and placed in 2 sets of heparinized bottles. One set was used immediately for hematological analysis while samples in the second set were centrifuged at 3500 rpm (Uniscope model SM 902B Bench centrifuge, Surgifriend Medicals, England) for 10 min each in order to obtain plasma. Plasma samples were collected and kept at -20°C until required.

Hematological analysis

The hematological indices namely red blood cell (RBC) counts and white blood cell (WBC) counts were estimated by visual counting improved by Neubauer counting chambers. Hemoglobin (Hb) and packed cell volume (PCV) were determined using cyanomethemoglobin and microhematocrit methods respectively (Dacie and Lewis, 1997).

Lipid peroxidation assay

Lipid peroxidation was determined spectrophotometrically by thiobarbituric acid reactive substances (TBARS) method as described by Varshney and Kale (1990). Results were expressed in terms of malondialdehyde (MDA) formed per mg protein.

Reduced glutathione

Reduced glutathione concentration in the blood was determined using the method of Jollow et al. (1974).

Statistical analysis

The data obtained were subjected to standard statistical analysis of variance (ANOVA) using the procedure of SAS (SAS Inst. Inc. 1999). Treatment means were compared using the Duncan procedure of the same software. The significance level was set at $P < 0.05$.

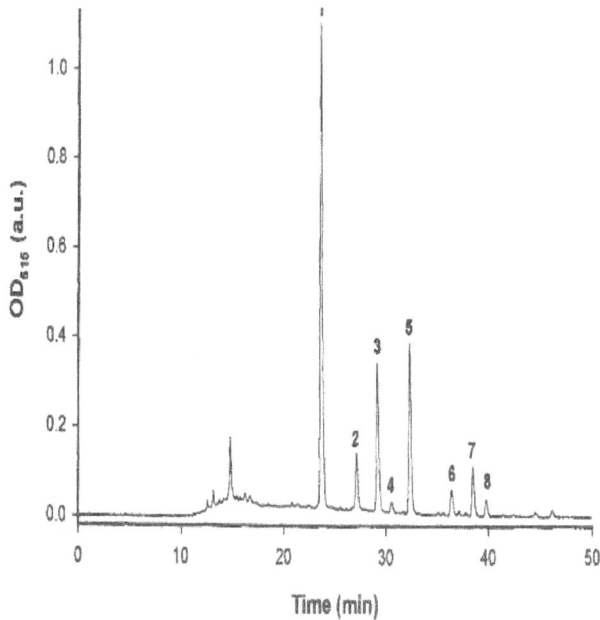

Figure 1. HPLC chromatogram of *H. sabdariffa anthocyanins.*

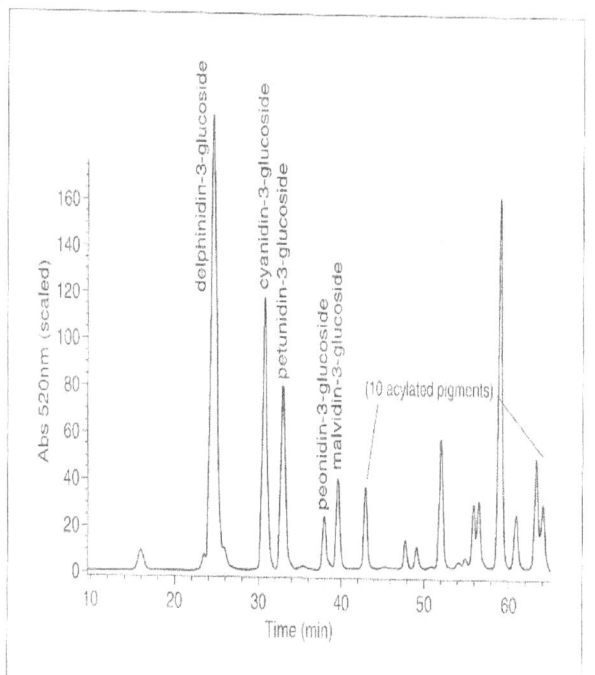

Figure 2. HPLC chromatogram of standard anthocyanins.

RESULTS

Figure 1 shows the result displayed on a multiwavelength detector used to monitor the absorbance of the effluent of *H. sabdariffa* extract at 515 nm. The peaks on the graph indicate the different anthocyanins present in the *H. sabdariffa* extract in form of their glucosides. The antho-

cyanins were identified by extrapolating from the graph shown in Figure 2 which is the HPLC chromatogram of known anthocyanins. The result showed that *H. sabdariffa calyces* contained several anthocyanins but the predominant ones were delphinidin-3-monoglucoside and cyanidin-3-monoglucoside.

Table 1 shows the effect of administering 2, 4-DNPH and *H. sabdariffa* anthocyanins on the levels of reduced glutathione (GSH) and malondialdehyde (MDA) of the studied rabbits. Intraperitoneal treatment of rabbits in group 2 with 2, 4-DNPH (28 mg/kg body weight) signify-cantly ($p < 0.05$) reduced blood content of GSH but increased the MDA level when compared to the control, Group 1.However, treatment of rabbits (Group 3) with anthocyanin extract alone led to significant ($P < 0.05$) increase in blood content of GSH and a decrease in MDA level relative to the values obtained for group 2. Fur-thermore, treatment of rabbits (group 4) with anthocyanin extract prior to 2, 4-DNPH intoxication maintained at normal, the levels of both MDA and GSH in relation to control.

The changes in red blood cell (RBC) and white blood cell (WBC) counts, packed cell volume (PCV) and hemo-globin (Hb) concentration of rabbits due to the effect of 2, 4-DNPH and *Hibiscus* anthocyanin extract are presented in Table 2. Treatment with 2, 4-DNPH significantly ($P < 0.05$) reduced rabbit blood RBC, PCV, Hb values but caused an increase in WBC counts compared to the control (Group 1). The RBC and WBC counts and Hb concentration of rabbits that received the anthocyanin extract alone (Group 3) and those pretreated with the extract before 2, 4-DNPH administration (Group 4) did not show any significant ($p < 0.05$) difference when com-pared to the control. However, the PCV of rabbits that were treated with the extract alone (Group 3) was sig-nificantly ($p < 0.05$) increased while the PCV of their counterparts that received the extract before 2, 4-DNPH administration (Group 4) did not show any significant ($p < 0.05$) difference when compared to the control.

DISCUSSION

Recently, much attention has focused on the protective biochemical functions of naturally occurring antioxidants in biological systems, and on the mechanisms of their action. Phenolic compounds, which are widely distributed in plants, were considered to play an important role as dietary antioxidants for the prevention of oxidative da-mage in living systems (Wang et al., 2000; Stanner et al., 2004).

Anthocyanins are phenolic compounds, and their anti-oxidant effects were investigated in this study. This study demonstrated that *H. sabdariffa* anthocyanins exhibited antioxidant bioactivity in intact cells and *in vivo* systems. As the integrity of cellular membranes is critical to normal cell function, the peroxidative decomposition of mem-brane lipids is an implication of chemical-induced toxicity.

Table 1. Effect of 2, 4-Dinitrophenylhydrazine and *Hibiscus* anthocyanin on the levels of reduced glutathione (GSH) and malondialdehyde (MDA) of rabbits.

Rabbit Group	Treatment	GSH Concentration (nmol per g protein)	MDA (µmol per mg protein)
1	2.5 ml H$_2$O /kg bd. wt. (control)	24.84 ± 1.33	1.38 ± 0.02
2	28 mg DNPH /kg bd. wt.	16.39[a] ± 0.92	8.50[a] ± 0.64
3	100 mg AN /kg bd. wt.	26.27[b] ± 0.59	1.07[b] ± 0.09
4	100 mg AN /kg bd. wt.+28 mg DNPH /kg bd. wt.	23.53[c] ± 0.26	1.44 ± 0.19

Results are means of 5 determinations ±SEM. Values carrying notations are significantly (p<0.05) different from control (Group 1). AN; - anthocyanin extract.

Table 2. Effect of 2,4-Dinitrophenylhydrazine and *Hibiscus* anthocyanin on the levels of red blood cells (RBC), white blood cells (WBC), packed cell volume (PCV) and hemoglobin (Hb) of rabbits.

Rabbit Group	Treatment	RBC (Counts/µL)×10^6	WBC (Counts/µL) ×10^3	PCV (%)	Hb concentration (g/dl)
1	2.5 ml H$_2$O /kg bd. wt. (control)	6.55 ± 0.71	7.4 ± 0.21	34.33 ± 0.88	11.27 ± 0.17
2	28 mg DNPH /kg bd. wt.	3.88[a] ± 0.40	11.47[a] ± 0.47	26.67[a] ± 1.76	7.30[a] ± 0.35
3	100 mg AN /kg bd. wt.	6.81 ± 0.34	6.60 ± 0.92	35.33[b] ± 1.53	11.93 ± 0.79
4	100 mg AN /kg bd. wt.+28 mg DNPH /kg bd. wt.	5.47 ± 0.21	7.60 ± 0.67	33.33 ± 1.76	11.27± 0.15

Results are means of 5 determinations ±SEM. Values carrying notations are significantly (p<0.05) different from control (Group 1).

In the present study, 2, 4-DNPH was shown to enhance lipid peroxidation, cytotoxicity in other words in animal systems. This is in consonance with our earlier reports (Ologundudu and Obi, 2005; Ologundudu et al., 2006a, b).

Evidence from a number of studies *in vitro* and *in vivo* suggests that phenylhydrazine and its derivatives interact with hemoglobin and cytochrome P-450 in an oxidation reaction, resulting in the generation of destructive free radicals, which are responsible for subsequent hemotoxicity (Itano et al., 1975; Jain and Hochstein, 1979; Maples et al., 1988). The results of the present study showed that *Hibiscus* anthocyanin extract effectively protect the blood from the oxidative damage caused by 2, 4-DNPH.

It is well established that reduced glutathione (GSH), the most important biomolecule protecting against chemically induced cytotoxicity, can participate in the elimination of reactive intermediates by conjugation or by direct free radical quenching. This study showed that 2, 4-DNPH caused a significant reduction in GSH levels in the blood and that a high dose of *Hibiscus* anthocyanins blocked the phenomenon effectively.

The effect of the anthocyanin extract on the hematological parameters in the animals treated with 2, 4-DNPH was also assessed (Table 2). The observed increase in the PCV, RBC counts and Hb concentration in the animals treated with the extract could be explained by the reduced loss of blood cells to lipid peroxidation as a re-

sult of the antioxidative properties of the pigments or the erythropoietic potencies already established for antioxidant molecules (Heda and Bhatia, 1986). The reduced PCV, RBC, Hb levels for animals treated with DNPH is a confirmation of the previously established hemotoxic properties exhibited by phenylhydrazine and its derivatives (Maples et al., 1988; Patil et al., 2000). Anthocyanins are known to induce the renal secretion of erythropoietin, the most important signal for differentiation and multiplication of the pluripotent stem cells involved in blood cell formation as reported by (Heda and Bhatia, 1986; Kaur and Kapoor, 2005).

Hemoglobin is a natural constituent of red blood cells and biochemically adapted to carry oxygen in the lungs and deposit it at the tissues for oxidative metabolism. Besides this function, it has been characterized to also play major role in physiologic carbon dioxide removal and acid-base balance. Therefore, an increased production of hemoglobin is an advantage to an organism. This metabolic status can only be ensured by decrease in red blood cell destruction or increased red blood cell production (Ponka, 1997).

White blood cells form part of the immune system in animals working against invading pathogens. The significant increase in WBC counts of the DNPH-treated animals when compared with the control is due to the ability of DNPH to act as hapten, thereby stimulating the production of plasma-cell derivatives of β-cells, thus accounting for the increased WBC levels. The proliferation of WBC by induced maturation of lymphocytes to matured WBC is the first stage of cell-mediated defense in the body in response to the presence of protein antigens and xenobiotics. Pretreatment of animals with anthocyanin extract prior to DNPH intoxication showed a feedback effect, with an observed significant reduction in WBC level. These results attest to the basis for the treatment of leukemia using anthocyanins in folk medicine, although, the biochemical mechanism is still unknown (Kaur and Kapoor, 2005).

REFERENCES

Dacie JV, Lewis M (1997). Blood counts. Practical hematology, 5th edition (1) Churchill Livingstone, New York pp. 20-40.

Davies K (1995). Oxidative stress: The paradox of aerobic life. Biochem. Soc. Symp. 6: 1-31.

Duke JA (1985). Proximate analysis. Handbook of medicinal herbs. 7th edition, Livingstone Group Ltd. Edinburgh pp. 228-229.

Harman D (1984). Free radical theory of aging: The free radical diseases. Age 7: 111-131.

Heda GL, Bhatia AL (1986). Hemocytometrical changes in Swiss albino mice after intrauterine low level HTO exposure. Proc. Asian Reg. Conf. Med. Phys. p. 390.

Hong V, Wrolstad RE (1990a). Use of HPLC separation/photodiode detection for characterization of anthocyanins. J. Agric. Food Chem. 38: 708-715.

Itano H, Hiraro K, Hosokawa K (1975). Mechanism of induction of hemolytic anemia by phenylhydrazine. Nature 256: 665-667.

Jain S, Hochstein P (1979). Generation of superoxide radicals by hydrazine: Its role in phenylhydrazine-induced hemolytic anemia. Biochem. Biophys. Acta, 586: 128-136.

Jollow DJ, Mitchel JR, Zampaghonic A, Gillette JR (1974). Bromobenzene-induced live necrosis; protective role of glutathione and evidence for 3, 4-bromobenzeneoxide as the hepatotoxic metabolite. Pharmacol. 11: 151-169.

Kaur C, Kapoor HC (2005). Antioxidant activity of some fruits in Indian diet. In ISHS Acta Hort. p. 696.

Lagenhoven P, Smith M, Letchame W, Simon J (2001). Hibiscus agrobusiness in sustainable national Africa plant products (ASNAPP). HIB-FS.

Maduka HCC, Okoye ZSC, Eje A (2003). The influence of Sacoglottis gabonensis stem bark extract and its isolate bergenin, Nigerian alcoholic additive, on the metabolic and hematological side effects of 2, 4-dinitrophenylhydrazine-induced tissue damage. Vasc. Pharmacol. 39: 317-324.

Maples K, Jordan S, Mason R (1988). In vivo rats hemoglobin free radical formation following phenylhydrazine administration. Mol. Pharmacol. 33: 344-350.

Ologundudu A, Obi FO (2005). Prevention of 2,4-dinitrophenylhydrazine-induced tissue damage in rabbits by orally administered decoction of dried flower of Hibiscus sabdariffa L J. Med. Sci. 5(3): 208-211.

Ologundudu A, Lawal AO, Adesina OG, Obi FO (2006a). Effect of ethanolic extract of Hibiscus sabdariffa L on 2,4- dinitrophenyl-hydrazine-induced changes in blood parameters in rabbits. Global J. Pure Appl. Sci. 12(3): 335-338.

Ologundudu A, Lawal AO, Adesina OG, Obi FO (2006b). Effect of ethanolic extract of Hibiscus sabdariffa L on 2, 4-dinitrophenyl-hydrazine-induced low glucose level and high malondialdehyde levels in rabbit brain and liver. Global J. Pure Appl. Sci. 12(4): 525-529.

Patil S, Kansase A, Kulkarni PH (2000). Antianaemic properties of ayurvedic drugs, raktavardhak, punarnavasav and navayas Louh in albino rats during phenylhydrazine induced hemolytic anaemia. Ind. J. Exp. Biol. 38: 253-257.

Pietta PG (2000). Flavonoids as antioxidants. J. Nat. Prod. 63 (7): 1035-1042.

Ponka P (1997). Tissue-specific regulation of iron metabolism and heme synthesis: distinct control mechanisms in erythroid cells. Blood 89:1–25.

Robinson T (1975). The organic constituents of higher plants. Their Chemistry and Interrelationships. 3rd edition. Cordus press, North Amherst. pp 190-220.

SAS Institute Inc. (1999). SAS/STAT User's Guide. Version 8 for Windows. SAS Institute Inc., SAS Campus Drive, Cary, North Carolina, USA.

Sies H (1997). Oxidative stress: Oxidants and antioxidants. Exp. Physiol. 82(2): 291-295.

Stanner SA, Hughes J, Kelly CN, Butriss J (2004). A review of the epidemiological evidence for the antioxidant hypothesis. Pub. Health Nutr. 7(3): 407-422.

Truswell AS (1992). ABC of nutrition, 2nd ed. Tavisteek Square Inc. London, pp. 50-93.

Tseng TH, Kao ES, Chu FP, Lin-Wa HW, Wang CJ (1997). Protective effect of dried flower extracts of Hibiscus sabdariffa L against oxidative stress in rat primary hepatocytes. Food Chem. Toxicol. 35(12): 1159-1164.

Vashney R, Kale RK (1990). Effect of calmodulin antagonist on radiation induced lipid peroxidation in microsomes. Int. J. Rad. Biol. 58: 733-743.

Vertuani S, Augusti A, Maufredina S (2004). The antioxidants and prooxidants network: An overview. Curr. Pharm. Des. 10(14): 1677-1694.

Wang CJ, Wang JM, Lin WL, Chu CY, Chou FP, Tseng TH (2000) Protective effect of Hibiscus anthocyanins against tert-butyl hydroperoxide-induced hepatic toxicity in rats. Food Chem. Toxicol. 38: 411-416.

Wolff SP, Garner A, Dean RT (1986). Free radicals, lipids and protein degradation. TIBS 11: 27-31

Cytoplasmic, peroxisomal and mitochondrial membrane phospholipid alteration in 3¹-diamino-azo-benzene (3¹-DAB)-induced hepatocellular carcinoma

Omotuyi I. O.[1], Ologundudu A.[1*], Okugbo T. O.[2], Oluyemi K. A.[3] and Omitade, O. E.[4]

[1]Department of Biochemistry, Adekunle Ajasin University, Akungba-Akoko, Ondo State, Nigeria.
[2]Department of Basic Sciences (Biochemistry option), Benson Idahosa University, G.R.A. Benin City, Edo State, Nigeria.
[3]Department of Biology/Biotechnology, William Paterson University, Wayne, New Jersey, USA.
[4]Department of Biochemistry, University of Port-Harcourt, Port-Harcourt, Rivers State, Nigeria.

Alteration in the cytoplasmic, mitochondrial and peroxisomal membrane phospholipid has been investigated in 3¹- diamino-azo-benzene (DAB) -induced hepatocellular carcinoma in rats. The cytoplasmic membrane of tumour group shows 27.5, 19.7, 38.7 and 34.7% increase in phosphatidylserine, phosphatidylcholine, phosphatidylethanolamine, and sphingomyelin respectively compared with the control group, mitochondrial membrane shows 46.4, 67.2, 64, and 73.4% decrease in phosphatidylserine, phosphatidylcholine, phosphatidyletholamine, and sphingomyelin respectively in tumour group compared with the control. Peroxisomal membrane phosphatidylserine, phosphatidylcholine, phosphatidylethanolamine and sphingomyelin composition of the tumour decreased by 20.0, 44.4, 79.2 and 78.6% compared with the control. The altered phospholipid is associated with altered cellular enzyme distribution. The soluble fraction of the cytoplasm shows specific activities of catalase and D-amino acid oxidase in soluble fraction of tumour group are 23.3 and 23.0% lower respectively compared to that of the control while Succinate dehydrogenase in tumour group is 42.2% higher than control group, the mitochondrial fraction shows specific activities of catalase, Succinate dehydrogenase and D-amino acid oxidase are 33.3, 33.3, and 58.3% higher respectively in control than the tumour group. The specific activities of catalase, Succinate dehydrogenase and D-amino acid oxidase are 61.5, 50.3, and 56.7% higher respectively in control compared with the tumour group. In the cytoplasmic, mitochondrial and the peroxisomal fractions of the control group, the phospholipid/protein ratio increased by 18.8, 35.7 and 48.6% respectively above the tumour group. These alterations might have implications on the metabolism, functions, morphology, metastasis and apoptosis in transformed cells.

Key words: Tumour, phospholipids, enzymes, phospholipid/protein ratio.

INTRODUCTION

The biochemical events that lead to the transformation of normal cell into tumour or cancer cells are complex (Liotta, 1992). Although these events are initialized by factors which alter the normal sequence of genetic material of cells (Flaks and Flaks, 1982; Devita et al., 1996; Lowry et al., 1951), the downstream consequences of such genetic alteration have not been fully elucidated (Devita et al., 1996). Some of the changes well documented in transformed cells include: uncontrolled cell division (Liotta, 1992), decreased requirement for growth factors, loss of contact (Devita et al., 1996), inhibition on growth, changes in cellular morphology and surface properties of transformed cells (Solomon et al., 1991; Folkman and Moscona, 1978; Morris, 1963) The integration of all these properties confers on cancerous cells their metastatic ability (Liotta, 1992; Solomon et al., 1991). Also, there are research evidences establishing

*Corresponding author: E-mail: oluologundudu@yahoo.com.

changes in intracellular composition of cells sequel to transformation (Morris, 1963). Organelles with reduced distribution have been reported in tumour cells, well documented ones include: mitochondria, lysosomes (Morris, 1963) and peroxisomes (Flaks and Flaks, 1982). To corroborate this claim, reduced activities of biomarker enzymes have been observed in these organelles in tumour and cancerous tissue samples (Rechcigl and Wollman, 1963). Also, recent findings about the involvement of lysosomes and peroxisomes in addition to the already discovered mitochondria (Chacon and Acosta, 1991; Kagan et al.,1992; Schulze-osthoff et al., 1992) in apoptotic cell death has raised questions about the exact biochemical reasons underlying the decreased distribution of these organelles in cancerous and tumour tissues.

This research is aimed at investigating distribution of marker enzymes and membrane phospholipids alteration in selected organelles (peroxisomes, mitochondria and cytoplasmic fraction) associated with 3^1-diamino-azo-benzene (3^1-DAB)-induced hepatocellular carcinoma in rats.

MATERIALS AND METHODS

Experimental animals

Female Sprague rats maintained in a closed colony of two rats per cage were fed rat cubes *ad libitum*. The water of the tumour group (experimental) contained 0.06% 3^1-diamino-azo-benzene (3^1-DAB). This treatment lasted for twenty four (24) weeks.

Tissue fractionation

All animals were killed under light anesthesia (chloroform). The liver was removed and rinsed in phosphate buffer saline solution (pH 7.2). The organ was then homogenized via up-down strokes of Teflon pestle rotated at 120rpm in a smooth glass homogenizer in a solution containing 0.25M sucrose solution containing 0.01M tris-HCL and 1mM $MgCl_2$ (pH 7.5) maintained at 4°C. Differential centrifugation procedure was carried out at 4°C on serval S-2 centrifuge equipped with SS-34 rotor for isolating mitochondrial and peroxisomal fractions as described previously (Ghosh et al., 1986)

Membrane phospholipids

The membrane phospholipids were extracted using the method described in Schulze-osthoff et al. (1992). Total phospholipids were estimated using the phospho-vanillin method as described in Wang et al. (1989). Individual phospholipids were first resolved on silica H gel plate, developed in chloroform/methanol/water/acetic acid (65:25:4:1) and run alongside with standard phospholipids. Each identified class of phospholipid was scrapped off the plate and dissolved in chloroform/ methanol (2:1). After decantation, the phospholipid was then estimated using the phospho-vanilin method.

Enzyme activity and protein estimation

Protein estimation was performed using method described by Lowry et al. (1951). The specific activities of catalase, Succinate

dehydrogenase, and D-amino acid oxidase were determined as biomarkers and to estimate the distribution of these enzymes in the cytoplasmic (whole liver or soluble fractions as indicated), peroxisomal (peroxisomes) and mitochondrial fractions of normal and tumour cells using the methods described by Folkman and Moscona (1978).

RESULTS

I. For the whole liver: The tumour group shows 27.5, 19.7, 38.7, and 34.7% increase for phosphatidylserine, phosphatidylcholine, phosphatidylethanolamine, and sphingomyelin respectively compared with the normal group.

II. For peroxisomal fraction: When compared with the normal group, phosphatidylserine, phosphatidylcholine, phosphatidylethanolamine and sphingomyelin composition of the tumour peroxisomal fraction decrease by 20.0, 44.4, 79.2 and 78.6% respectively.

III. For mitochondrial fraction: There is 46.4, 67.2, 64 and 73.4% decrease in phosphatidylserine, phosphatidylcholine, phosphatidyletholamine and sphingomyelin respectively in tumour group compared with the normal.

IV. Soluble fraction: The specific activities of catalase and D-amino acid oxidase in soluble fraction of tumour group are 23.3 and 23.0% lower respectively compared to that of the normal group. While the specific activity of succinate dehydrogenase in tumour cells is 42.2% higher than soluble fraction of normal group compared to the normal group.

V. Mitochondrial fraction: the specific activities of catalase, succinate dehydrogenase and D-amino acid oxidase are 33.3, 33.3 and 58.3% higher respectively in normal group compared with the tumour group.

VI. Peroxisomal fraction: the specific activities of catalase, succinate dehydrogenase and D-amino acid oxidase are 61.5, 50.3 and 56.7% higher respectively in normal group compared with the tumour group.

VII. Whole liver: The phospholipid: protein ratio of the normal group is 18.8% higher than that of the tumour group

VIII. Mitochondrial fraction: The phospholipid: protein ratio of the normal group is 35.7% higher than that of the tumour group

IX. Peroxisomal fraction: The phospholipid: protein ratio of the normal group is 48.6% higher than that of the tumour group.

DISCUSSION

The data obtained for phospholipids component of membrane in a given cell and its organelles is only valid when such membranes are isolated and purified free of other components of the cell possessing phospholipids. The purity of the hepatic cytoplasmic, peroxisomal and mitochondrial membranes was ascertained by the specific activities of the marker enzymes (urate oxidase

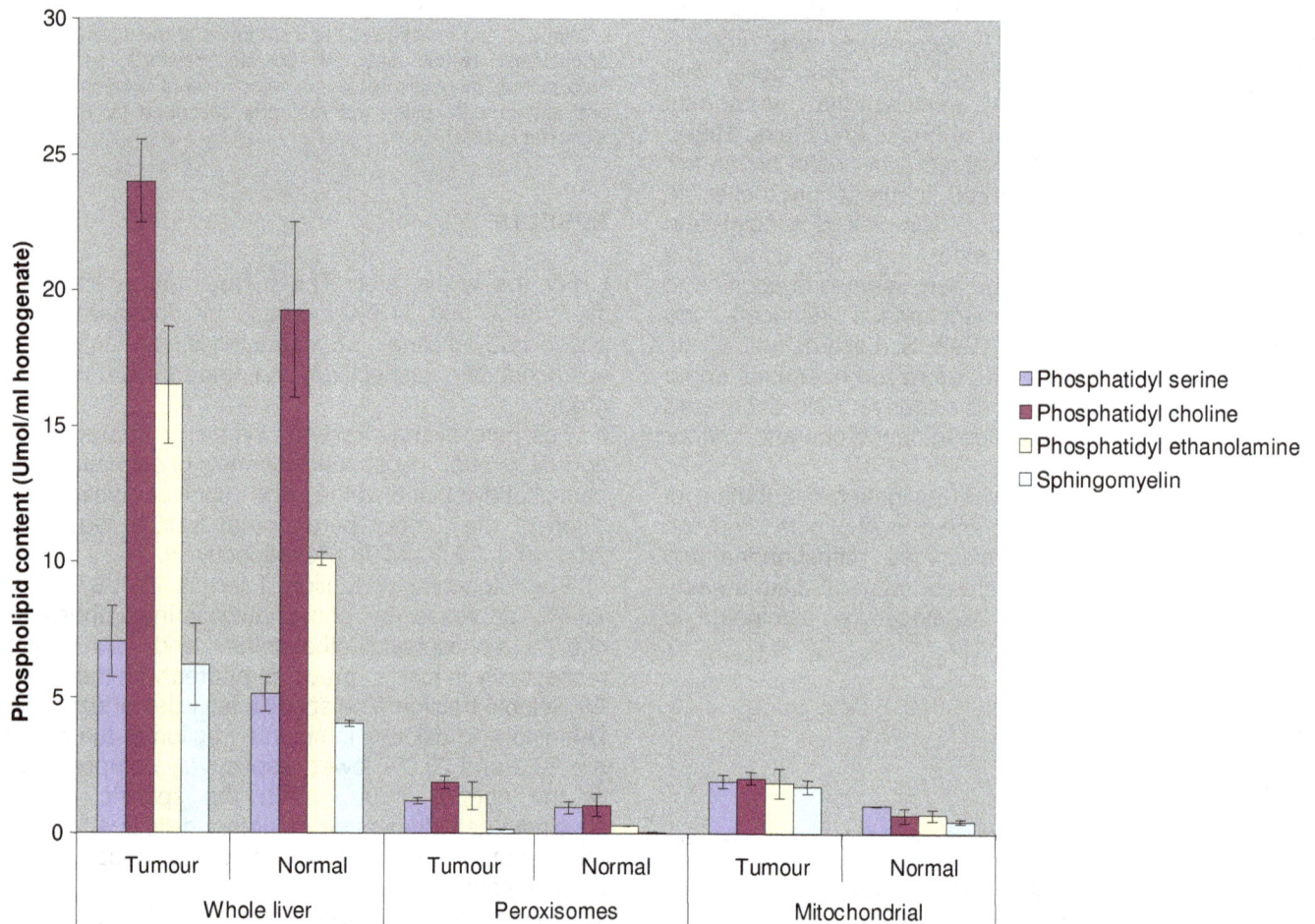

Figure 1. Shows the membrane phospholipid distribution in whole liver and liver peroxisomes and mitochondrial in normal and tumour groups.

for peroxisomal fraction, succinate-cytochrome C-dehydrogenase for mitochondrial fraction, data not shown) in their respective fractions.

The phospholipids of the whole liver and selected organelles (peroxisomes and mitochondria) of normal and 3^1-diamino-azo-benzene (3^1-DAB) treated rats were investigated (Figure 1). The data show that for the whole liver, the tumour group showed 27.5, 19.7, 38.7 and 34.7% increase for phosphatidylserine, phosphatidyl-choline, phosphatidylethanolamine, and sphingomyelin respectively compared with the normal group.

Biochemically, cells synthesize phospholipids for maintenance of membrane integrity during turnover, expansion of cytoplasmic volume in cell growth, cell division and differentiation.

Therefore, the quantitative pattern of phospholipids reported in the data possibly explains the source of phospholipids needed for the formation of membranes of daughter cells in the rapidly dividing hepatocytes of tumour group.

Also, the individual class of phospholipids showed unique quantitative pattern, which can improve our general understanding of phospholipid metabolism and membrane formation in tumour cells. Sphingomyelin has the highest percentage compared to other classes of phospholipids estimated. In nervous tissues this finding would not be unusual, considering its role as the major lipid of the myelin sheath. In tumour cells of hepatic origin, this can be explained however, as a deliberate biochemical strategy to reduce the concentration of ceramide in the cell. Ceramide is the immediate precursor of sphingomyelin in a single step involving transfer of phospho-choline from phosphatidylcholine to ceramide to form 1, 2-diacylglycerol and sphingomyelin respectively (Perry et al., 2000). This process shifts the equilibrium from ceramide to sphingomyelin. Ceramide and its catabolite sphingosine, not sphingomyelin are implicated in linking cell surface receptors, such as tumor necrosis factor alpha (TNF-α) and the Fas receptor, to apoptotic pathways. To maintain this equilibrium, the synthesis of phosphatidylcholine is increased to 38.7% above the normal cells. This process may also be accompanied by the decrease in sphingomyelinase activity in tumour cells.

The data also reveal that phosphatidylcholine has the lowest percentage. This contrasts the published work of Podo et al. (2007) that reported high phosphatidylcholine in breast and ovarian cancer. It has been biochemically proven that phosphatidylcholine is the donor of phosphocholine during the synthesis of sphingomyelin. Therefore, an initial increase in the intracellular concentration of phosphatidylcholine is followed by sphingomyelin synthesis. This is consistent with reported increase in phosphocholine-containing compounds in tumour cells by Glunde et al. (2004). While ceramide induces apoptosis, sphingomyelin induces cell proliferation which aids tumorigenesis (Perry et al., 2000; Radin, 2003).

Peroxisomes have been reported to participate in reactions which involved oxidation of D and L-amino acid, metabolism of long chain fatty acids and most recently, induction of apoptosis. When compared with the normal group, phosphatidylserine, phosphatidylcholine, phosphatidylethanolamine and sphingomyelin composition of the tumour peroxisomal fraction decrease by 20.0, 44.4, 79.2 and 78.6% respectively. To explain this finding, two assumptions would be made: assuming there is no significant difference in the number of peroxisomes of the tumour and the normal group respectively, this means that the peroxisomes of the normal group is more stable and dynamic compared to the tumour group (since the peroxisomes of the tumour group has lowered phospholipid distribution). The alternative assumption would be that given that the peroxisomes in the tumour and the normal groups have the same phospholipid composition, but the data indicate quantitative difference in the number and distribution of peroxisomes with the normal group having more peroxisomes compared with the tumour group. We agree that both assumptions may be pivotal to tumourigenesis depending on the nature of the carcinogen and the stage of its progression. Also, the role of peroxisomes in the apoptosis during tumourigenesis might be the biochemical force driving the alteration of their cellular distribution and modification of their phospholipid composition in tumour groups. Sphingomyelin in tumour peroxisomes was 78.6% lower than the estimated value for the normal group. The probable explanation for this is likely repression of key enzymes in synthesis of sphingosine, ceramide and sphingomyelin. Sphingosine, which is the immediate precursor of ceramide, has been reported to release cytochrome C and activate caspase (Radin, 2003), though the mechanism by which this occurs is still unknown. While ceramide strongly activates the c-Jun N-terminal kinase signaling pathway (JNK/SAPK pathway) leading to the phosphorylation of the N-terminus of c-Jun, c-Jun and c-Fos then heterodimerize to form AP-1, which is capable of inducing apoptosis. It can therefore be concluded that the lowered sphingomyelin in the peroxisomes is aimed at reducing its intracellular composition and subsequently nullifying its role in apoptosis (Radin, 2003; von Haefen et al., 2002;

Kroesen et al., 2001).

Mitochondrion is the power-house of the cells due to their ATP-generating ability. This metabolic feat depends on the electron transport chain and the oxidative phosphorylation which is also dependent on the integrity of mitochondrial membranes. Recently, other membrane-dependent processes that determine the life-span of a cell have been traced to mitochondrial membrane integrity and function (Kroesen et al. 2001; Zhuang et al., 2000). These processes include: maintenance of low cytosolic calcium concentration (Chacon and Acosta, 1991), regulation of production of highly reactive oxygen species and the regulation of the release of cytochrome C in response to tumour necrosis factor (Kagan et al., 1992; Ghosh et al., 1986; von Haefen et al., 2002). Although, early reports claimed that the mitochondrial ATP generation proceeds at such a low rate such that it does not meet the metabolic demands of a fast dividing cells Hinkle et al., 1991), studies have also confirmed that mitochondria have reduced occurrence in tumour cells (Morris, 1963). Reduced enzyme activities and altered structures have also been well documented (Morris, 1963). The alteration in the structure and the composition of mitochondria can be traced to the alteration of the cell membrane and this can be traced to altered transportation of protein from the cytoplasm to the mitochondrial matrix sequel to possible altered mitochondrial phospholipid/protein composition. Our data show 46.4, 67.2, 64 and 73.4% decrease in phosphatidylserine, phosphatidylcholine, phosphatidyletholamine, and sphingomyelin respectively in tumour group compared with the normal. The reduced incorporation of phospholipids in the mitochondria might be biochemically tailored by the tumour cells to reduce mitochondrial functions thereby avoiding apoptosis (Kroesen et al., 2001; Zhuang et al., 2000). Sphingomyelin has the lowest percentage distribution (73.4%) which is consistent with our previous findings in peroxisomes. This can be ascertained that tumour cells inhibit the pathway for the synthesis of sphingosine in order to inhibit apoptosis (Perry et al., 2000).

Figure 2 shows the distribution of the key enzymes in the soluble fraction of the liver, mitochondria and peroxisomes.

For the soluble fraction of the liver, the specific activities of catalase and D-amino acid oxidase in soluble fraction of tumour group are 23.3 and 23.0% lower respectively compared to that of the normal group while the specific activity of succinate dehydrogenase in tumour cells is 42.2% higher than soluble fraction of normal group compared to the normal group.

Catalase is a peroxisomal enzyme, its synthesis takes place in the cytosol in form of apocatalase. Peroxisomes have a single membrane (Perry et al., 2000), such that under agitation, the membrane breaks and catalase appears in the soluble fraction during fractionation. The activity of this enzyme depends on availability of iron III

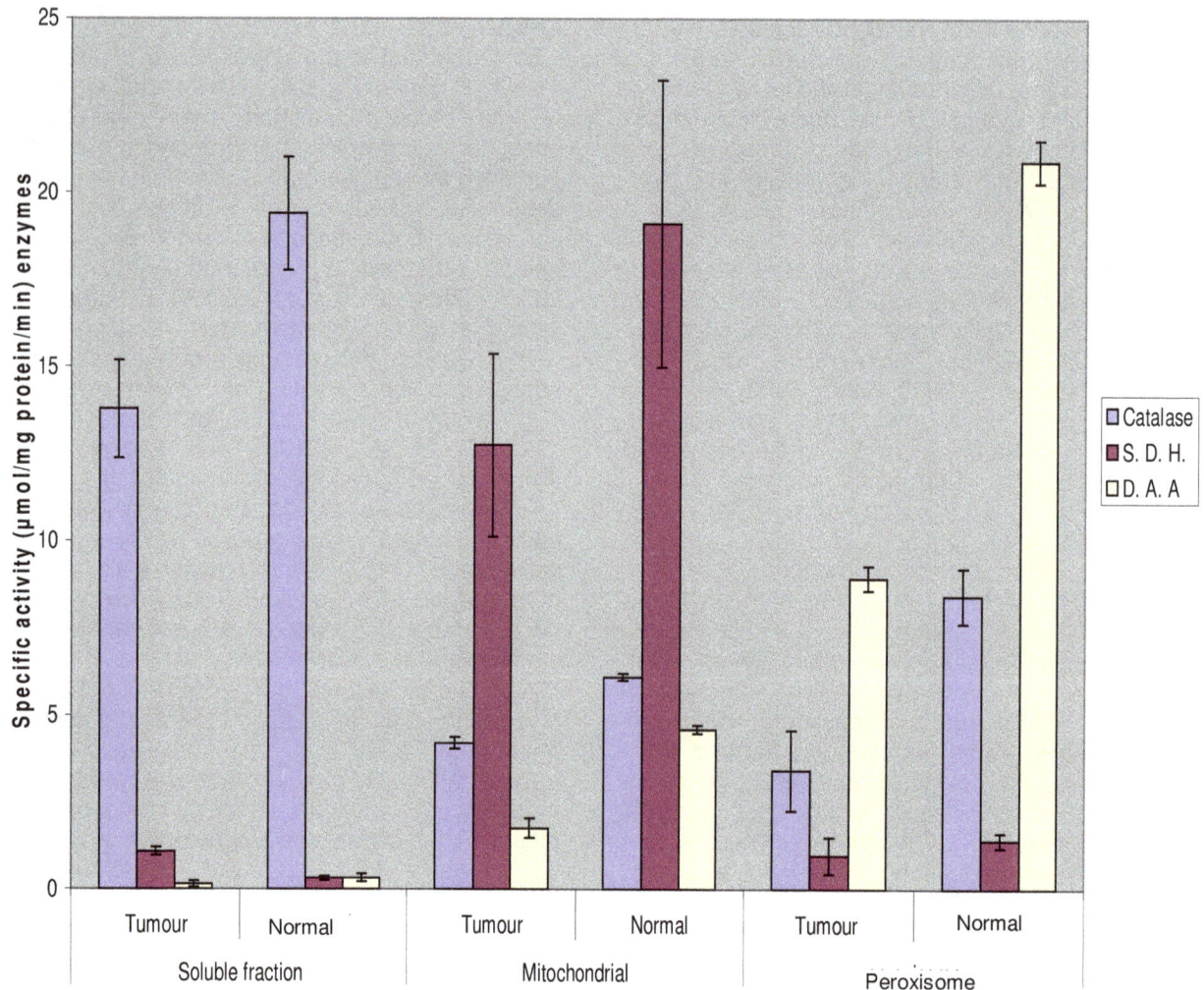

Figure 2. Shows the specific activities(μmol/mg protein/min) of marker enzymes in fractions of normal and transformed liver of rats.

ion (De-Hoop and Ab, 1992). Its lowered activity in the soluble fraction can then be traced to three factors: reduced synthesis in tumour cells, reduced transportation of apocatalase into the peroxisomes for conversion into catalase and the unavailability of iron III ion for incorporation into apocatalase to form catalase (Perry et al., 2000; De-Hoop and Ab, 1992), the crescendo of events that summarizes into lowered specific activity of catalase in tumour cells might be more complex. Orally administered 3^1-DAB chelates iron and other metals, interacts with cells of duodenum, and proximal jejunum-the portion of the small intestine responsible for iron absorption. This interaction possibly lowers their iron-absorbing capacity (Brittenham, 2000; Hoelzer, 1997), which in turn reduces the availability of iron in the liver for incorporation into apocatalase (De-Hoop and Ab, 1992). In fast dividing cells, there is an absolute necessity for metabolic economy. Metabolic emphasis would rather be on angiogenesis and erythropoiesis which require iron rather than catalase formation (Dupont et al., 1998).

Another plausible explanation is the progressive loss of liver function associated with hepatocellular carcinoma, there is progressive reduction in the synthesis and secretion of proteins involved in the absorption, transportation and storage of iron. For D-amino acid oxidase, its synthesis can be repressed due to its function as generator of hydrogen peroxide-a well documented oxidant (Konno and Yasumura, 1992; Katagiri et al., 1991), capable of inducing cell injury and death. Also, tumour cells might direct FAD prosthetic group to other enzymes of anabolic importance geared towards cell division rather than activating D-amino acid oxidase whose activity may be lethal to the cell especially deficient in catalase (Konno and Yasumura, 1992).

The high specific activity of succinate dehydrogenase in the soluble fraction of the tumour group is a strong indication of a reduced stability of membrane in the mitochondria of tumour cells (Kagan et al., 1992), since the membrane is double and they are less likely to respond to agitation by breaking as compared with

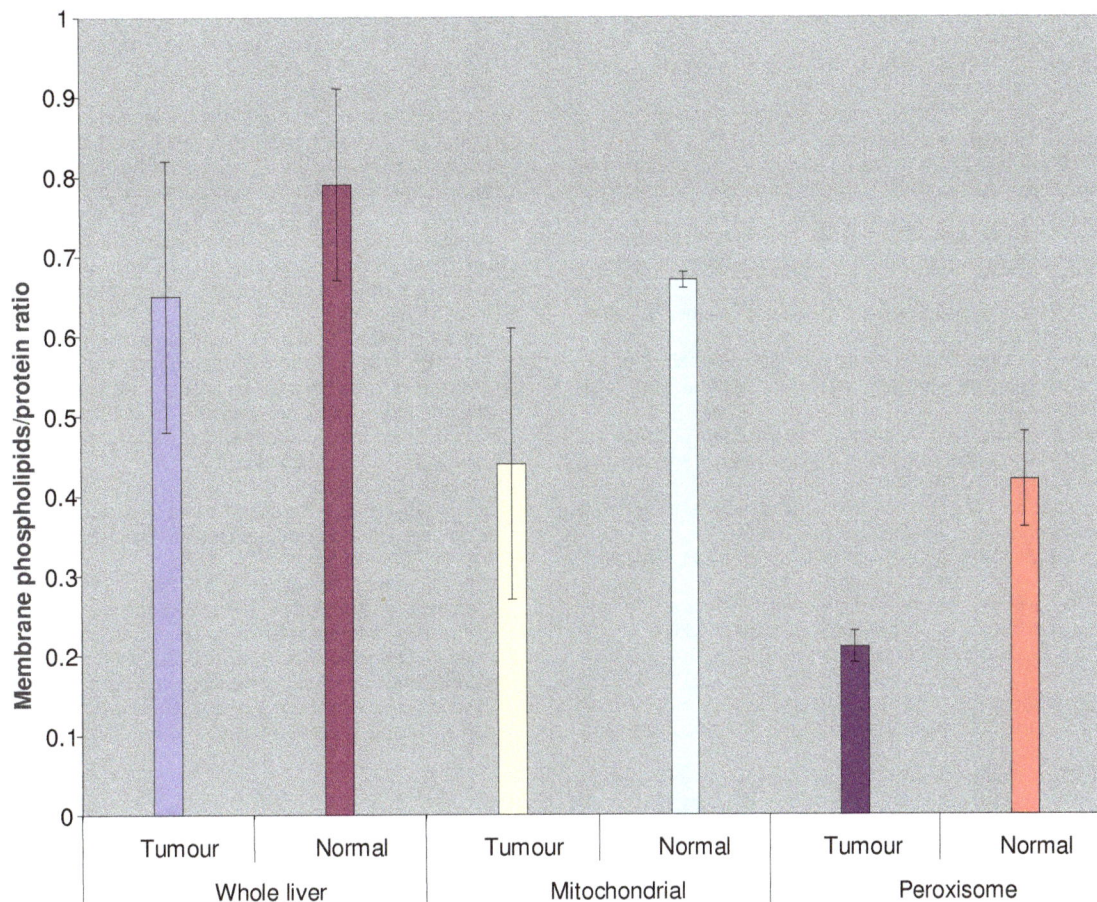

Figure 3. Shows total phosphol lipids to protein ratio in the whole liver, liver peroxisomal and mitochondrial fraction of normal and tumour group rats.

peroxisomal membrane. This finding agrees with the data generated for the mitochondrial membrane phospholipid distribution in tumour cells (Figure 1). It is well reported that permeability of mitochondria is increased with progression of tumour (Schulse-Osthoff et al., 1992).

Peroxisomal fraction of the tumour cells also exhibited 56.7% loss of D-amino acid oxidase specific activity compared with the normal group. This can be accounted for by assuming reduced distribution of peroxisomes, tumour-specific channeling of energy into metabolic processes aimed at cell division rather than expression of peroxisomal proteins which in turn might participate in death via apoptosis and hydrogen peroxide induced cell death.

The phospholipid/protein ratio for whole liver, mitochondria, and peroxisomes of the normal group is 18.8, 35.7 and 48.6% higher than that of the tumour group. This can be explained as the result of reduced phospholipid distribution in the membranes (Figure 3).

In conclusion, although the biochemical event that ultimately leads to tumourigenesis starts from the alteration of cellular DNA, the transformed cells and their organelles have registered biochemical foot-prints during

the entire process. Altered membrane phospholipid distribution, key enzyme distribution and the phospholipid-to-protein ratio in the whole liver, mitochondria and the peroxisomal fraction are some of such foot-prints with dire implication on the metabolisms, cell/tissue functions, morphology and metastasis in tumour and cancerous cells.

REFERENCES

Brittenham GM (2000). Disorders of iron metabolism: Iron deficiency and overload. In: hematology: Basic principle and practice. 3rd ed. Churchill livingstone, pp 397–428.

Chacon E, Acosta D (1991). Mitochondrial regulation of superoxide by calcium ion: an alternative mechanism for cardiotoxicity of doxorubicin. Toxicol. Appl. Pharmacol., 107(1): 117-128.

De-Hoop M, Ab G (1992). Import of proteins into peroxisomes and other microbodies. Biochem. J., 286: 657-669.

Devita VT, Hellman S, Rosenberg S (1996). Cancer: Principle and practice of oncology, 5th ed. Lippincott, pp. 2586-2598.

Dupont E, Riviere M, Latreille J (1998). Neovastat: An inhibitor of angiogenesis with anti-cancer activity. Presented at the American Association of Cancer Research Special Conference on Angiogenesis and Cancer, Orlando, FL, January 24-28.

Flaks A, Flaks B (1982). 3-Methyl cholanthrene inhibition of hepatocellular carcinogenesis in rats due to 31- methyl-4-dimethyl amino benzene or 2-acetyl aminoflourene: a comparative study. Carcinogenesis, 3(9): 981-991.

Folkman J, Moscona A (1978). Role of cell shape in growth control. Nature, 273: 345-349.

Ghosh M, Mridul K, Hajra AK (1986). A rapid method for the isolation of peroxisomes from rat liver. Anal. Biochem., 159(1): 169-174.

Glunde K, Jie C, Bdujwalla Z (2004). Molecular causes of the aberrant choline phospholipid metabolism in breast cancer. Cancer Res., 64(12): 4270-4276.

Hinkle P, Kumar A, Resetar A, Harris D (1991). Mechanistic stoichiometry of mitochondrial oxidative phosphorylation. Biochem., 30: 3576.

Hoelzer D (1997). Hematopoietic growth factors-not whether but when and where. N. Engl. J. Med., 336: 1822.

Kagan B, Baldwin L, Munoz D, Wisnieski J (1992). Formation of ion-permeable channels by tumour necrosis factor-alpha. Sci., 225(5050): 1427-1430.

Katagiri M, Tojo H, Horiike K, Yamano T (1991). Immunochemical relationship of D-amino acid oxidase in various tissues and animals. Comp. Biochem. Physiol., 99B: 345-350.

Konno R, Yasumura YD (1992). Amino-acid oxidase and its physiological function. Int. J. Biochem., 24: 519-524.

Kroesen B, Pettus B, Luberto C, Busman M, Sietsma H, de Leij L, Hannun L (2001). Induction of apoptosis through β-cell receptor cross-linking occurs via de novo generated C16-ceramide and involves mitochondria. J. Biol. Chem., 276(17): 13606-13614.

Liotta L (1992). Cancer cell invasion and metastasis. Sci. Am., 266: 54-63.

Lowry OH, Rosebrough NJ, Farr AL, Randall RJ (1951). Protein measurement with the Folin phenol reagent. J. Biol. Chem., 193: 265-275.

Morris HP (1963). Some growth, morphology and biochemical characteristics of hepatoma 5123 and other new transplantable hepatomas. Progr. Exptl. Tumour Res., 3: 371-411.

Perry DK, Carton J, Shah AK, Meredith F, Uhlinger DJ, Hannun YA (2000). Serine palmitoyltransferase regulates de novo ceramide generation during etoposide-induced apoptosis. J. Biol. Chem., 275(12): 9078-9084.

Podo F, Sardanelli F, Lorio E, Canesse R, Carinelli G, Fausto A, Canevari S (2007). Abnormal choline phospholipid metabolism in breast and ovary cancer. Molecular bases for noninvasive imaging approaches. Curr. Med. Imaging Rev., 3: 123-137.

Radin SN (2003). Killing tumours by ceramide-induced apoptosis: a critique of available drugs. Biochem. J. 371 (2) 243-256.

Rechcigl M Jn., Wollman S (1963). Effect of hypohysectomy on the lowering of organ catalase by a transplanted tumour. J. Nat. Cancer Inst., 31: 651-669.

Schulze-osthoff K, Bakker C, Vanhaesebroeck B, Beyaert R, Jacob W, Fiers W (1992). Cytotoxic activity of TNF is mediated by early damage of mitochondrial function: evidence for the involvement of mitochondrial radical generation. J. Biol. Chem., 267(8): 5317-5323.

Solomon E, Borrow J, Goddard D (1991). Chromosome aberration and cancer. Sci., 254: 1153-1160.

von Haefen C, Wieder T, Gillissen B, Stärck L, Graupner V, Dörken B, Daniel P (2002). Ceramide induces mitochondrial activation and apoptosis via a bax-dependent pathway in human carcinoma cells, 21(25): 4009-4019.

Wang H, Elwell H, Oberley W, Goedde IV (1989). Manganese superoxide dismutase is essential for cellular resistance to cytotoxicity of tumour necrosis factor. Cell, 58(5): 923-931

Zhuang S, Demirs J, Kochevar I (2000). P38 mitogen-activated protein kinase mediates bid cleavage, mitochondrial dysfunction, and caspase-3 activation during apoptosis induced by singlet oxygen but not by hydrogen peroxide. J. Biol. Chem., 275(34): 25939 – 25948.

The effects of mannan-oligosaccharides on cecal microbial populations, blood parameters, immune response and performance of broiler chicks under controlled condition

Saeed Khalaji*, Mojtaba Zaghari and Somaye Nezafati

Department of Animal Science, College of Agriculture and Natural Resource, University of Tehran, Karaj31587-11167, Iran.

A study was conducted to evaluate the effects of mannan-oligosaccharides (MOS) on cecal microbial populations, immune responses to phytohaemagglutinin-P (PHA-P) and sheep red blood cell (SRBC) and performance (weight gain and feed conversion ratio) of broiler chicks under strict controlled condition. Sixty four day-old male broiler chicks were randomly assigned to 12 battery cages pens of 4 chicks each and were fed from 1 to 42 days of age. Two basal diets were formulated for starter (1 to 21 days) and grower (22 to 42 days) period and a graded level of MOS (0.5, 1 and 1.5 g/kg) were added to basal diet to make diets 2 to 4. Body weight gain and feed conversion ratio were measured at 21, 35 and 42 days of age. Immune response to PHA-P was measured at 35 and to SRBC at 28 and 42 days of age. Cecal contents were assayed for *lactobacilli* and *Escherichia coli* at 42 days of age. Plasma triglyceride, cholesterol and HDL concentration was measured at 42 days of age. Body weight gain and feed conversion ratio did not differ among dietary treatments. Carcass, breast, tight, gizzard, duodenum, jejunum and ileum relative weight and duodenum, jejunum and ileum relative length were not affected by treatments. MOS increased immune response to PHA-P and SRBC numerically. Plasma triglyceride and HDL concentration did not differ among the treatments but plasma cholesterol concentration decreased linearly by adding MOS to the basal diet (P<0.05). Results showed no significant differences in *lactobacilli* and *E. coli* content of ceca among the treatments. Results of this experiment showed that MOS could affect the immune response and performance and as well as plasma cholesterol regardless of strict controlled condition.

Key words: Mannan-oligosaccharides, broiler chicks, performance, immune response, plasma factors.

INTRODUCTION

The subtherapeutic usage of antibiotic growth promoters (AGP) is under intense scientific and public scrutiny because their use has been linked to the development of antibiotic-resistant pathogenic bacteria, which pose a threat to human health (Smith et al., 2003). At the same time, food safety remains a major public health concern worldwide. Prebiotics are defined as "a nondigestible food ingredient that beneficially affects the host by selectively stimulating the growth and/or activity of one or a limited number of bacteria in the colon" (Gibson and Roberfroid, 1995). Mannose-based carbohydrates occur naturally in many products, such as yeast cell walls or different gums, which are available at reasonable prices. The observation on the effects of mannanoligosaccharides (MOS) on broiler performance is controversial. Hooge (2004) reported that growth performance and feed efficiency were improved in birds fed MOS compared with those fed basal diet. Based on the published research, it appears that MOS have variable effects on broiler performance. These may be attributed to differences in

*Corresponding author. E- mail: saeed.khlj@gmail.com.

the type of product, experimental conditions, diet formulation, and health status of the birds. It is reported that (Sims et al., 2004; Hooge, 2004) most beneficial additives are most effective under stress and disease conditions. Supplementation of poultry diets with MOS results in improved production in terms of body weight gain and feed conversion (Parks et al., 2001), partly due to its hypothesized nutrient sparing effect but primarily due to its influence on nutrient utilization in the gastrointestinal tract (Kumprecht et al., 1997; Savage et al., 1997; Sonmez and Eren, 1999). In addition, previous reports suggest that MOS supplementation resulted in significant improvement in antibody responses in broiler and layers (Cotter et al., 2000; Raju and Devegowda, 2002). Most of the experiments have investigated the effects of MOS on performance and gut microflora, but there is no information about the effects of these products on blood parameters such as cholesterol. On the other hand almost all of the experiments in this field were done at the commercial or challenging condition where the effects of MOS are so clear. The hypothesis tested in this study was that mannanoligosaccharides (MOS) from yeast cell walls even so are able to improve broiler performance, decrease cholesterol concentrations of plasma and stimulate immune response in broiler chicks under strict controlled conditions.

MATERIALS AND METHODS

Sixty four 1-day old male Ross 308 broilers were obtained from a commercial local hatchery and grown over a 42 days experimental period. The chicks were housed in thermostatically controlled batteries with raised wire floors in an environmentally controlled building. At hatch, chicks were weighed and randomly allotted to 4 pens of 4 chicks per treatment so that each pen of chicks had a similar initial weight and weight distribution. Throughout the study, the birds were brooded following standard temperature regimens, which gradually decreased from 32 to 24°C, and under a 23L: 1D cycle. The 4 experimental diets included control diet and control diet with addition of graded levels (0.05, 0.1 and 0.15%) of TechnoMOS (Biochem®) as a commercial source of mannan oligosaccharide. The ingredient composition and nutrient content of the diets are shown in Table 1. Body weights and feed intakes were measured at the end of Weeks 3, 5 and 7. Weight gain, feed conversion ratio and mortality were calculated for each pen replicate. No mortality occurred in any of the experimental groups.

Immune measurements

At the 21 and 35 days of age 0.1 ml/kg body weight of 0.5% sheep red blood cell was injected into brachial vein of two chicks per replicate. Seven days after each injection, blood was collected in nonheparinized tubes by puncturing the brachial vein. Serum was isolated and stored at –20°C. Individual serum samples were analyzed for antibody responses against SRBC by ELISA technique using commercial kits, and the plates were read at 405 nm on an ELISA reader. Two chicks per pen were randomly chosen to evaluate a cutaneous basophil hypersensitivity test to phytohaemagglutinin-P (PHA-P) on day 36. At 36 days of age, the right toe web thickness was measured with a constant tension caliper before injection of 100 µg of PHA-P suspended in 0.10 ml of

sterile PBS. Twenty-four hours after the injection, the toe web was measured again. Relative swelling and toe thickness were indicators of the cellular immune response. A saline control was not used due to minor differences observed between the PHA-P and the saline-injected toe web (Corrier et al., 1990).

Carcass analysis and blood parameter

At the end of experimental period, 2 chicks with weight closest to the mean average of each replicate were selected randomly and blood samples were collected in heparinized tubes by puncturing the brachial vein to measuring the cholesterol, triglyceride and HDL, then these chicks were slaughtered by neck cutter and the weights of breast, thigh, abdominal fat, gizzard, duodenum, jejunum and ileum, also the length of duodenum, jejunum and ileum were measured.

Cecal microbial populations

At 42 days, 2 chicks from each pen replicate were slaughtered by neck cutter for the extraction of cecal contents. The cecal contents from ceca of each bird were pooled together for serial dilution. Microbial populations were determined by serial dilution (10^{-4} to 10^{-7}) of cecal samples in anaerobic diluents before inoculation onto Petri dishes of sterile agar as described by Bryant and Burkey (1953). Lactobacilli were grown on Rogosa SL agar, and E. coli were grown on EMB agar. Plates for Lactobacillus was incubated anaerobically (73% N: 20% CO2: 7% H2) at 37°C. E. coli were incubated aerobically at 37°C. Plates were counted between 24 and 48 h after inoculation. Colony forming units (CFU) were defined as being distinct colonies measuring at least 1 mm in diameter.

Statistical analysis

Data were analyzed by one way ANOVA using the general linear model (GLM) procedure of SAS software (SAS Institute, 2003) with pen as the experimental unit for performance parameters and bird as the experimental unit for microbiological parameters. Differences between treatment means were tested using Duncan multiple comparison test, and statistical significance was declared at a probability of $p < 0.05$. Microbiological concentrations were subject to log10 transformation before analysis.

RESULTS

Dietary treatments did not alter growth performance or feed conversion ratio (Table 2). MOS had no significant effect on relative weight of breast, thigh, abdominal fat and gizzard. Neither relative weight nor relative length of duodenum, jejunum and ileum were affected by addition of MOS (Table 3). The effect of MOS on immune response was significant and MOS increased linearly cutaneous basophil hypersensitivity test to phytohaemagglutinin-P ($p < 0.05$) but immune response against the SRBC was not significant and immune response increased numerically by addition of MOS compared to the basal diet (Table 4). There was no significant effect of MOS on plasma triglyceride and HDL concentration at 42 days of age (Table 4). Supplementation of the basal diet with MOS significantly reduced the plasma cholesterol concentration at 42 days

Table 1. Nutritional composition of basal diets %.

Ingredients	0 - 21 days of age	22 - 42 days of age
Wheat grain	10	10
Corn grain	39.3	49.28
Soybean meal (44%)	41.63	31.01
Soybean oil	5	6.07
Dicalcium phosphate	2.07	1.73
Oyster shell	0.86	0.76
Common salt	0.36	0.36
Vitamin mix[1]	0.25	0.25
Mineral mix[1]	0.25	0.25
DL-Met	0.22	0.22
L-Lysine	0.06	0.07
Calculated analysis		
CP (%)	23	19.2
ME (kcal/kg)	3010	3200
Digestible Lys (%)	1.19	0.96
Digestible Met (%)	0.52	0.49
Digestible Met + cystine (%)	0.86	0.77

[1]Vitamin and mineral mix supplied the following per kilogram of diet: vitamin A, 11,000 IU; vitamin D3, 1,800 IU; vitamin E, 11 mg; vitamin K3, 2 mg; vitamin B2, 5.7 mg; vitamin B6, 2 mg; vitamin B12, 0.024 mg; nicotinic acid, 28 mg; folic acid 0.5 mg; pantothenic acid, 12 mg; choline chloride, 250 mg; Mn, 100 mg; Zn, 65 mg; Cu, 5 mg; Se, 0.22 mg; I, 0.5 mg; and Co, 0.5 mg.

Table 2. Effects of mannanoligosaccharide on body weight and feed conversion of broiler chicks.

Age	Treatments				SEM[1]	p value
	Control	MOS				
		0.05%	0.1%	0.15%		
BW (g)						
Day 21	784.00	814.88	791.31	830.63	9.54	0.30
Day 35	1974.38	1947.50	1995.00	2029.38	22.35	0.66
Day 42	2485.31	2436.25	2497.19	2517.69	15.97	0.34
Feed conversion						
Day 21	1.300	1.290	1.320	1.270	0.017	0.72
Day 35	1.586	1.555	1.558	1.552	0.014	0.84
Day 42	1.674	1.659	1.652	1.643	0.012	0.86

[1]-Standard error of mean.

of age ($p<0.05$). There was no effect of dietary treatment on cecal lactobacilli and *E. coli* populations.

DISCUSSION AND CONCLUSION

Body weight gain and feed conversion ratio were not affected by MOS. In studies with broilers fed with MOS, virginiamycin, a combination of MOS and virginiamycin, or an antibiotic free diet, Waldroup et al. (2003) reported no improvement in growth performance and feed efficiency. However, based on a metanalysis of 44 research trials with broilers, Hooge (2004) concluded that birds fed MOS showed improved growth performance and feed efficiency compared with antibiotic free diet. It is reported that MOS and most beneficial additives (Hooge, 2004) are most effective under disease and stress conditions, such as extremes of ambient temperature, crowding, and poor management, which are invariably present in commercial broiler production. The present study was conducted under good hygienic conditions (new experimental facility, strict biosecurity measures,

Table 3. Effect of MOS on relative weight of different parts of gut and relative length of duodenum, jejunum and ileum (g/100 g BW).

Treatments	%Weight								%Length		
	Carcass	Breast	Thigh	Abdominal fat	Gizzard	Duodenum	Jejunum	Ileum	Duodenum	Jejunum	Ileum
Control	72.975	24.913	19.890	1.335	1.141	0.376	0.939	0.822	1.039	2.884	3.191
0.05% MOS	72.269	25.228	19.540	1.258	1.049	0.346	0.914	0.705	1.125	2.926	3.029
0.1% MOS	72.488	25.628	19.440	1.468	1.109	0.388	1.004	0.798	1.157	2.779	2.779
0.15% MOS	72.974	25.140	19.412	1.761	1.122	0.382	1.034	0.841	1.121	2.897	2.973
SEM	0.272	0.195	0.11	0.078	0.027	0.009	0.02	0.019	0.022	0.055	0.046
p value	0.8	0.67	0.43	0.09	0.7	0.47	0.11	0.3	0.26	0.83	0.21

Table 4. Effect of MOS on cutaneous basophil hypersensitivity test to PHA-P, immune response against SRBC and on blood cholesterol, triglyceride and HDL with cecal microbial populations.

Treatments	Toe web thickness (mm)	SRBC 28 day	SRBC 42 d	Triglyceride (mg/dl)	HDL(mg/dl)	Cholesterol (mg/dl)	Lactobacilli	Escherichia coli
							log10 cfu/g of DM	
Control	0.995[b]	4.13	7.75	97.25	80.625	135.13[a]	9.11	6.04
0.05 % MOS	1.685[ab]	5.38	8.125	122.13	82.625	118.00[ab]	8.92	5.93
0.1 % MOS	1.890[a]	4.88	7.75	120.75	77.125	112.50	8.52	6.15
0.15 % MOS	2.045[a]	4.88	8.125	83.38	79.25	108.75[b]	8.51	6.34
SEM	0.157	0.285	0.362	7.607	2.122	3.634	0.21	0.11
P Value	0.06	0.53	0.97	0.2	0.86	0.02	0.16	0.1

no access to feces, clean litter, good ventilation, and low stocking density), thus implying minimum bacterial challenge. Under such conditions, the chicks may not have required any feed additive for maximum productive response. Relative weight of different parts of gut and relative length of small intestine were not influenced by supplementation of MOS. Ferket et al. (2002) reported that intestinal weight and crypt depth were similar when turkeys were fed MOS or an antibiotic-free diet; however, muscularis thickness was significantly reduced. Increases in gut mass are associated with inflammation following bacterial infection (Walton, 1988), and this notion is supported by the observation that germ-free birds have thinner muscularis mucosa than conventional birds (Gordon and Bruckner-Kardoss, 1961). The reason for *the lack of an effect of the feed additives on gut parameters may be that under the* conditions of this experiment the pathogen load in the gut was low. Controlling the growth of intestinal microflora is important for improving the well being of the host. Many bacteria compete with the host for nutrients within the gastrointestinal tract, elicit an immune response that causes a reduction in appetite and an increase in muscle catabolism to maintain the immune response, cause disease, and reduce nutrient absorption in the intestine (Bedford, 2000). There was no significant effect of MOS on ceca population in the condition of this experiment. Spring et al. (2000) also reported no effect of MOS on lactobacilli populations in the ceca of broilers. In studies with turkeys, Fairchild et al. (2001) reported that intestinal populations of lactobacilli and bifidobacteria did not differ by supplementing of MOS. Factors contributing to variability in the effects of MOS on population of beneficial bacteria in the gut may include differences in experimental conditions, diet formulation, seasonal effects, and health status of the flock. For example, Coon et al. (1990) reported any effects of oligosaccharide on cecal microbe populations with soybean meal diet, because soybean meal contains approximately

6% raffinose plus stachyose, the lack of response in cecal microbial numbers may have resulted because the dietary soybean meal already provided a large amount of indigestible oligosaccharides. Regardless of the strict controlled conditions, results of this experiment showed that MOS can improve the immune response of broiler chicks. Results of research have shown that MOS has immunomodulatory properties (MacDonald, 1995; Savage et al., 1996; Cotter, 1997; Cotter et al., 2002). The yeast cell wall has powerful antigenic stimulating properties, and it is well established that this property is a characteristic of the mannan chain (Ballou, 1970). MOS had no effect on plasma triglyceride and HDL concentration but significantly reduced cholesterol. Results of many experiments have shown that prebiotics can reduce plasma cholesterol by different ways.

REFERENCES

Ballou CE (1970). A study of the immunochemistry of three yeast mannans. J. Biol. Chem., 245: 1197-1203.

Bedford M (2000). Removal of antibiotic growth promoters from poultry diets: Implications and strategies to minimize subsequent problems. World's Poult. Sci. J., 56: 347-365.

Bryant MP, Burkey LA (1953). Cultural methods and some characteristics of some of the more numerous groups of bacteria in the bovine rumen. J. Dairy Sci., 36: 205-217.

Coon CN, Leske KL, Akavanichan O, Cheng TK (1990). Effect of oligosaccharide-free soybean meal on true metabolizable energy and fiber digestion in adult roosters. Poult. Sci., 69: 787-793.

Corrier DE, DeLoach JR (1990). Interdigital skin test for evaluation of delayed hypersensitivity of cutaneous basophil hypersensitivity in young chickens. Am. J. Vet. Res., 51: 950-954.

Cotter PF (1997). Modulation of the immune response: Current perceptions and future prospects with an example from poultry. In Biotechnology in the Feed Industry. T. P. Lyons and K.A. Jacques, ed. Nottingham University Press, Nottingham, pp. 195-204

Cotter PF, Malzone A, Paluch B, Lilburn MS, Sefton AE (2000). Modulation of humoral immunity in commercial laying hens by a dietary prebiotic. Poult. Sci., 79(Suppl. 1): 38 (Abstr.).

Cotter PF, Sefton AE, Lilburn MS (2002). Manipulating the immune system of layers and breeders: Novel applications for mannan oligosaccharides. In Nutritional Biotechnology in the Feed and food Industries. T. P. Lyons and K. A. Jacques, ed. Nottingham University Press, Nottingham, pp. 21-28.

Fairchild AS, Grimes JL, Jones FT, Wineland MJ, Edens FW, Sefton AE (2001). Effects of hen age, Bio- Mos, and Flavomycin on poult susceptibility to oral Escherichia coli challenge. Poult. Sci., 80: 562-571.

Ferket PR, Parks CW, Grimes JL (2002). Benefits of dietary antibiotic and mannan oligosaccharide supplementation for poultry. http://www.feedinfo.com/files/multi 2002-ferket.pdf Accessed Jun.

Gibson GR, Roberfroid MB (1995). Dietary modulation of the human colonic microbiotica: Introducing the concept of prebiotica. J. Clin. Nutr., 125: 1404-1412.

Gordon HA, Bruckner-Kardoss E (1961). Effect of normal microflora on intestinal surface area. Am. J. Physiol., 201: 175-178.

Hooge DM (2004). Meta-analysis of broiler chicken pen trials evaluating dietary mannan oligosaccharides, 1993–2003. Int. J. Poult. Sci., 3: 163-174.

Kumprecht I, Zobac P, Siske V, Sefton AE, Spring P (1997). Effects of dietary mannanoligosaccharide level on performance and nutrient utilization of broilers. Poult. Sci., 76(Suppl. 1):132. (Abstr.).

MacDonald F (1995). Use of immunostimulants in agricultural applications. In Biotechnology in the Feed Industry. T. P. Lyons and K.A. Jacques, ed. Nottingham University Press, Nottingham, pp. 97-103

Parks CW, Grimes JL, Ferket PR, Fairchild AS (2001). The effect of mannanoligosaccharides, bambermycins, and virginiamycin on performance of large white male market turkeys. Poult. Sci., 80: 718-723.

Raju MVLN, Devegowda G (2002). Esterified-Glucomannan in broiler chicken diets-contaminated with aflatoxin, ochratoxin and T-2 toxin: Evaluation of its binding ability (in vitro) and efficacy as immunomodulator. Asian- Aust. J. Anim. Sci., 15: 1051-1056.

Savage TF, Cotter PF, Zakrzewska EI (1996). The effect of feeding mannanoligosaccharide on immunoglobulins, plasma IgG and bile IgA of wrolstadMWmale turkeys. Poult. Sci., 75(Suppl. 1): 143 (Abstr.).

Savage TF, Zakrzewska EI, Andreasen JR (1997). The effects of feeding mannan oligosaccharide supplemented diets to poults on performance and morphology of small intestine. Poult. Sci., 76(Suppl. 1): 139 (Abstr.).

Sims MD, Dawson KA, Newman KE, Spring P, Hooge DM (2004). Effects of mannanoligosaccharide, bacitracin methylene disalicyclate, or both on live performance and intestinal microbiology of turkeys. Poult. Sci., 83: 1148-1154.

Smith DL, Johnson JA, Harris AD, Furuno JP, Perencevich EN, Morris JG (2003). Assessing risks for a pre-emergent pathogen: Virginiamycin use and the emergence of streptogramin resistance in Enterococcus faecium. Lancet Infect. Dis., 3: 241-249.

Sonmez G, Eren M (1999). Effects of supplementation of zinc bacitracin, mannanoligosaccharide and probiotic into the broiler feed on morphology of the small intestine. Vet. Fak. Derg. Uludag Univ., 18: 125-138.

Spring P, Wenk C, Dawson KA, Newman KE (2000). The effects of dietary mannanoligosaccharides on cecal parameters and the concentrations of enteric bacteria in the ceca of Salmonella-challenged broiler chicks. Poult. Sci., 79: 205-211.

Waldroup PW, Fritts CA, Yan F (2003). Utilization of Bio-Mos mannan oligosaccharide and Bioplex copper in broiler diets. Int. J. Poult. Sci., 2: 44-52.

Walton JR (1988). The modes of action and safety aspects of growth promoting agents. In Proc. Maryland Nutr. Conf. Univ. Maryland, College Park, pp. 92-97.

20

Dyslipidemic and atherogenic effects of academic stress

Adekunle Adeniran S.

Department of Biochemistry, Ladoke Akintola University of Technology, Ogbomoso Oyo State, Nigeria.
E-mail: kunleniran@yahoo.com.

Evidence suggests that there is strong relationship between stress and development of cardiovascular disorders. Here, the relationship between academic stress and selected traditional markers of cardiovascular disorder such as lipids and lipoprotein profile and apoproteins were investigated. Eighty apparently healthy male and female students participated in the study. Plasma concentration of selected biochemical parameters such as total cholesterol, triglyceride, high density lipoproteins, Apoproteins A and B were determined before academic activities and one hour to an examination. Student's 't' test was used to compare the values before and after stress periods. Results showed significant elevations in total cholesterol, triglyceride, Apo A and B while there was reduction in mean concentration of high density lipoprotein cholesterol (HDL-C) during the intense academic period when compared with when there was no academic activity. These results suggest that intense academic activity may cause stress which may affect plasma lipids and lipoproteins.

Key words: Dyslipidemia, stress, atherosclerosis, apoproteins.

INTRODUCTION

Link between stress and development of cardiovascular disorder has been reported by previous workers (Brent et al., 2010). The signs of stress may be cognitive, emotional, physical or behavioral. There are two (2) types of stress which include the acute and the chronic stress. The acute stress affects an organism or an individual in the short term and the chronic stress affects an organism or an individual over a longer term. The relationship between stress, heart disease and sudden death has been recognized since antiquity. The incidence of heart attacks and sudden death has been shown to increase significantly following the acute stress.

Studies have shown that stressful situations resulted in increased levels of serum lipids that is, total cholesterol, low density lipoproteins, triglyceride and reduction in high density lipoproteins (Feroza et al., 2008; Stoney, et al., 2002). Furthermore, excessive elevations in the serum concentrations of these lipid profiles are traditional predictors of cardiovascular diseases. Therefore, this study aimed at determining the possible effects of academic stress occasioned by end of semester examination on traditional risk factors of cardiovascular disorders among apparently healthy male and female

students. In addition, effect of academic stress on apoproteins such as Apoprotein A and B was assessed in this study. The assessment of apoproteins was included based on studies that Apo B in LDL-C particles act as ligand for low density lipoprotein cholesterol (LDL-C) receptors on various cells. Informally, it 'opens doors to cells allowing cholesterol to be deposited into cells. Through a mechanism that is not fully understood, high level of Apo B can lead to plaques that cause vascular disease (atheriosclerosis), leading to heart disease. And most importantly, through series of studies, considerable evidence have emerged that levels of Apo B are a better indicator of heart disease risk than total cholesterol or LDL-C (Walldius and Jungner, 2006; Yusuf et al., 2004).

METHODS

Sample collection and storage

Eighty apparently healthy male and female students (mean age 21 years, mean weight 60.8 kg, mean height 1.70 m) were recruited from the Faculty of Basic Medical Sciences of Ladoke Akintola University of Technology, Ogbomoso, Oyo State. The inclusion

Table 1. Effects of academic stress on risk factors of atherogenicity.

Parameter	Before stress	During stress
Total cholesterol (mg/dl)	157 ± 53.29	263.5 ± 54.02[#]
Triglyceride (mg/dl)	162.9± 46.18	366.5± 216.57[#]
HDL-C (mg/dl)	358.8± 216.57	216.57± 144.18[#]
Apo A (mg/dl)	117.39 ± 37.08	182.55 ± 31.83[#]
Apo B (mg/dl)	81.6± 41.69	126.44 ± 38.78[#]
TC/ HDL	0.44± 0.43	1.22 ± 1.26[#]
Apo B/ Apo A	0.70 ± 0.59	0.69 ± 0.25

Significant at [#]$p<0.05$ when compared with values before stress.

criteria are: no history of hypertension, should not be suffering from any cardiovascular disorder such as atherosclerosis, stroke, and hypertension etc. All were normotensive, non-diabetic, without renal or liver dysfunction and with body mass index of less than 30 kg/m^2. Ten milliliters of veinous blood sample was collected from each of the volunteers in the second week of resumption when they have not undergone stress. The blood samples were collected into a heparinized bottle. Physical parameters of the students such as height, weight, hip and waist ratio were measured. Another 10 mm of veinous blood were collected 1 h before the commencement of a first semester examination when they appeared stress. Plasma was separated and used for analysis of total cholesterol; triglyceride, high density lipoprotein cholesterol, Apoproteins A and B. Participants were clearly informed of the procedure in English language since all of them understand and speak English fluently and they gave their consent. The approval of the University Ethical Committee was sought and gotten to be able to carry out this study.

Determination of total cholesterol

Total cholesterol was determined using enzymatic method described by Allain et al. (1974). Cholesterol esterase hydrolyses cholesterol esters to free cholesterol. The free cholesterol produced is oxidized by cholesterol oxidase to cholesten-4-ene-3-one with simultaneous production of hydrogen peroxide which oxidatively couples with 4-aminoantipyrine and phenol in the presence of peroxidase to yield chromogen with maximum absorption at wavelength 510 nm, The colour intensity is proportional to the cholesterol concentration.

Determination of triglycerides

Triglyceride was determined using enzymatic method described by Buccolo and David (1973). Triglycerides are hydrolyzed by lipases to yield glycerol and fatty acids. The glycerol produced is oxidized to dihydroxylacetone phosphatase with the production of hydrogen peroxide which oxidatively couples with 4-aminophenazone and 4-chlorophenol to produce a chromogen referred to as quinoneimine. The reaction is catalyzed by peroxidase. The degree of absorbance of the chromogen is directly proportional to the concentration of triglyceride measured at 505 nm.

Determination of high density lipoprotein

The precipitation method by Assmann et al. (1983) was used to determine HDL-cholesterol. The addition of phosphotungistic acid in the presence of magnesium ions precipitates quantitatively low density lipoprotein; very low density lipoprotein and chylomicron

fractions from whole plasma, leaving the HDL fraction in the supernatant. The cholesterol in the HDL which remains in the supernatant after centrifugation is estimated using the enzymatic method of Allain et al. (1974).

Determination of Apoproteins A and B

Apoproteins A and B were determined using EasyRID human plasma proteins quantitative determination by radial immunodiffusion kit from Via Scozia, Zona Industriale Roseto d.A. (TE), Italy. This allows the determination of human plasma proteins in radial immunodiffusion. The antigen (protein) inoculated in the well of the plate, diffuses radially in the agarose gel, reacting with specific antibodies incorporated in agarose and forming immune complexes visible as precipitin rings. Diameter of precipitin ring is directly proportional to the concentration of the relevant protein in the sample (Sniderman et al., 1975).

Statistical analysis

Quantitative data were presented as mean ± SD. Triglyceride, total cholesterol, phospholipids, high density lipoprotein and apoproteins between the two periods were compared using student's 't' test. A value of $p<0.05$ was considered statistically significant.

RESULTS

Table 1 shows the effect of stress on the biochemical parameters and atherogenic indices. There were significant ($p<0.05$) elevations in mean serum concentration of total cholesterol, triglyceride, Apo A, Apo B, and total cholesterol-HDL-C ratio during stress when compared with corresponding parameters before stress. There were decreased in mean serum concentrations of high density lipoprotein cholesterol HDL-C and Apo B-Apo A ratio during stress when compared with the corresponding parameters before stress.

DISCUSSION

There were significant increases in the levels of certain lipids (that is, total cholesterol, triglycerides) in the volunteers during the period of stress when compared with period when there was no stress. A consequent reduction in the level of HDL-cholesterol was observed during the period of stress. There were also significant increases in the levels of Apolipoprotein A and B during the period of stress.

One way that stress might influence lipid concentrations is through stress induced hormonal changes that affect lipid metabolism. For example, stress induces sympathetic nervous system activity leading to concomitant increases in catecholamine (epinephrine and nor-epinephrine) and glucocorticoids (cortisol). These increases have resultant effects on metabolism of the lipoproteins. For example, catecholamines can directly stimulate adipose tissue to release free fatty acids into circulation through the process of lipolysis. Epinephrine induced-increase in free fatty acids may be the result of increased blood flow through adipose tissue or stimulation of adipose-β2 adrenoreceptors. In either of the cases, the accumulation of circulating free fatty acids can trigger the production of triglyceride-rich very low-density lipoprotein (VLDL), which will eventually result in increased concentrations of circulating LDL-cholesterol (Brindley et al., 1993).

Like epinephrine, nor-epinephrine influences lipoprotein metabolism. Increased nor-epinephrine activity stimulates adipose β adrenergic receptors which may result in diminished lipoprotein lipase activity, subsequent decline in triglyceride clearance, lower concentrations of HDL-C and increased level of LDL-C. This may be responsible for elevated triglyceride, total cholesterol, and reduced HDL-C observed in this study. Cortisol is another hormone that may be affected by stress. It also has profound influence on the mobilization of lipids and lipid metabolism through activation of the hypothalamic-pituitary-adrenocortical (HPA) axis. Moreover, cortisol and free fatty acids stimulate the secretion of VLDL, increase the synthesis of hepatic triglycerides, inhibit insulin secretion, and increase insulin insensitivity in the tissues (Bjorntop, 2000). These activities may delay LDL clearance from the blood, result of which will be elevated serum lipids level as observed in this study.

A high level of stress promotes an increase in the levels of VLDL-cholesterol and LDL-cholesterol which apolipoprotein B forms a significant part of. It will also lead to the reduction in the levels of HDL-cholesterol (good cholesterol) which contains a significant amount of apolipoprotein A-I or apolipoprotein A-II. This condition is known as dyslipidaemia (Adiels et al., 2006). Hence increase in stress tends to cause an increase in apolipoprotein B concentration and a reduction in the concentration of apolipoprotein A, an observation which is consistent with the findings of this study. Series of studies have shown that Apo B in LDL-C particles act as ligand for LDL-C receptors on various cells. Informally, it 'opens' doors to cells allowing cholesterol to be deposited into cells. Through a mechanism that is not fully understood, high level of ApoB can lead to plaques that

cause vascular disease (atheriosclerosis), leading to heart disease. And most importantly, through series of studies, considerable evidence have emerged that level of Apo B is a better indicator of heart disease risk than total cholesterol or LDL-C (Walldius and Jungner, 2006; Yusuf et al., 2004). Results of elevated Apo B concentrations in this study further corroborated the association between stress and risk factors of cardiovascular disorder.

The principal hypothesis to support the relationship between stress and cardiovascular diseases is that the stress-induced activation of hypothalamic and sympatho-hormonal regions occurs repeatedly over time, leading to cardiovascular adjustments that increase hypertension risk. The incidence of heart attacks and sudden death have been shown to increase significantly following the acute stress due to natural disasters like hurricanes, earthquakes and tsunamis. Coronary heart disease is much more common in individuals subjected to chronic stress. The source of stress in this instance was exposure to examination condition. The effect of stress was assessed one hour to the examination. During this period, the students were impatient and have excited pulse. The excited pulse due to examination stress is an indication that there is association between stress and heart beat. Excessive beat of the heart may result in cardiovascular disorder.

This view is also supported by the results of the blood biochemical parameters. There were significant elevations ($P<0.05$) of total cholesterol and triglyceride concentrations during the examination period (stress condition) when compared with the corresponding values before the period. Elevated serum lipids are some of the biochemical markers of cardiovascular diseases such as hypertension, stroke, artheriosclerosis etc. Studies have shown that elevated concentration of triglyceride is a risk factor for cardiovascular disease for both men and women in the general population (John and Melissa, 1996). Several studies detected a trend towards prominent role of lipid levels in the pathogenesis of cardiovascular diseases (Schoofs et al., 2004; Meier et al., 2000; Wang et al., 2000). The pathological background of the artherosclerosis of coronary arteries is the formation of artherosclerotic plaque due to elevated artery cholesterol which additionally induces other cardiovascular diseases (Valentaviciene et al., 2005).

The elevations in the serum concentrations of total cholesterol, triglycerides, total cholesterol- HDL ratio and reduction in serum HDL-C is suggestive of artherogenic tendency of stress. One of the predictive markers of artherogenesis is elevated levels of total cholesterol, triglyceride, LDL-C and reduction in HDL-C. The findings in this study are consistent with this well established theory. Prolonged stress results in inability of the body to convert cholesterol into needed products such as hormones, a situation that may results in accumulation of cholesterol. This may be a reason for the observed elevation of serum total cholesterol in this study. In a study carried out on female monkeys at Wake forest University by Alice in 2009, he observed that individual monkey suffering from high levels of stress have higher levels of visceral fat in their bodies. This suggests a possible cause and effect link where stress promotes accumulation of visceral fat which in turn causes hormonal and metabolic changes that contribute to heart diseases and other health problems.

In this study, significantly ($P<0.05$) low serum HDL-C concentration was observed during stress when compared with before stress. This may be an indication that stress reduces serum concentration HDL-C that is, that is ability to mop-up excess cholesterol in the blood. In epidemiological studies, high density lipoprotein cholesterol (HDL-C) has also been identified as a strong and independent inverse predictor of cardiovascular events. Higher level of HDL-C predicts longevity (Zuliani et al., 2010). The anti-atherogenic or protective mechanisms of HDL-C are multiple. Most well known is its role in reverse cholesterol transport, by which excess cellular cholesterol is returned to the liver for excretion in the bile (Rothblat and Phillips, 2010). HDL-C has other important roles and these include decreasing inflammation, preventing endothelial cell apoptosis and improving endothelial function. The reduction in HDL-C observed during stress condition in this study may suggest that prolonged and sustained stress may reduce capacity of HDL-C to moderate cholesterol efflux and perform other functions.

REFERENCES

Allain CC, Poon LS, Chan CSG (1974). Enzymatic determination of total serum cholesterol. Clin. Chem., 20: 470-475.

Assmann G, Schriewer H, Schmitz G, Hagele EO (1983). Quantification of HDL-C by precipitation with phosphotungstic acid and MgCl$_2$. Clin. Chem., 29(12): 2026-2030.

Brent TM, Susan KR, Micheal GZ, Milos M, Roland VK, Joel ED, Paul JM, Thomas LP, Mathew AA, Sonia A, Igor G (2010). Stress of caring for someone with demential may impair endothelial function. J. Am Coll. Cardiol., 55(23): 2599-2606.

Buccolo G, David H (1973). Quantitative determination of serum triglyceride by the use of enzymes. Clin. Chem., 19: 476-482.

Feroza HW, Muhammad SM, Allah NM, Muhammad HSW, Syed AT, Javel I (2008). Estimation and correlation of stress and cholesterol levels in college teachers and housewives of Hyderabad-Pakistan. J. Pak. Med. Assoc., 15(1): 15-18.

John EH, Melissa AA (1996). Plasma triglyceride is a risk factor for cardiovascular disease. Eur. J. Cardiovasc. Prev. Rehabil., 3(2): 213-219.

Meier CR, Schwinger RG, Kraenzlin ME, Schlegel B, Jickk H (2000). HMB-COA reductase inhibitors and the risk of fractures. JAMA, 283: 3205-3210.

Rothblat GH, Phillips MC (2010). HDE and Cardiovaslar mortality in the elderly. Curr. Opin. Lipodol., 21: 117-126

Schoofs MW, Sturkenboom MC, Vander KM, Hofman A, Pols HA, Sticker BH (2004). HMG COA reeducates inhibitors and the risks of vertebral fracture. J. Bone Miner. Res., 19: 1525-1530.

Sniderman A, Feng B, Jerry M (1975). Determination of B protein of low density lipoprotein directly in plasma. J. Lipid Res., 16: 465-467.

Stoney CM, West SG, Hughes JW, Lentino LM, Finney ML, Falko J, Bausserman L (2002). Acute psychological stress reduces plasma triglyceride clearance. Psychophysiology, 39(1): 80-85.

Valentaviciene G, Paipaliene P, Nedzelskiene I, Zilinskas J, Anuseviciene OV (2005). The relationship between blood serum lipids and periodontal condition. Stomatologiya, 8: 96-100.

Walldius G, Jungner I (2006). The apo B/apo A-I ratio; a new predictor of fatal stroke, myocardial infarction and other ischaemic diseases-stronger than LDL and lipid ratios. Atherosclerosis, 7: 468.

Wang PS, Solomon DH, Mogun H, Avorin J (2000). HMG COA reductase inhibitors and the risk of lip fractures in elderly patients. JAMA, 283(24): 3211-3216.

Yusuf S, Hawken S, Ounpuu S, Dans T, Avezum A, Lanas F, McQueen M, Budaj A, Pais P, Varigos J, Lisheng L (2004). Effect of potentially modifiable risk factors associated with myocardial infarction in 52 countries: case-control study. Lancet, 364: 937-952.

Zuliani G, Cavalieri M, Galvani M, Volpato S, Cherubim A, Bandinelli S, Corsi M, Lauretain I, Guranik J, Fellin R, Ferrucci L (2010). Relationship between low levels of high density impoprotein cholesterol and dementis in the elderly. The inchianti study. J. Garentol. Series A, 65A: 559-564.

21

Radio-protective chelating agents against DNA oxidative damage

M. H. Awwad[2], Samy A. Abd El-Azim[1], F. A. M. Marzouk[3]*, E. A. El-Ghany[3], and M. A. Barakat[1]

[1]Faculty of Pharmacy, Cairo University Cairo, Egypt.
[2]Faculty of Science, Benha University Benha, Egypt.
[3]Labelled Compounds Department, Hot Laboratory Center, Atomic Energy Authority. P. O. Box 13759, Cairo, Egypt.

The present study aims to investigate the strong neuro and radio-protective activity of chelating agents through inhibition of metal catalyzed Fenton reaction and aims to explore the potential of PCR-RFLP as biomarkers. Polymorphisms of rat brain angiotensin II subtype 2 receptor gene (AT2RG) were investigated as a biomarker for the effect of 6 Gy irradiation with or without prior carnosine and dimer-captosuccinic acid (DMSA) injection alone or in combination. AT2RG was separated into purified form ($\approx$2950 bp) using PCR magnification followed by digestion of AT2RG with restriction endonucleases which gave different polymorphism profiles which illustrated the following, strong antioxidant activity of carnosine alone and moderate antioxidant activity of DMSA. Maximum protection was achieved by the dual action of both carnosine and DMSA. Antioxidants prevented oxidative stress induced mutations of AT2RG that confer stability to its function such as neuroprotection and blood pressure osmoregulation. PCR-RFLP analysis explored the possibility of using of AT2RG polymorphisms restricted by Xbal, Acc65I restriction endonucleases as characteristic of new biomarkers for non radiation exposed individuals.

Key words: New biomarkers, PCR-RFLP, radio-protective, AT2RGen, chelating agents.

INTRODUCTION

The steadily increasing uses of advanced technologies are paralleled by increasing sources of oxidants such as ionizing radiation. Radiation effects in exposed tissues are the consequence of damage to various critical cell systems. Brain is the one of the highest sensitive organ for radiation induced oxidative stress because of its high O_2 utilization rate, the brain accounts only for a small percentage of the body weight, but consumes about 20% of basal O_2 metabolism which leaks reactive oxygen species as a by product of oxidative metabolism. Having higher metabolism rate, the developing brain are more susceptible to oxidative stress lesions than the mature one. Many of these lesions induced by ionizing radiation are chemically similar to those induced as by-product of oxidative metabolism. Radiation and oxidative impaired metabolism in the brain can lead to excess extracellular

glutamate levels which lead to a rising of intracellular Ca^{2+} to pathological levels (Halliwell, 2001; Dai et al., 2007). In addition to its high content of free ferrous ions and autoxidisable neurotransmitters (L-DOPA, epinephrines, encephalins and endorphins) leak free radicals in the brain (Halliwell, 2001; Halliwell and Gutteridge, 1999). Antioxidant defenses in the brain are modest in action and not all antioxidants can penetrate blood brain barrier (Colton et al., 1996; Sherki et al., 2001).

Brain is enriched with iron-containing proteins and irradiation damage to brain tissue readily releases iron and copper ions in forms that are capable of catalyzing such free radical formation, lipid peroxidation, and autoxidation of neurotransmitters (Spencer et al., 1994). Iron chelators were found to alter metal iron catalyzed oxidation. Iron atoms that are close to DNA can interact with highly diffusible, but weak free radical H_2O_2 in a metal-catalyzed Fenton-type reaction to produce OH$^-$ mediated DNA damage, strand breaks and DNA mutations (Henle and Linn, 1997).Chelating agents may

---

*Corresponding author. E-mail: fawzymarzouk@yahoo.com.

(a)   (b)

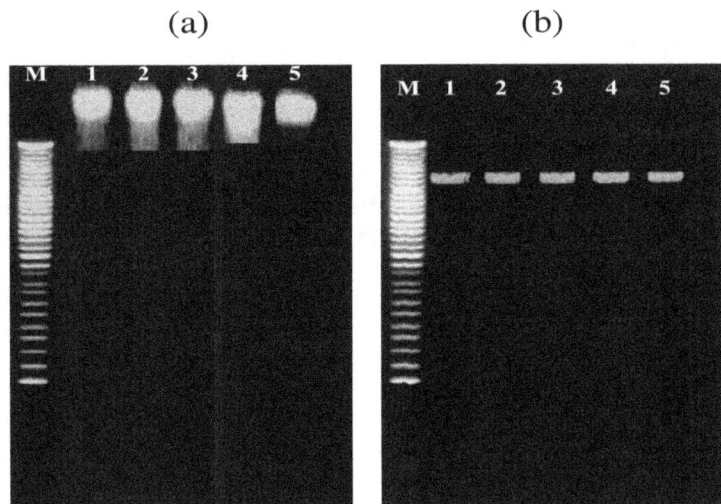

Figure 1. Lane M represents DNA marker; 1 to 5 represent 1 kb ladder lanes in (a) DNA genome (b) AT2R Gene in the studied groups.

protect DNA from hydroxyl radical and prevent enzymes damage which are responsible for DNA enzymatic repair system.

The brain owns a complete rennin-angiotensin system, including both AT_1 and AT_2-receptor subtypes (Reineke et al., 2006; Gasparo et al., 2000; Steckelings et al., 1992). But under pathological conditions such as stroke and oxidative stress as irradiation (Li et al., 2005), increased expression of the AT_2-receptor in brain tissue has been noticed with the involvement possibility of this receptor in maintaining anti-inflammatory functions and neuroprotection of the central nervous system (Li et al., 2005; Schulz and Heusch, 2006). In addition, polymorphisms in the rennin-angiotensin system genes were shown to be associated with immunoglobulin A and diabetic nephropathy, an increased risk of myocardial infarction, and the response to treatment with ACE inhibitors (Steckelings et al., 2005).

This study aims is to evaluate the antioxidant neuro-protective and radio-protective properties of chelating agents as antioxidants such as DMSA and carnosine alone or in combination against irradiation exposure. Also the present study aims is to explore new biomarkers for different radiation exposed groups using PCR-RFLP analysis of AT2R gene and to evaluate possible oxidative stress related pathological conditions for AT2R gene.

MATERIALS AND METHODS

Animals and irradiation protocol

In this study thirty mature male albino rats (*Rattus rattus*) aged approximately 3 months and weighing 100±10 g were used. Rats were bred in the animal house of the Atomic Energy Authority. The animals were kept in the same conditions living and nourishment. All experiments were carried out with the permission of the Egyptian Atomic Energy Authority. The animals were divided into five groups

each group consists of six rats. The first group (1) injected with saline as a negative control group while the second were injected with saline before irradiation by one hour as a positive control one (2). The third group was injected with dose equivalent to 200 mg/kg of carnosine before irradiation by one hour (3). The fourth group was injected with both DMSA and carnosine before irradiation by one hour (4). The fifth group was injected with 50 mg/kg of DMSA before irradiation by one hour also (5). The last four groups were γ-irradiated with 6 Gy. Whole body irradiation was performed with ^{60}Co - γ-cell at dose rate of 2 Gy/min. All animals were sacrificed and dissected after one day of irradiation and whole brain were taken for DNA extraction.

DNA preparation

The whole brain was homogenized then one gram from each sample of the five groups was mixed in 500 µl of isotonic saline solution and centrifuged at 5,000 rpm for 5 min. The cell pellet was re-suspended in 500 µl of UNSET (Lysis solution; 8 M urea, 2% sodium dodecyl sulfate, 0.15 M NaCl, 0.001 M EDTA, 0.1 M Tris pH 7.5). The aqueous lysate was repeatedly extracted with 500 µl Phenol-chloroform-isoamyl alcohol several times until protein interface disappeared to separate the organic and aqueous phases. To precipitate the nucleic acid, iced absolute ethanol was added (2:1 v/v), and left to incubate at -20°C for 24 to 48 h. The nucleic acids were recovered by centrifugation at 5,000 rpm for 15 min. The pellet was dried and then re-suspended in 40 µl of sterile H_2O (Awwad, 2003). 1 µl of the resuspend pellets was checked by gel electrophoresis for the presence of DNA (Figure 1a).

Preparation of 0.8% agarose gel

0.8 gm of agarose (sigma) was dissolved in 100 ml TAE buffer (242 gm tris, 3.72 gm EDTA, 700 ml H_2O, 57 ml of glacial acetic acid and the volume brought to 5 L).

Gene amplification and purification using the standard polymerase chain reaction (PCR)

Polymerase chain reaction amplification: A fragment of

angiotensin II subtype two receptor gene of approximately 2950 base pairs length was amplified using the primers:, 5' TTTGGTATGCATTAAGCCTTTTCT 3'; as a forward primer and , 5' GAATTCATTTCCGACATATGCT 3' as a reverse primer. Primers were designed from an alignment of the angiotensin II subtype two receptor gene sequences of rat. The standard PCR reaction mixture was done as mentioned (Kissing et al., 1989). The standard polymerase chain reaction program for amplification of angiotensin II subtype two receptor gene was: 30-53 cycles each cycle included three steps; 1 min, 94℃; 2-3 min, 45℃; and 3 min, 72℃. Glass milk DNA purification was used to purify the gene from the agarose gel. 1 μl of the resuspend pellet was checked by agarose gel electrophoresis for the presence of angiotensin II subtype two receptor gene (Figure 1b) (Kissing et al., 1989).

Restriction fragment length polymorphism (RFLP) protocol

Restriction enzymes were used in this study; these are DraI, AvaI and Acc65I (Toyobo Biochemicals); XbaI and BstXII (Boehringer-Mannheim) and BsmI and DraI (Sigma).Restriction endonucleases were used to digest the AT₂RG of all groups. The digestion was performed for 3.5 h at 37ºC, and the products were evaluated on 2% TAE-agarose gels, stained with ethidium bromide, and bands were detected upon ultraviolet transillumination and photographed (35 mm Kodak Film, England). RFLP profiles were obtained according to the following steps: 1- The restriction buffer (5 μl) was transferred to the Labeled 0.5 μl tubes, 2- 5 μl of each PCR product was added to the labeled tubes. 2- The tubes was centrifuged and placed in water bath (37℃) for 1.5 h. 3- The tubes were placed on an ice after digestion (Vidigal et al., 1998).

RESULTS

The study dealt with antioxidant activity donated by chelating agents of both carnosine and DMSA alone or in combination against DNA irradiation induced oxidative damage. Radio protective and neuroprotective activities of both chelators were estimated by using PCR-RFLP analysis of AT2RG polymorphic fragments as biomarkers.

The resulting fragments of rat brain of different groups, which have been separated by agarose electrophoresis as photos enable us to identify their different alleles. DNA genome has a high molecular weight about, so it remained in the field of the agarose gel electrophoresis (Figure 1).

The anti-inflammatory angiotensin II subtype II receptor gene (AT2RG) was separated into a purified fragment form by using PCR. The size of AT₂RG was found to be approximately 2950 bp for rats (Figure 1b).

BsmI restriction endonuclease fragmented AT2RG into approximately identical four restriction cuts (200,600,650 and 1500 bp; Figure 2. DraI restriction enzyme cut AT₂RG for all groups into five restriction fragment (150, 300, 350,600 and 1550 bp; Figure 3). XbaI restriction endonuclease isolated the non-irradiated group in quite different profile from other groups.

XbaI restriction endonuclease clustered the five groups of rat brain AT₂RG for all groups into two groups, group 1 in the first and group 2, 3, 4 and 5 in the second cluster

(Figure 4). XbaI restriction enzyme digested AT2RG for group A into thee restriction cuts with unique one (400, 1150 and 1400 bp; lane 1). The rest for other groups of rat brain AT2RG were restricted into five bands (50, 400, 450,650 and 1400 bp; lanes 2, 3, 4 and 5).

As shown in Figure 5 AvaI restriction enzyme differentiated group 2 into five restriction fragments with two unique ones (100, 250, 600, 700 and 1300 bp ; lane 2). The same enzyme restricted AT2R Gene of groups 3 and 5 into four restriction fragments (350, 600, 700 and 1300 bp; lanes 3 and 5). Groups 1 and 4 were fragmented into three bands by the action of the same enzyme (600, 700 and 1650 bp; lanes 1 and 4).

BstXI restriction endonuclease clustered AT₂R Gene groups into three groups (Figure 6). Group 2 of AT2R Gene isolated into five bands with two characteristic ones (50, 200, 800, 850 and 1050 bp; lane 2).groups 1 and 4 showed three restriction fragments (200, 1000 and 1750 bp; lanes 1 and 4) while groups 3 and 5 showed four restriction fragments (200,850,900 and 1000 bp; lanes 3 and 5).

Rat brain of group A was differentiated when its AT₂RG was digested by Acc65I restriction endonuclease. As shown in Figure 7 Acc65I restriction enzyme cuts groups 2, 3 and 5 into four similar bands (250, 500, 1000 and 1200 bp; lanes 2, 3 and 5). Acc65I restriction endonuclease digested the AT₂RGene of group A into two different bands with characteristic one (500 and 2450 bp; lane 1). The same enzyme restricted AT2R Gene of groups 4 into three restriction fragments (250, 500 and 2200 bp; lanes 1 and 4). So some antioxidant activity was observed by the dual action of carnosine and DMSA (group D) that concluded from less number of mutations with the respect of group 2 (irradiated one). Individual treatment of either carnosine or DMSA (groups 3 and 5) showed quite similar profile with the respect of group 2 (irradiated one) by using Acc65I enzyme which means no detectable antioxidant activity by using this enzyme.

DISCUSSION

Reactive oxygen species are formed as a by pass products during cellular metabolism or during stress conditions. Also reactive oxygen species were used to propagate through auto-oxidation steps using transition metal as a catalyst. Irradiation causes oxidative stress which in turn increases free iron that leads to oxidative stress propagation and exaggeration. It is well known that the brain is one of the most sensitive organs for oxidative stress because of its high metabolic rate and high iron content. So chelating agents considerably have potent antioxidant property on the brain especially which can penetrate blood brain barrier.

Since biochemical studies have mainly focused in the last few years on oxidation-reduction biochemistry using molecular tools and chelating agents considerably have potent antioxidant property on the brain especially that

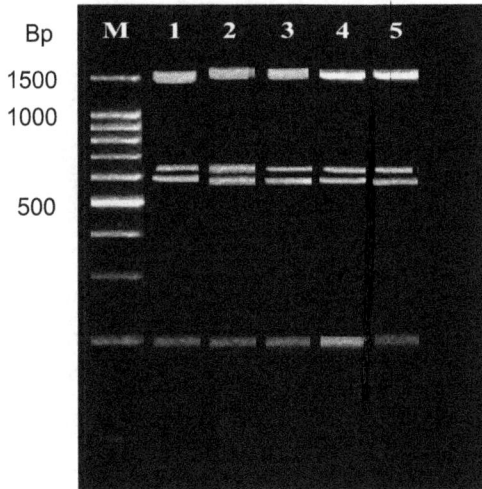

Figure 2. 1 kb ladder in lane M. Lanes 1,2,3,4 and 5 represent the length of AT$_2$R Gene fragments, which restricted with the *BsmI* endonuclease in the rat groups.

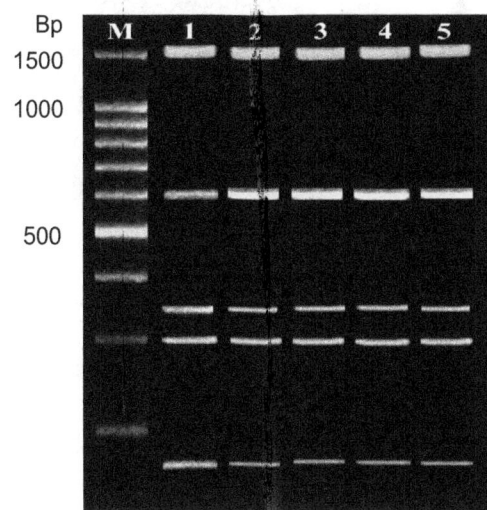

Figure 3. Lane M represents 1 kb ladder; lanes 1 to 5 represent the length of AT2R Gene fragments, which restricted with the endonuclease DraI in the studied groups.

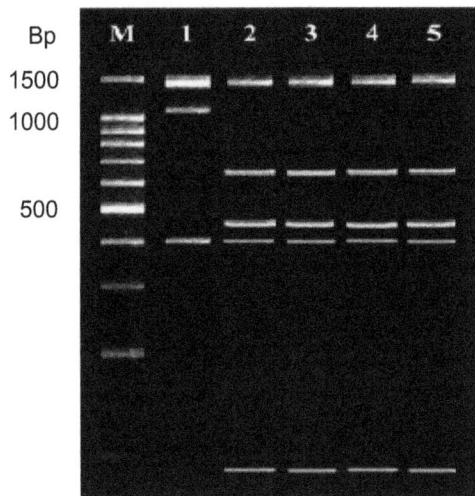

Figure 4. 1 kb ladder in lane M; Lanes 1, 2, 3, 4 and 5 represent the length of AT2R Gene fragments, which restricted with the XbaI endonuclease in the rat groups.

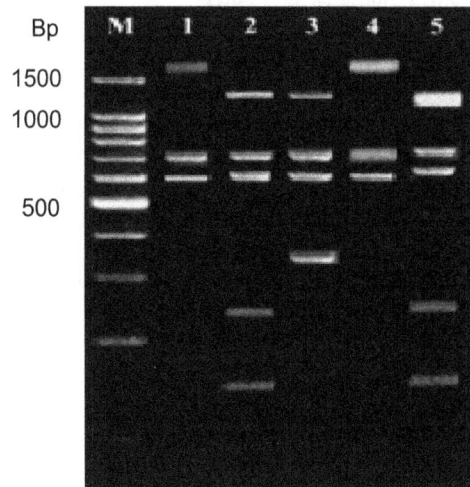

Figure 5. 1 kb ladder in lane M; Lanes 1, 2, 3, 4 and 5 represent the length of AT2R Gene fragments, which restricted with the AvaI endonuclease in the rat groups.

can penetrate blood brain barrier, molecular techniques were used to evaluate oxidative stress and to be as a diagnostic tool for certain stress condition (Collins et al., 1997; Dib, 1996; Dainiak, 2002).

So the present study aims to investigate the neuro-protective and radio-protective properties of either carnosine or DMSA alone or in combination as an antioxidant chelating agents after 6 Gy of gamma irradiation. Also this aims to explore new biomarkers for different groups using PCR-RFLP analysis of AT2RG.

In the present work it was found that polyphylogenecity between all groups which are diagnostic biomarkers for

these groups by DNA analysis. The data obtained revealed that normal control group has a characteristic polymorphic bands with *Acc65I* and *XbaI* restriction endonucleases .While irradiated rats group showed a diagnostic polymorphic markers with *AvaI* and *BstXII* restriction enzymes respectively. Also the polymorphism of all irradiated groups showed three characteristic bands of due to gene induction. All irradiated groups treated with AOs have diagnostic cuts by using *Acc65I* and *BstXII* restriction endonucleases respectively. So DNA analysis of the AT$_2$R gene polymorphism for different groups showed high specificity as a diagnostic tool that is

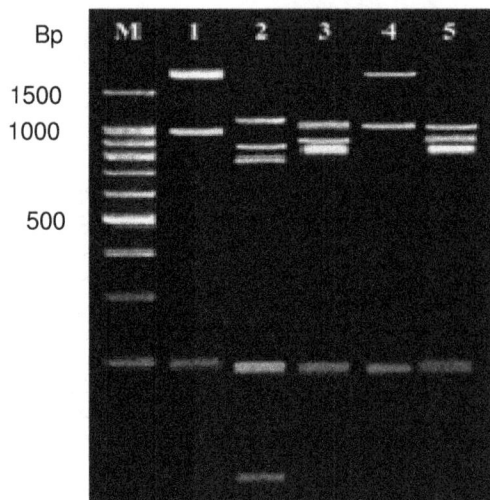

Figure 6. Lane M represents 1kb ladder lanes 1 to 5 represent the length of AT2R Gene fragments, which restricted with the endonuclease BstXI in the studied groups.

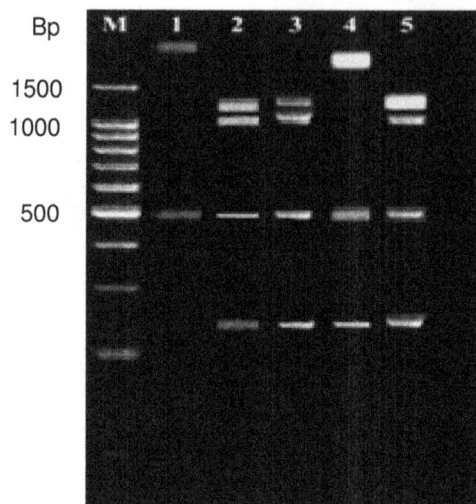

Figure 7. Lane M represents 1kb ladder lanes 1 to 5 represent the length of AT2R Gene fragments, which restricted with the endonuclease Acc65I in the studied groups.

confirmed with Collins et al. (1997).

One of the important roles of chelators is to detoxify metal ions and prevent poisoning. Iron chelation therapy leads to low cytosolic iron concentrations that facilitate resistance by protecting proteins, more than DNA, from IR-induced oxidative damage. By decreasing iron content in some bacteria species by three folds may lead to increasing radio-resistance up to 2000 times than human lethal dose. *D. radiodurans* resistance was attributed to DNA protection from hydroxyl radical and also attributed to enzymes protection which involved in base DNA repair system. Radiation induced DNA mutations can undergo enzymatic repair involving base excision repair or recombination (Coleman and Stevenson, 1996; Oleinick, 1990; Miller et al., 1995).

Ionizing radiation as oxidative stress inducer has sufficient energy to eject an electron from molecules, with the critical target being DNA to give DNA radical (Oleinick, 1990; Powell and McMillan, 1990). Ionizing radiation deposits energy in a bio-distribution model that gives closely spaced damage area, termed "locally multiply damaged sites (Coleman and Stevenson, 1996). Several types of DNA base damage and cross-links are produced that giving rise to both single and double-strand breaks (Coleman and Stevenson, 1996; Oleinick, 1990; Powell and McMillan, 1990). When an electron is ejected, an unstable DNA radical (DNA°) is produced, which is exposed to rapid biochemical reactions, and results in the formation of potentially lethal single and double strand breaks in the DNA. The DNA lesion can be biochemically repaired by reducing species (such as thiol -SH containing groups such as DMSA and other AOs species such as carnosine) or be made more permanent by combining with oxygen species. The accumulation of unrepaired double strand break correlates closely with

the loss of cell integrity (Coleman and Stevenson, 1996; Powell and McMillan, 1990).

From oxidative stress point of view AT$_2$R gene fragments showed high variations among normal control, irradiated (positive control) and all irradiated rat groups treated with AOs according to the difference of the profile obtained with *XbaI, AvaI, Acc65I* and *BstXII* restriction endonucleases. On the other hand *BsmI* and *DraI* restriction endonucleases did not clarify any difference between all rat groups based on the similar of profiles obtained with these restriction endonucleases. From all previous, the digestion of AT$_2$R gene for all groups by different restriction enzymes showed polyphylogenetic relationship between AT$_2$R structures for all groups.

Irradiated control rats which exposed to 6 Gy of γ-irradiation showed different AT$_2$R gene cuts from normal control group in all restriction endonucleases except for *BsmI* and *DraI* ones which indicates DNA fragmentation induced oxidative stress. Results obtained with *BsmI* and *DraI* restriction endonucleases did not clarify the difference between all rat groups of AT$_2$R gene based on the similarity of profiles obtained with these restriction endonucleases so if we exclude them we will find the followings: Carnosine treated group before γ-irradiation exposure showed the same cuts with two restriction enzyme endonucleases (*StyI and Bst*EII) for normal control group and two similar restriction cuts profile of *Acc65I and XbaI* restriction enzymes with irradiated control group but there are four different PCR/RFLPs profiles from non irradiated control rat but with less number of mutations in relation to irradiated control with *AvaI* and *BstXII* restriction endonucleases. All of that indicates the strong antioxidant activity of carnosine but not enough to completely block radiation induced oxidative stress which compatible with published data (Dizdaroglu, 1994; Halliwell,

1998).

Irradiated carnosine and DMSA treated rats gave the same restriction cuts for two enzymes which are AvaI, and BstXII restriction endonucleases in compared to normal control group of rats. It was found that one similar restriction cut profile by using XbaI restriction enzyme with irradiated control group and three different ones in relation to normal control by Acc65I, restriction endonuclease with less number of mutations compared to irradiated control one. So it can be concluded that maximum protection can be assessed through dual action of both DMSA and carnosine investigated by better gene profile which is confirmed with Kannan and Flora who declared that the dual action of DMSA with other AOs (Vit. E) are more effective (Kannan and Flora, 2004).

Irradiated DMSA treated rats exhibited similar profile with StyI restriction enzyme only when compared to normal control rats in other study for us under processing. In contrast to the above more similarity with irradiated control rats were shown with gene cuts by Acc65I, XbaI and BstEII restriction endonucleases. In addition there are four different PCR/RFLPs profiles from non irradiated control group with less number of mutations in relation to irradiated control one with PstXII and AvaI restriction endonucleases. This in turn means the moderate antioxidant activity of DMSA through free iron chelation and chelation of iron released through oxidative stress. Also DMSA may gives its antioxidant effect through reduction of formed DNA radical because of its content of free thiols (Powell and McMillan, 1990; Kannan and Flora, 2004).

DMSA and carnosine are tetra-chelating agents that mean every two atoms of free iron require three molecules of either chelating agents in stoichiometric chelation. In this study these chelating agents were found to alter iron mediated DNA oxidative damage leading to improvement in the picture of PCR-RFLP profile. According to PCR/RFLPs profiles Carnosine alone may have stronger antioxidant property than DMSA because it may exceed iron chelation by forming zinc and copper chelates that may have SOD-mimic action and its alkaline buffering action. But the dual action of both carnosine and DMSA may have maximum brain protection through free metal chelation, SOD-mimic action and direct reducing power for both chelates.

Angiotensin II receptor were found to have two subtypes of receptors with antagonistic action to each other, angiotensin II subtype 1 receptor (AT1R) over expression increases Na/water retention, inflammation, proliferation, vasoconstriction and oxidative stress while the AT2R expression have the reverse action so it represents one of the body protective mechanism. AT1R blocker increases expression of AT2R with modulating action on blood lipids and possible role of antioxidant activity (Steckelings et al., 2005;, Steckelings et al., 1992; Baykal et al., 2003). AT1R blockers screening for many patients showed possible lowering action on incidence

rate of diabetes mellitus. In the present study antioxidants may prevent or decrease AT2R gene mutation induced oxidative stress that elicits beneficial effect on vascular system (Li et al., 2005; Schulz and Heusch, 2006). Thus antioxidants may decrease incidence rate of diabetes mellitus, hypertension and hyperlipidaemia. Antioxidants may have a beneficial effect on glucose level, blood pressure and total lipid profiles which may be through inhibition of AT2R gene mutation as a possible mechanism.

Conclusion

PCR-RFLP biomarkers were found to be an advanced, simple and rapid tool for evaluation of radio-protecting and antioxidant activity of chelating agents alone or in combination against oxidative stress irradiation induced. Also it may be used as a new biomarker for the diagnosis of radiation exposure at different conditions with respect to antioxidant treatment. PCR-RFLP analysis cleared that the possibility of using AT2RG polymorphisms fragmented by XbaI Acc65I restriction enzymes as a characteristic new biomarkers for non radiation exposed individuals. Carnosine was shown to have a stronger antioxidant activity than that of DMSA. Maximum protection was achieved by the dual action of both chelating agents' carnosine and DMSA. Chelating agents prevented oxidative stress induced mutations of AT2RG that confer stability to its function such as neuroprotection and blood pressure osmoregulation.

REFERENCES

Awwad MH (2003). Molecular identification of *Biomphalaria alexandrina* and *Bulinus tranctus* using PCR-RFLP of actin gene. J. Egypt. Acad. Soc. Environ.Dev., 3(1): 39-52.

Baykal Y, Yilmaz M, Celik T, Gok F, Rehber H, Akay C,Kocar H (2003). Effect of antihypertensive agents, alpha receptor blockers, beta blockers, angiotensin converting enzyme inhibitors, angiotensin receptor blockers and calcium channel blockers on oxidative stress. J. Hypertens., 21(6): 1207-1211.

Coleman CN, Stevenson AM (1996). Advances in Cellular and Molecular Radiation Oncology. Urol. Oncol., 2: 3-13.

Collins FS, Guyer MS, Chakravarti A (1997). Variations on a theme cataloging human DNA sequence variations. Science, 278: 1580-1581.

Colton C, Wilt S, Gilbert D (1996). Species differences in the generation of reactive oxygen species by microglia. Mol. Chem. Neuropathol., 28: 15-20.

Dai X, Sun Y, Jiang Z (2007). Protective Effects of Vitamin E against Oxidative Damage Induced by Abeta (1-40) Cu (II) Complexes. Acta Biochim. Biophys. Sin. (Shanghai), February, 39(2): 123-130.

Dainiak N (2002). Hematologic consequences of exposure to ionizing radiation. Exp. Hematol., 30: 513-528.

Dib C (1996). A comprehensive genetic map of the human genome based on 5264 microsatellites. Nature, 380: 152-154.

Dizdaroglu M (1994). Chemical determination of oxidative DNA damage by gas chromatography-mass spectrometry. Methods Enzymol., 234: 3-16.

Gasparo M, Catt K, Inagami T, Wright JW, Unger T (2000). International Union of Pharmacology. XXIII. The angiotensin II receptors. Pharmacol. Rev., 52: 415-472.

Halliwell B (1998). Can oxidative DNA damage be used as a biomarker of cancer risk in humans? Problems, resolutions and preliminary results from nutritional supplementation studies. Free Radicals Res., 29: 469-486.

Halliwell B (2001). Role of Free Radicals in the Neurodegenerative Diseases: Therapeutic Implications for Antioxidant Treatment. Drugs Aging, 18(9): 685-716.

Halliwell B, Gutteridge JMC (1999). Free radicals in biology and medicine. 3rd ed. Oxford: University Press.

Henle ES, Linn S (1997). Formation, prevention, and repair of DNA damage by iron/hydrogen peroxide. J. Biol. Chem., 272: 19095-19098.

Kannan GM, Flora SJ (2004). Chronic arsenic poisoning in the rat: Treatment with combined administration of succimers and an antioxidant. Ecotoxicol. Environ. Saf., 58(1): 37-43.

Kissing B, Croom H, Martin A, McIntosh C, McMillan W, Palumpi S (1989). The simple fools guide to PCR version1.0. Department of Zoology, University of Hawaii, Honolulu.

Lees-Miller SP (1995). Godbout R, Chan DW, Absence of p350 subunit of DNA-activated protein kinase from a radiosensitive human cell line. Science, 267: 1183-1185.

Li J, Culman J, Hortnagl H, Zhao Y, Gerova N, Timm M (2005). Angiotensin AT_2 receptor protects against cerebral ischemia-induced neuronal injury. FASEB J., 19: 617-619.

Oleinick NL(1990). Ionizing radiation damage to DNA: Molecular aspects. Rad. Res., 124: 1-6.

Powell S, McMillan TJ (1990). DNA damage and repair following treatment with ionizing radiation. Radiother. Oncol., 19: 95-108.

Reineke C, Steckelings UM, Unger T (2006). Angiotensin receptor blockers and cerebral protection in stroke. J. Hypertens., 24: 116-119.

Schulz R, Heusch G (2006). Angiotensin II type 1 receptors in cerebral ischaemia-reperfusion: initiation of inflammation. J. Hypertens., 24: 124-128.

Sherki Y, Melamed E, Offen D (2001). Oxidative stress induced-neurodegenerative diseases: The need for antioxidants that penetrate the blood brain barrier. Neuropharmacology, 40: 959-975.

Spencer JP, Jenner A, Aruoma OI (1994). Intense oxidative DNA damage promoted by L-dopa and its metabolites: Implications for neurodegenerative disease. FEBS Lett., 353: 246-250.

Steckelings UM, Bottari SP, Unger T (1992). Angiotensin receptor subtypes in the brain. Trends Pharmacol. Sci., 13: 365-368.

Steckelings UM, Kaschina E, Unger T (2005). The AT2 receptor - A matter of love and hate. Peptides, 26: 1401-1409.

Vidigal T, Dias E, Spattz LN, Pires R, Simpson A, Carvalho O (1998). Genetic variability and identification of the intermediate snail host of Schistosoma Mansoni. Mem. Inst. Oswald Cruz., 93: 103-110.

22

Assessment of current iodine status of pregnant women in a suburban area of Imo State Nigeria, twelve years after universal salt iodization

Cosmas O. Ujowundu*, Agwu I. Ukoha, Comfort N. Agha, Ngwu Nwachukwu and Kalu O. Igwe

Biochemistry Department, Federal University of Technology, Owerri, Imo State, Nigeria.

Three hundred and two pregnant women participated in this study. Our results showed that the range, mean and median urinary iodine excretion (UIE) were 28.1 to 218.1, 152.09 ± 41.65 and 163.1 µg/l, respectively. The range, mean and median TSH concentration were 0.7 - 5.9, 1.4 ± 0.7 and 1.3 µIU/ml, respectively. Our results showed that none of the women have severe iodine deficiency, 2% had moderate iodine deficiency, 12% had mild iodine deficiency, while 80% had optimal iodine nutrition and 6% have more than adequate. We observed a progressive and significant (P = 0.0009) decrease in the mean UIE from the 1st to the 3rd trimester. We observed also that 95% of the pregnant women had TSH concentration within the normal range. The TSH values between the three trimesters showed no significant difference (P = 1.20). The Urinary Iodine Excretion and Thyroid Stimulating Hormone concentration values suggest that iodine deficiency has been eliminated as a public health problem in Orlu. The progressive decrease in the median UIE from the 1st to the 3rd trimester should be addressed to meet the increased demand of iodine as a result of the pregnancy.

Key words: Urinary iodine, thyroid stimulating hormone, iodine deficiency, hyperthyroidism, pregnant women, Orlu, Nigeria.

INTRODUCTION

Iodine is an essential trace element necessary for the synthesis of thyroid hormones (Delange, 1994; Hetzel and Maberly, 1989). These hormones promote growth and development of bone, muscle, height and weight and maintain the stabilization of energy and material metabolism (Fuge and Johnson, 1986). The thyroid hormones are also vital for growth and development of all organs especially the brain, reproductive organs, nerves, skins, nails and teeth (Fisher and Delange, 1998). Deficiency of iodine resulting from inadequate dietary intake is related to a spectrum of diseases collectively known as iodine deficiency disorders (IDD) (Hetzel, 1983). IDD can be corrected by re-supplying iodine in the diet (Delange, 2000). The impact of IDD is enormous and it affects all the stages of life (Hetzel, 1983; ICCIDD/UNICEF/WHO, 2001). Iodine deficiency disorders are primarily the result of inadequate amounts of iodine in soil, water and food as well as consumption of foods rich in goitrogenic substances (Aston and Brazier, 1979; Sharma et al., 1999; Ene-Obong, 2001). Iodine deficiency in the fetus is the result of iodine deficiency in the mother. The consequence of iodine deficiency during pregnancy is impaired synthesis of thyroid hormones by the mother and the fetus. An insufficient supply of thyroid hormones to the developing brain may result in mental retardation (Morreale et al., 2004; Auso et al., 2004; Koibuchi and Chin, 2000; Delange, 2001). It has been established through some experimental evidence that the varying manifestation of IDD in fetus could be as a result of low thyroxine level in the blood of the iodine deficient mother and the lower the level of thyroid hormone of the pregnant women, the greater the threat to the fetus development (WHO, 1996). These children also have a greater occurrence of congenital abnormalities, lower birth weight and lower mortality rate as indicated by higher perinatal and infant mortality (US Foods and Nutrition Board, 2001). The commonest manifestation of iodine deficiency is goitre, which occurs when the iodine level of the blood is low; the cells of the thyroid gland enlarge in an attempt to trap as many particles of iodine

*Corresponding author. E-mail: ujowundu@yahoo.com.

as possible. Sometimes the gland enlarges until it is visible (Hamilton et al., 1998; Chatterjea and Rana, 2004). This gland enlargement is caused by an increased production of thyroid stimulating hormone (TSH) (Ubom, 1991).

Inadequate dietary iodine leads to reduced synthesis of thyroid hormones (T3 and T4). Lower level of T4 in the blood stimulates the pituitary gland to secrete TSH to fulfill the production of thyroid gland hormones. It is important not to over consume iodine as it has a relatively narrow range of intakes that reliably support good thyroid function. Consumption of an excessive amount of iodized salt or seaweeds could readily result to complex disruptive effect on the thyroid and may cause either hyperthyroidism or hypothyroidism in susceptible individuals, as well increasing the risk of thyroid cancer. A large percentage of the world population is at a risk of IDD (Delange and Hetzel, 2003). Several parts of Nigeria have been before now identified with goiter endemicity and hence labeled the "goitre belt" (Nwokolo and Ekpechi, 1966; Olurin, 1975; Isichie et al., 1987; Ubom, 1991). In 1993 a national goitre rate of 20% was reported and 20 million Nigerians were estimated to be affected by IDD (UNICEF, 1993). The Participatory Information Collection Study (1993), using thyroid hormone concentrations as indicators of iodine status reported an iodine deficiency prevalence of 65.6% in South-East, 41% in the South-West, 43% in the North- West of Nigeria. As part of the strategies to reduce the prevalence of IDD in Nigeria, the Universal Salt Iodization (USI) Programme was introduced in 1995. The update from the report of the Nigeria Demographic and Health Survey (NDHS, 2003) showed that almost all Nigerian households (97.3%) consumed adequately iodized salt, while about 1.7% consumed uniodized salt. This study was used to evaluate the iodine nutrition in Orlu suburban area of Nigeria using pregnant women as a case study after several years of availability and consumption of iodized salt.

MATERIALS AND METHODS

Experimental design

This study is a hospital based study conducted amongst pregnant women attending the antenatal clinic of Imo State University Teaching Hospital, a government hospital, in Orlu senatorial district of Imo State South-East, Nigeria. Three hundred and two pregnant women (mean age 33 years) participated in the study. All subjects were volunteers and were selected from among pregnant women visiting the antenatal clinic. All subjects provided written informed consent in accordance with the ethical standards of the local ethical committee. Selection criteria were the absence of chronic disease such as thyroid disease, diabetes mellitus, anemia, hypertension and coronary artery. Casual urine samples were obtained from the three hundred and two pregnant women, labelled and immediately preserved in a cooler with ice chips. Blood samples were collected from two hundred and nine of the women in standardized conditions to reduce sources of pre-analytical variation. Venipuncture was performed after an overnight fast. All blood samples were collected by experienced medical laboratory scientists using conventional

venipuncture. The blood samples were allowed to clot before separation by centrifugation at 3000 g for 15 min. All serum samples were stored frozen at refrigerated until testing at the end of the collection period. The urine and blood samples obtained from the women were used for the determination of urinary iodine excretion (UIE) and thyroid stimulating hormone (TSH).

Determination of urinary iodine excretion

Measurement of urinary iodine is the most common method to monitor dietary iodine intake (Fray et al., 1973). This makes urinary iodine a good biochemical marker for control of iodine deficiency disorders. The iodine in the urine is measured by a modification of the traditional colorimetric method of Sandell and Kolthoff (1937). This was done using the Ammonium Persulfate Method as described by Pino et al. (1996). Urine was digested with ammonium persulphate. The iodine in the urine samples catalyses the reduction of ceric ammonium sulphate (yellow colour) to the cerous form (colourless) in the presence of arsenious acid. The degree of reduction in colour intensity of the yellow ceric ammonium sulphate is proportional to the iodine content in the urine sample. This method was applied to all urine samples.

Measurement of the TSH serum concentration

Serum TSH concentration was measured by enzyme-linked immunosorbent assay using commercial kits (Syntron Bioresearch, Inc. Carlsbad, CA - USA). The normal range of TSH concentration determined with this kit was 0.5 - 4.10.

Statistical analysis

The data obtained were subjected to statistical analysis using the statistical software package, Statistical Analysis Software (SAS). The mean, Median and range of the data was determined. Results will be considered significant when $P<0.05$ at 95% confidence.

RESULTS

Urinary iodine excretion (UIE)

Urinary iodine excretion (UIE) of 302 pregnant women was used to monitor the effect of Universal Salt Iodization in Nigeria. The results obtained from the 1st to the 3rd trimester showed that the mean and median UIE were 152.09 ± 41.65 and 163.1 µg/l, respectively. The UIE ranged from 28.1 to 218.1 µg/l. The results showed that none of the women have severe (<20 µg/l) iodine deficiency, 5 (2%) have moderate (20 - 49 µg/l) iodine deficiency, 37 (12%) have mild (50 - 99 µg/l) iodine deficiency, while 242 (80%) have optimal (100 - 199 µg/l) iodine nutrition range and 18 (6%) have more than adequate (200 - 299). The UIE for the three trimesters varied significantly ($P = 0.0009$). The UIE between the 1st trimester and the 2nd trimester varied slightly ($P = 0.046$), while UIE between the 1st and 3rd trimester varied significantly ($P = 0.0001$). There was no significant difference between UIE in the 2nd and 3rd trimester ($P = 0.06$). From the result, a progressive decrease in the

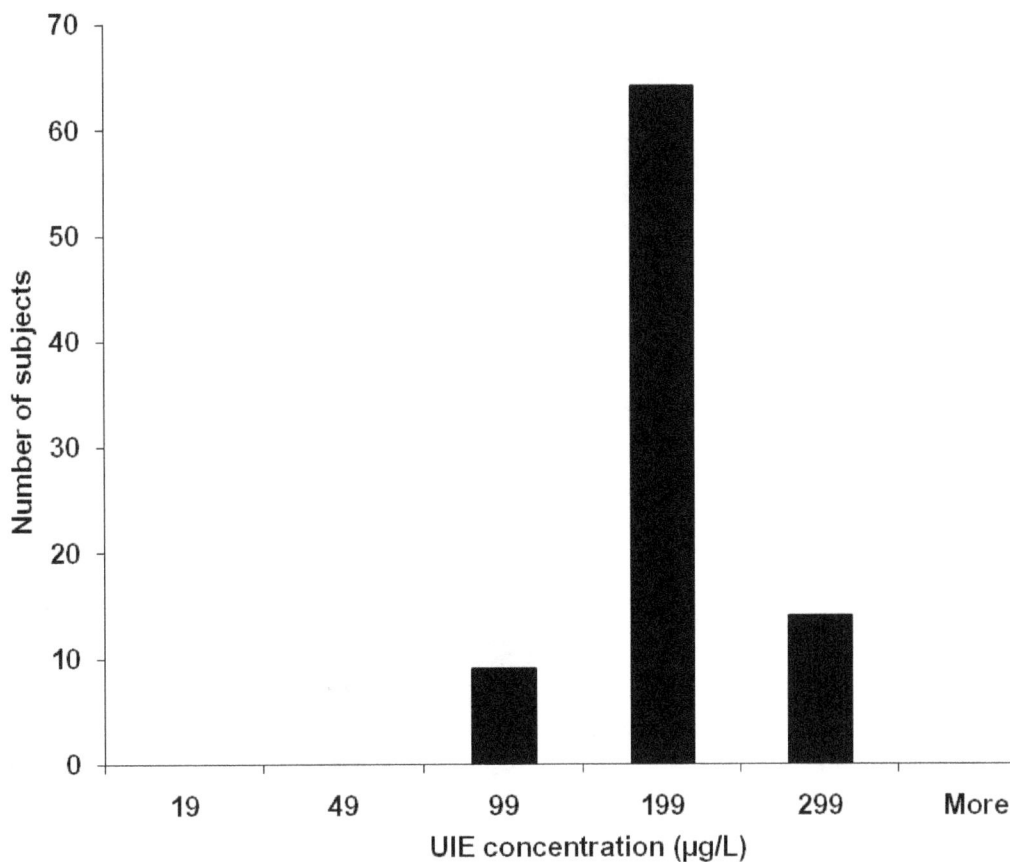

Figure 1. Distribution of UIE concentration values of 87 pregnant women in 1st trimester of pregnancy.

mean UIE from the 1st to the 3rd trimester was observed and the decrease was significant (P = 0.0009). The results obtained from each trimester are presented in Figures 1, 2 and 3.

The result of UIE of 87 pregnant women in their first trimester is shown in Figure 1. The result showed that none of the women have severe iodine deficiency but 9 (10%) have mild iodine deficiency, 64 (74%) have optimal iodine nutrition and 14 (16%) have more than adequate. The results from the pregnant women in the first trimester gave a mean UIE of 164.22 ± 40.56 µg/l, median UIE of 173.1 µg/l and the UIE ranged from 53.1 to 218.1 µg/l.

The result UIE of 112 pregnant women in their 2nd trimester is shown in Figure 2. The results showed that none of the women have severe iodine deficiency, 2 (2%) have moderate iodine deficiency, 17 (15%) have mild iodine deficiency. The result also showed that, 90 (80%) have optimal iodine nutrition and 3 (3%) have more than adequate. From the result, the mean UIE was 152.22 ± 42.80 µg/l, median UIE was 170.6 µg/l. The UIE ranged from 48.1 to 205.6 µg/l.

The result of UIE of 103 pregnant women in their 3rd trimester is shown in Figure 3. The results showed that none of the women have severe iodine deficiency, 3 (3%) have moderate iodine deficiency. Also, 11 (11%) have mild iodine deficiency, 88 (85%) have optimal iodine

nutrition and 1 (1%) have more than adequate. From the result, the mean UIE was 141.69 ± 38.78 µg/l, median UIE was 150.6 µg/l. The UIE ranged from 28.1 to 200.6 µg/l.

Thyroid Stimulating Hormone

The result of TSH concentration measurements, of 58 pregnant women in their 1st trimester of pregnancy are showed in Figure 4. The result showed that 7 (12 %) of the pregnant women had TSH values between 0.0 to 0.5 µIU/ml (Hyperthyroid status). A total of 50 (86%) of the women had TSH level between 0.6 to 4.10 µIU/ml (normal thyroid status). Also the result showed that 1(2%) of the pregnant women had TSH value > 4.10 µIU/ml (Hypothyroid). From the result, the mean and median TSH concentration values were 1.11 ± 0.76 and 1.07 µIU/ml, respectively and the TSH values ranged from 0.39 to 6.00 µIU/ml.

The result of TSH concentration measurements, of 82 pregnant women in their 2nd trimester of pregnancy are showed in Figure 5. The result showed that 2(2%) of the pregnant women had TSH values between 0.0 to 0.5 µIU/ml (Hyperthyroid status). A total of 80 (98%) of the pregnant women had TSH level between 0.6 to 4.10 µIU/ml

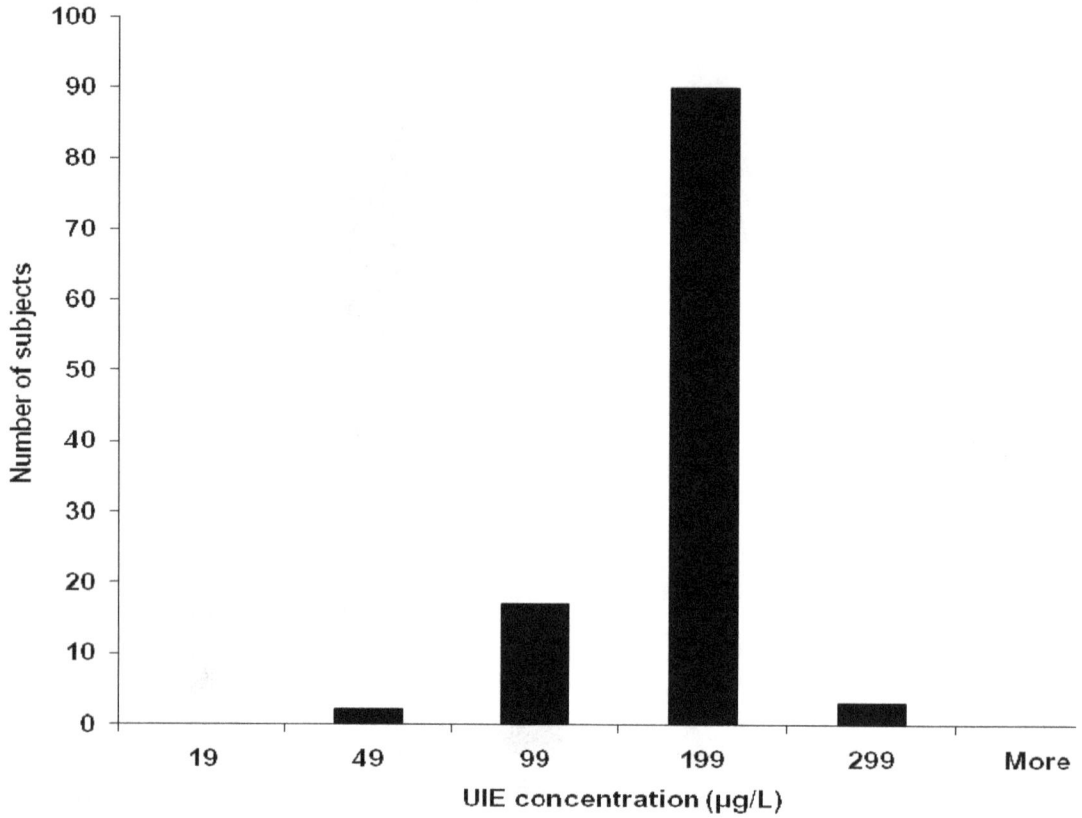

Figure 2. Distribution of UIE concentration values of 112 pregnant women in 2nd trimester of pregnancy.

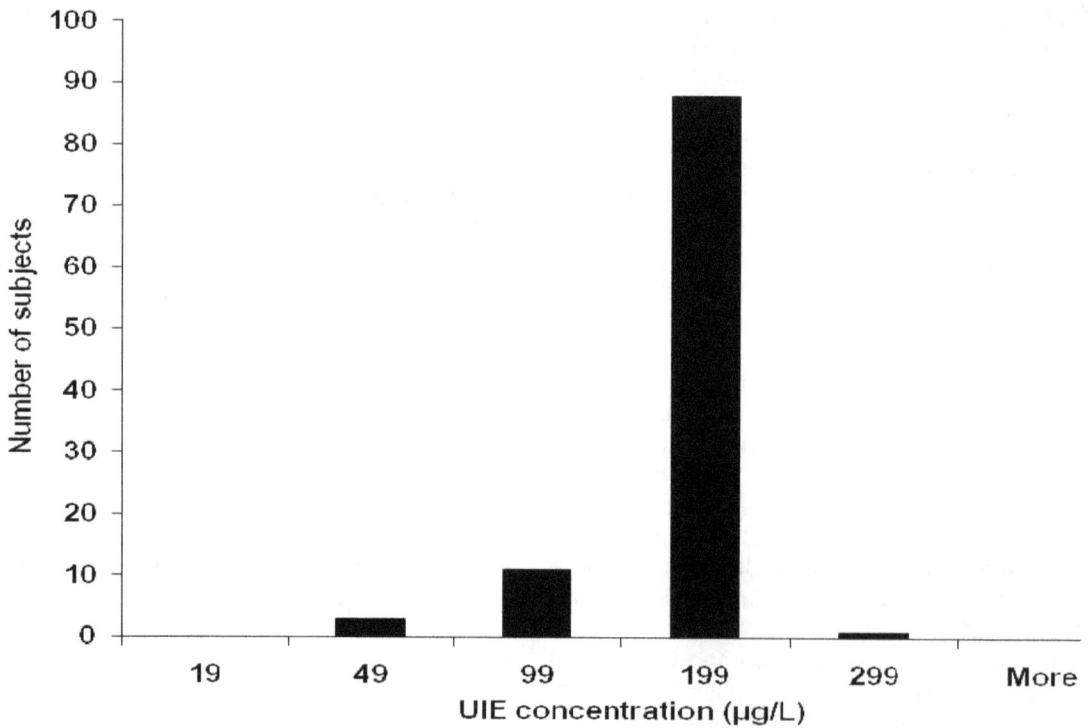

Figure 3. Distribution of UIE concentration values of 103 pregnant women in 3rd trimester of pregnancy.

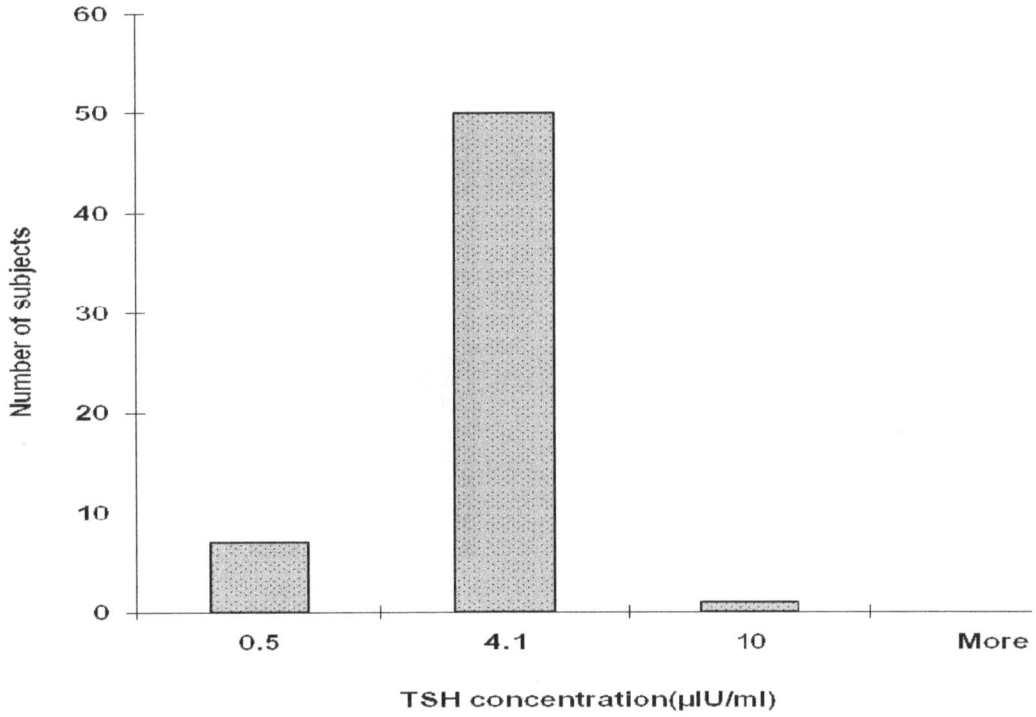

Figure 4. TSH concentration distribution of 58 pregnant women in 1st trimester of pregnancy.

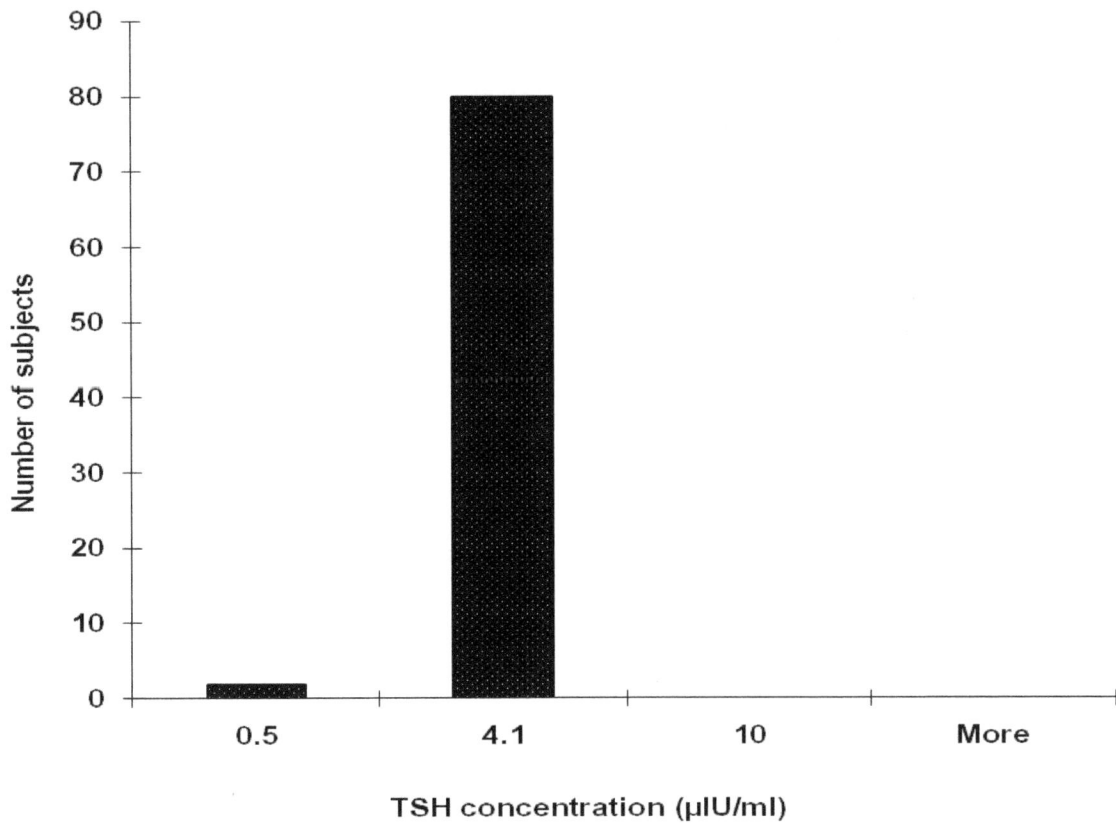

Figure 5. TSH concentration distribution of 82 pregnant women in 2nd trimester of pregnancy.

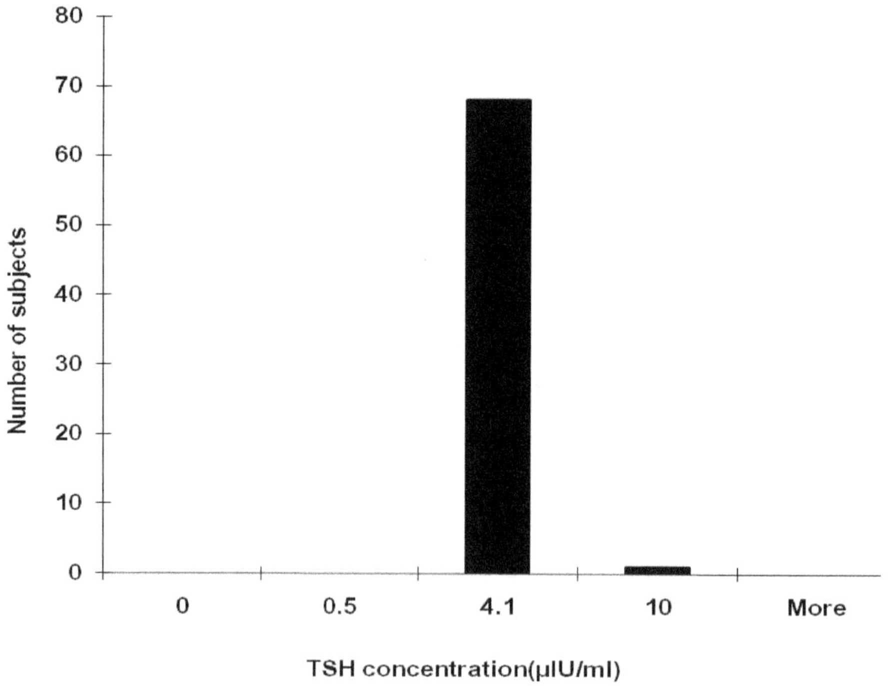

Figure 6. TSH concentration distribution of 69 pregnant women in 3rd trimester of pregnancy.

(normal thyroid status). The result also showed that none of the pregnant women had TSH value > 4.10 µIU/ml (Hypothyroid condition). From the result, the mean and median TSH concentration values were 1.39 ± 0.53 and 1.39 µIU/ml, respectively. The TSH values ranged from 0.47 to 3.09 µIU/ml.

The result of TSH concentration measurements, of 69 pregnant women in their 3rd trimester of pregnancy are showed in Figure 6. The result showed that none of the pregnant women had TSH values between 0.0 to 0.5 µIU/ml (Hyperthyroid status), while 68 (99%) of the pregnant women had TSH level between 0.6 to 4.10 µIU/ml (normal thyroid status). The result also showed that 1(1%) of the pregnant women have TSH value > 4.10 µIU/ml (Hypothyroid condition). From the result, the mean and median TSH concentration values were 1.7 ± 0.8 and 1.62 µIU/ml, respectively. The TSH values ranged from 0.78 to 5.92 µIU/ml

DISCUSSION

The results obtained showed a progressive decrease in the median UIE from the first to the third trimester. This decrease can be attributed to the increased demand of iodine as a result of the pregnancy (Aboul-Khair et al., 1964). Increased demand for micronutrients predisposes women of reproductive age, pregnant and lactating women to micronutrient deficiencies (McGuire, 1993; ICCCIDD/UNICEF/WHO, 2001). This result also suggests that the quantity of iodized salt consumed is not

meeting the increased demand for iodine. Increased iodine requirement during pregnancy is to provide for the needs of the fetus and to compensate for increased loss of iodine. It has being reported that intake between 200 - 250 µg/day, for pregnant women is very important to prevent goiter development and to keep serum levels of T4 and T3 stable (ICCIDD/WHO/UNICEF, 2001; Glinoer et al., 1995). The UIE value results showed that 80% of the pregnant women had optimal iodine nutrition. Also, our results showed that 5 (2%) of 302 pregnant women had moderate (20 - 49 µg/l) iodine deficiency and 37 (12%) had mild (50 - 99 µg/l) iodine deficiency. These results suggested that 42 pregnant women (14%) did not consume the necessary amount of iodine.

The results of TSH measurements showed that over 90% was observed to have serum TSH within normal suggesting the optimal iodine nutrition, in which most of them were was able to keep serum levels of T4 and T3 stable.

The number of the women (80%) with UIE values corresponding to the optimal nutrition is in line with the result of the TSH in which over 90% was observed to have serum TSH within normal, the normal range which suggests good thyroid function status and an adequate dietary intake of iodide. The results of the TSH for women consuming adequate iodine is in line with the requirement by WHO/UNICEF/ICCIDD (1994) and the report of Delange (1999). The result of the TSH showed that the mean concentration of the TSH in the three trimesters increased gradually as pregnancy progressed, however variation of the TSH concentration between the three trimesters was not significant (P = 1.20). The value of the mean serum TSH concentration of the pregnant women suggests that, they had normal thyroid function status. The result also showed that only 4% of the women had serum TSH concentration below normal (hyper-thyroidism), this justifies the 6% of women observed to have UIE values above optimal. These findings are in line with that obtained in Australia where lowest TSH was associated with UIE values of between 200 - 300 µg/l (Buchinger et al., 1996).

The results obtained in this study, in measuring USI effect on UIE and TSH concentration, of pregnant women indicate that iodine deficiency was eliminated as a public health problem in Orlu. Also, with these results and the results obtained in previous studies in measuring the effect of USI on iodide nutrition, Nigeria can be said to be succeeding in, mitigating the large-scale losses of brain-power and productivity caused by Iodine Deficiency Disorders. The challenge faced today by the Nigerian government, salt industries, the communities and other stakeholders in the IDD elimination drive will be to sustain these achievements and consolidate the gains of the universal salt iodization through certification and monitoring at all levels as a routine by all concerned.

REFERENCES

Aboul-Khair SA, Turnbull AC, Hytten FE (1964). The physiological changes in thyroid function during pregnancy. Clin. Sci. 27: 195-207.

Aston SR, Brazier PH (1979). Endemic goiter, the factors controlling iodine deficiency in soils. Sci. Total Environ. 11: 99-104.

Auso E, Lavado-Autric R, Cuevas E, Escbar del Rey F, Morreale de Escobar G, Berbel PA (2004). moderate and transient deficiency of maternal thyroid function at the beginning of fetal neocoticogenesis alters neuronal migration. Endocrinol., 145: 4037-4047.

Buchinger W, Lorenz-Wawschinek O, Binter G, Langsteger W, Bonelli R, Eber O (1996). Relation between serum thyrotropin and thyroglobulin with urinary iodine excretion. In: The Thyroid and Iodine. Nauman J, Glinoer D, Braverman LE, Hostalek U, eds. pp. 189-190. Stuttgart, Germany, Schattauer.

Chartterjea MM, Rana S (2004). Medical Biochemistry, 6th edition Oxford Press London pp. 700-750.

Delange F (1994). The disorders induced by iodine deficiency. Thyroids 4: 107.

Delange F (1999). Neonatal thyroid screening as a monitoring tool for the control of iodine deficiency. Acta. Padiat. Scand. Suppl. 432: 21-24.

Delange F (2000). Iodine deficiency. In The thyroid. A fundamental and clinical text. Braverman LE, Utiger RD, editors. Philadelphia: Lippincott publ. pp. 295-316.

Delange F (2001). Iodine deficiency as a cause of brain damage. Postgrad. Med. J. 77: 217-220.

Delange F, Hetzel BS (2003). The iodine deficiency disorders. In: The Thyroid and its Diseases. Basle karger pp. 324-44.

Ene-Obong HN (2001). Substances occurring in foods. In: Eating Right (Nutrition Guide). University of Calaber Press, Nigeria p. 48.

Fisher DA, Delange F (1998). Thyroid hormone and iodine requirements in man during brain development. In: Stanbury J.B., Delange F., Dunn J.T. and Pandav C.S. eds. Iodine in pregnancy. New Delhi, Oxford University Press pp. 1–33.

Fray HM, Rosenlund B, Torgersen JB (1973). Value of single urine specimens in estimation of 24 h urine iodine excretion. Acta. Endocrinol. 72: 287-92.

Fuge R, Johnson CC (1986). The geochemistry of iodine. Environ Geochem. Health, 8: 3l–54.

Glinoer D, de Nayer P, Delange F, Lemone M, Toppet V, Spehl M, Grün JP, Kinthaert J, Lejeune B (1995). A randomized trial for the treatment of mild iodine deficiency during pregnancy: Maternal and neonatal effects. J. Clin. Endocrin. Metab., 80: 258-269.

Hamilton EM, Whitney EN, Sizer FS (1998). Nutrition concepts and controversies. 5th edition west publishing company, Washington pp. 255 – 493.

Hetzel BS (1983). Iodine Deficiency Disorders (IDD) and their eradication. Lancet, 11: 1126-1129.

Hetzel BS (1989). The Story of Iodine Deficiency, 1st edition. Oxford Medical Publications, Great Britain p. 3.

Hetzel BS, Maberly GF (1986). Iodine. In: Mertz W, ed. Trace Elements in Human and Animal Nutrition, Vol. 2, 5th ed. New York: Academic Press pp.139–208.

ICCIDD/UCICEF/WHO (2001). Assessment of iodine deficiency disorders and monitoring their elimination. A guide for programme managers; WHO document WHO/NHD/01.1.

Isichie UP, Das SC, Egbuta JO, Banwo AI, Marimot I, Nagataki S (1987). Endemic goitre in plateau state, Nigeria; the possible aetiological factor and the establishment of endemic goitre map for the region. Proceedings of Nigeria/Japan Conferences, Jos pp. 78-81.

Koibuchi N, Chin WW (2000). Thyroid hormone action and brain development. Trends Endocrinol. Metab., 4: 123-128.

McGuire J (1993). Best practice in addressing micronutrient malnutrition. In: SCN News Geneva. No. 9.ACC/SCN pp. 1-10.

Morreale de Escobar G, Obregon M, Escobar del Rey F (2004). Role of thyroid hormone during early brain development. Eur. J. Endocrinol., 15: 25- 37.

Nigeria Demographic and Health Survey (2003). Infant feeding and children and women nutritional status. National Population Commission, Federal Republic of Nigeria pp. 160.

Nwokolo C, Ekpechi OL (1966). New foci of on the handling and storage of table salt. endemic goitre in Eastern Nigeria. Trans. Roy. Soc. Trop. Med. Hyg., 6: 97-108.

Olurin EO (1975). The fire of life (The Thyroid Gland). Inaugural lecture. University of Ibadan.

Pino S, Fang SL, Braverman LE (1996). Ammonium persulfate: A safe alternative oxidizing reagent for measuring urinary iodine .Clin. Chem. 42(2): 239-243

Sandell EB, Kolthoff I (1937). Micro determination of iodine by catalytic method. Mikrochim Acta, 1: 9-25.

Sharma SK, Chelleng PK, Gogoi S (1999). Iodine status of food and drinking water of a sub-Himalayan zone of India. Int. J. Food Sci. Nutr., 50: 95-98.

The Participatory Information Collection Study, 1993.

Ubom GA (1991). The Goitre-Soil-Water-Diet Relationship: Case Study in Plateau State, Nigeria. The Sci. the Total Environ. 107: 1-11.

UNICEF (1993). Regional goitre survey in school children, Nutrition Section, Lagos Nigeria.

US Food and Nutrition Board (2001). Dietary Reference Intakes. A report of the Institute of Medicine. pp 8-1 to 8-27, National Academy Press, Washington, DC.

WHO, UNICEF, ICCIDD (1994). Indicators for assessing Iodine Deficiency Disorders and their control through salt iodization. Geneva: WHO publ. WHO/NUT/94.6 pp. 1-55.

WHO (World Health Organisation) (1996). Trace elements in human nutrition and health. WHO, Geneva.

Trace metals and oxidative metabolic changes in malignant prostate cancer patients

Akiibinu M. O.[1]*, Ogundahunsi A. O.[1], Kareem O. I.[2], Adesiyan A. A.[3], Idonije B. O.[4]
and Adeniyi F. A. A.[5]

[1]Department of Chemical Pathology and Immunology, College of Health Sciences, Olabisi Onabanjo University, Ago-Iwoye, Ogun State, Nigeria.
[2]Department of Nuclear Medicine, University College Hospital, Ibadan, Nigeria.
[3]Department of Biomedical Sciences, Ladoke Akintola University of Technology, Ogbomosho, Oyo State, Nigeria.
[4]Department of Chemical Pathology, College of Medicine, Ambrose Alli University, Ekpoma, Edo State, Nigeria.
[5]Department of Chemical Pathology, College of Medicine, University of Ibadan, Ibadan, Nigeria.

Metabolic derangement commonly associated with prostate cancer (PCa) has not been well researched in Nigerian malignant PCa patients yet. The present study was designed to assess the levels of selected trace metals (Zn, Cu, Cr, Fe, Co, and Se) and markers of oxidative stress, such as total plasma peroxide (TPP), total antioxidant potential (TAP), oxidative stress index (OSI) and free malondialdehyde (MDA), in Nigerian patients with malignant (PCa).   Twenty three newly diagnosed patients with malignant PCa participated in this study. Inclusion criteria included prostate-specific antigen (PSA) levels in blood >20 µg/ml and clinical presentation. Thirty apparently healthy staffs of University College Hospital, Ibadan, Nigeria served as controls. There were significantly ($p < 0.05$) lower plasma levels of Zn, Fe, Cu, Co, Se and TAP in malignant PCa patients when compared with the controls. In contrast, there was no significant change in the plasma level of Cr when compared with the controls. Meanwhile, significantly ($p < 0.05$) higher levels of TPP, MDA and OSI were observed in malignant PCa patients when compared with the controls. Our results indicate trace metal deficiency and oxidative stress in malignant PCa patients. Since micronutrients' deficiencies play critical roles in oxidative stress, micronutrients supplementation might be used to overcome complications in malignant PCa patients.

Key words: Trace metals, oxidative metabolites, malignant prostate cancer.

INTRODUCTION

Prostate cancer (PCa) is the consequence of chromosomal aberration and pathological proliferation of cells of the prostate tissue (Anderson, 1985). It is the most common neoplasm in men and the second cause of cancer death worldwide (Segev and Native, 2006). Various factors associated with the chromosomal aberration include nutritional factor, genetics, immunologic exhaustion and infection (Nelson et al., 2004; Pathak et al., 2005).

Recent data revealed that chronic inflammation of the prostate gland and high free radical load contribute to DNA damage and genomic instability, which may facilitate subsequent progression of cancer cells (Nelson et al., 2004). It is also evident that oxidative stress provoked by toxins, dietary fat consumption, or high level of androgens is important etiologic factors in the development and progression of prostate cancer (Pathak et al., 2005). Increased free radical generation has been reported in cancer cells when compared with normal cells (Kumar et al., 2008). The free radicals generated further attack the DNA of other cells and increase the mutation rates, genome instability and apoptosis evasion (Sawa and Ohshima, 2006; Szabo and Ohshima, 1997).

The effect of free radicals on the DNA has been

---

*Corresponding author. E-mail: akiibinumoses@yahoo.com.

Abbreviations: PCa, prostate cancer; TPP, total plasma peroxide; TAP, total antioxidant potential; OSI, oxidative stress index; MDA, malondialdehyde; PSA, prostate-specific antigen.

Table 1. Levels (Mean ± SD) of Cu, Zn, Co, Se, Fe Cr, in malignant PCa patients and controls.

	N	Cu (µg/dL)	Co (µg/dL)	Zn (µg/dL)	Se (µg/L)	Fe (µg/dL)	Cr (µg/dL)
Pca patients	23	36.8 ±9.4	51.2±4.0	100.5±11.6	39.4 ± 7.8	61.5±10.0	29.6± 4.4
Controls	30	46.4±10.2	60.0±5.7	120.2±20.5	57.2±10.8	79.2±9.2	27.9±3.4
t, p values		2.4,<0.05*	6.6,< 0.01*	3.0,<0.02*	8.2, <0.01*	4.1,<0.01*	1.4,>0.05

N=number of subjects. *=significantly different from controls.

implicated in the pathophysiology of many cancers (Sawa and Ohshima, 2006). Ramanujam (2004) reported that the DNA of a cell undergoes about 10,000 free radical attacks each day; the effect of which may cause gene mutation in the absence of efficient antioxidant system. High free radical load can cause fragmentation, cross-linking, aggregation and ultimately denaturation of protein molecules, such as hormones and enzymes (Ramanujam, 2004). The synergistic effect of nutritional factors and oxidative metabolic changes may therefore play prominent roles in the progression of benign to malignant prostate cancer (Ramanujam, 2004). Ruffin and Rock (2001) stressed that dietary antioxidants such as carotenoids, vitamins C, E and Se play significant roles in DNA and cell maintenance and repair. Therefore, the present study was designed to assess the plasma levels of Zn, Fe, Cu, Co, Se, Cr, total antioxidant potential, total plasma peroxide, malondialdehyde and oxidative stress index in malignant PCa patients.

HUMAN SUBJECTS AND METHODS

Volunteer human subjects

A total number of twenty three (23) PCa patients volunteered to participate in this study. The PCa patients were recruited at the Medical Out Patient Department, University College Hospital (UCH), Ibadan, Nigeria. Recruitment criteria included the plasma levels of prostate specific antigen (PSA)≥20.0 µg/ml and the clinical assessment by the consultant in charge of the patients. Another 30 apparently healthy individuals (PSA ≤ 4.0 µg/ml), who were staffs of University College Hospital, Ibadan, served as controls.

Determination of trace metals

Trace metals' levels were determined by atomic absorption spectrophotometer (AAS), as described by Kaneko (1999).

Estimation of MDA

Level of lipid peroxidation was determined by measuring the formation MDA using the method of Varshney and Kale (1990). This procedure is based on the fact that malondiahydehyde (MDA) produced from the peroxidation of membrane fatty acid reacts with the chromogenic reagent 2-thiobarbituric acid (TBA) under acidic conditions to yield a pink–coloured complex measured spectrophotometrically at 532 nm. 1, 1, 3, 3-tetramethoxylpropane was used as standard.

Estimation of TAP

TAP was determined using the ferric reducing/ antioxidant power (FRAP) assay (Benzie and Strain, 1996; Harma et al., 2003). 1.5 ml of working pre-wormed 37 °C FRAP reagent (300 mM acetate buffer - pH 3.6, 10 mM 2,4,6- tripyridyl-s-triazine in 40 mM HCl and 20 mM $FeCl_3$ at ratio 10:1:1) was vortex mixed with 50 µl of test sample and standards. Absorbance was read at 593 nm against a reagent blank. The result was reported as µmol Trolox equiv./ L.

Estimation of TPP

Determination of total plasma peroxide levels made use of the reaction of ferrous-butylated hydroxytoluene-xylenol orange complex (F0X-2 reagent) with plasma hydrogen peroxide, which yields a colour complex that was measured spectrophotometrically at 560 mm. H_2O_2 was used as standard. 1.8 ml of F0X-2 reagent was mixed with 200 µ1 of plasma. This was incubated at room temperature for 30 m. 100 µM H_2O_2 was used as standard. The mixture was centrifuged and the supernatant separated for reading at 560 nm (Harma et al., 2003).

Determination of oxidative stress index (OSI)

OSI, an indicator of the degree of oxidative stress, is the percent ratio of the TPP to the TAP values (Harma et al., 2003).

Statistical analysis

The data are presented as Means ± Standard deviation. Student (t) test was used for comparison between groups. Results were considered significant when the p-values were less than 0.05.

RESULTS

As shown in Table 1, the mean levels of Cu, Zn, Co, Se and Fe decreased significantly (p<0.05) in malignant PCa patients when compared with the controls. However, plasma levels of Cr in malignant PCa patients (p>0.05) were similar to controls. Table 2 shows a significantly (p<0.05) lower level of TAP with significantly (p<0.05) higher levels of TPP, MDA and OSI in malignant PCa patients when compared with the controls.

DISCUSSION AND PERSPECTIVES

Cancer cells depend on the host for their nutrition, blood

Table 2. Levels (Mean ± SD) of markers of oxidative stress in malignant PCa patients and controls.

	N	TAP (µmol Trolox equiv/ L)	TPP (µmolH$_2$O$_2$/L)	OSI (%)	MDA (nMol/ml)
Pca patients	23	1025 ± 450	28.1 ± 8.0	2.7 ± 1.7	10.7 ± 1.6
Controls	30	1480 ± 610	12.4 ± 5.0	0.8 ± 0.6	6.2 ± 1.5
t, p values		3.1, <0.02*	6.9, <0.01*	5.1, <0.01*	10.5, <0.01*

N=number of subjects. *=significantly different from controls.

supply and supporting stroma. Due to their altered metabolism and high energetic demands, cancer cells ultimately override the entire system by diverting the body's nutritional materials for their own use (Bongaerts et al., 2006; Ferreira, 2010). Exhaustion of nutritional materials causes the deficiencies of several micronutrients commonly implicated in many cancers. In the present study, significantly lower plasma levels of more trace metals (Cu, Co, Zn, Se and Fe) were observed in malignant PCa patients. These significantly lower levels of the trace metals could be due to increased demand by the cancerous tissues and other associated metabolic dysfunctions commonly encountered in cancers.

Since micronutrients play important roles in the general metabolic activities in man, their deficiencies may have profound effect in various metabolic dysfunctions and cancer formation (Xia et al., 1999). Deficiencies of certain trace metals (that is, Zn and Fe) cause general low cellular immune response and decreased secretion of interferon-γ, tumor necrotic factor–α and interleukin–2 (Hopkins and Failla, 1999). Cobalt, in the presence of other factors has profound influence on erythropoiesis. It is an essential trace metal linking the four pyrrol rings of cobalamin for effective synthesis of red blood cells (Hall and Malia, 1984). Deficiency of cobalt, in the presence of other factors may therefore contribute to anaemia commonly encountered in malignant prostate cancer patients.

Trace metals are also important in the catalytic activities of major antioxidant enzymes and DNA repair, short-circuiting the generation of malignancy, tumor growth and cancer spread (Xia et al., 1999). Low cancer risk has been associated with adequate plasma levels of vitamins A, E and C, Cu, Zn, and Se (Simopoulos, 2004). The deficiencies of trace metals therefore contribute to various metabolic dysfunctions and cancer development (Xia et al., 1999). Zn is the most abundant trace metal in the cells, and increasing evidences emphasize its role in genetic stability and function. Zn is a component of many DNA repair proteins and it is also involved in a variety of general cellular functions, including cell signal transduction, transcription and replication. Xia et al., (1999) stressed that the loss of Zn from biological membranes could increase the susceptibility of such cells to oxidative damage and impaired cell functions. Ozmen et al. (2006) reported significantly lower levels of Se and Zn in plasma of PCa patients. Arinola and Marbel (2008) also reported significantly lower level of Se in all categories of PCA patients.

Micronutrients are essential constituents of antioxidant system (Ruffin and Rock, 2001). Certain vitamins (vitamins C, E and A) exert direct antioxidant effects while trace metals are integral parts of antioxidant enzymes. Zn, Cu and Mn are integral parts of superoxide dismutase. Se is an integral part of glutathione peroxidase and Fe is an integral part of catalase (Ruffin and Rock, 2001). The TAP is therefore an index of all classes of antioxidants. The malignant PCa patients recruited for this study demonstrated significantly lower level of TAP. Since low risk of cancers have been associated with adequate plasma levels of antioxidants (Simopoulos, 2004), lower level of TAP in our PCa patients could account for the development and progression of prostate malignancy. Also, in the report of Clark et al., (1996), significantly lower incidences of lung, colorectal and prostate cancers were observed in the group of people given Se therapy. The present result corroborates that of Akinloye et al. (2009) who reported significantly lower level of antioxidant vitamins in all categories of PCa patients. Since Cu, Zn, Se and Fe are integral parts of antioxidant enzymes, their significantly lower levels observed in this study could account for the lower level of TAP in the malignant PCa patients.

In this study, significantly higher levels of TPP, MDA and OSI were observed in malignant PCa patients, when compared with the controls. This result corroborates the findings of some previous workers who implicated oxidative stress in the oncogenesis and pathogenesis of several cancer cases. Thalmann et al. (2000) reported that in all the free radicals, expressing higher levels of hydrogen peroxide (H$_2$O$_2$) are more tumorigenic and metastatic. In another study, it was reported that higher level of H$_2$O$_2$ could be generated from cells that have mitochondrial DNA mutation during tumor development and progression (Bianchi et al., 2001; Mandavilli et al., 2002). These reports therefore support the present study where higher level of TPP was found in the malignant PCa patients. Kumar et al. (2008) observed increased free radical generation in cancer cells compared with normal cells. The free radicals generated attack the DNA and increase the mutation rates, genome instability and loss of apoptosis in cancer cells (Sawa et al., 2006; Szabo and Ohshima, 1997). Increased free radicals in prostate cancer cells and tissues correlated with increased potential of the prostate cancer cells and tissues (Lim et al., 2005). The high free radical load and oxidative stress are associated with the gelatinous transformation of the bone marrow which causes anaemia (Ghaffari, 2008) commonly found in malignant PCa patients. The elevated level of TPP and exhaustion of TAP observed in this study may account for the oxidant-antioxidant imbalance that caused the increased level of OSI in our PCa patients.

MDA is an index of lipid peroxidation, which was found significantly higher in the malignant PCa patients used for this study. Since excessive free radical generation causes macromolecular damage through lipid peroxidation and protein fragmentation, the significantly higher level of MDA observed in this study could be associated with the higher level of TPP in the patients. Our findings corroborate that of Osmen et al. (2006) who reported significantly higher level of MDA in patients with prostate cancer. Caliskan-can et al. (2008) also reported increased oxidative stress and higher levels of products of lipid peroxidation in lung cancer patients.

In conclusion, trace metal deficiency and oxidative stress are features of malignant PCa. Therefore, micronutrient supplements may be required as adjuvant therapy in the management of malignant PCa patients.

REFERENCES

Akinloye O, Adaramoye O, Kareem O (2009). Changes in antioxidant status and lipid peroxidation in Nigerian patients with prostate Carcinoma. Pol. Arch. Med. Wewn., 119 (9): 526-32.

Anderson JR (1985). Muirs Textbook of Pathology. 12th Ed. Edward Arnold. p. 14.36-14.38.

Arinola OG, Marbel CDA (2008). The serum levels of Trace metals in Nigerians males with different PSA values. Malaysian J. Med. Sci., 15(2): 39-42.

Benzie IFF, Strain JJ (1996). The ferric reducing ability of plasma (FRAP) as a measure of antioxidant power (the FRAP assay). Anal. Biochem., 239: 70-76.

Bianchi NO, Bianchi MS, Richard SM (2001). Mitochondrial genome instability in human cancers. Mutat. Res., 488:9–23.

Bongaerts GPA, van Hatteren HK, Verhagen CAM, Wagener DJT (2006). Cancer cachexia demonstrates the energetic impact of gluconeogenesis in human metabolism. Med. Hypotheses, 67(5): 1213-1222.

Caliskan-Can E, Firat H, Ardic S, Sinsek B, Torun M, Yardim- Akaydin S (2008). Increased levels of 8-hydioxyguanosine and its relationship with lipid peroxidation and antioxidant vitamins in lung cancer. Clin. Chem. Lab. Med., 46 (1): 107-12.

Clark LC, Combs GF Jr, Turnbull BW, Slate EH, Chalker DK, Chow J, Davis LS, Glover RA, Graham GF, Gross EG, Krongrad A, Lesher JL Jr, Park HK, Sanders BB Jr, Smith CL, Taylor JR (1996). Effect of selenium supplementation for cancer prevention in patients with carcinoma of skin. A randomized controlled trial. Nutritional Prevention of Cancer Study Group. J. Am. Med. Ass., 276(24): 1957-1963.

Ferreira LM (2010). Cancer metabolism: The Warburg effect today. Exp. Mol. Pathol., 89(3): 372-380.

Ghaffari S (2008). Oxidative stress in the regulation of normal and neoplastic. Redox Signal., 10(11): 1923.

Hall R, Malia RG (1984). In Textbook of Medical Laboratory Haematology 1st Ed. Butterworths, London.. p. 32.

Harma M, Harma M, Enel O (2003). Increased oxidative stress in patients with hydatidiform mole. Swiss Med. Wkly., 133: 563-566.

Hopkins RG, Failla ML (1999). Transcriptional regulation of interleukin-2 gene expression is impaired by copper deficiency in Jurkat human T lymphocytes. J. Nutr., 129: 596-601.

Kaneko JJ (1999). Clinical Biochem of Animal 4th edition. Kaneko JJ editor Academic press Inc. New York., p. 932.

Kumar B, Koul S, Khandrika L, Meacham RB, Koul HK (2008). Oxidative Stress Is Inherent in Prostate Cancer Cells and Is Required for Aggressive Phenotype. Cancer Res., 68: 1777.

Lim SD, Sun C, Lambeth JD, Marshall F, Amin M, Chung L, Petros JA, Arnold RS (2005). Increased Nox1 and hydrogen peroxide in prostate cancer. Prostate. 62: 200-7.

Mandavilli BS, Santos JH, Van Houten B (2002). Mitochondrial DNA repair and aging. Mutat. Res., 509: 127–51.

Nelson WG, De Marzo AM, DeWeese TL, Isaacs WB (2004). The role of inflammation in the pathogenesis of prostate cancer. J. Urol., 172: 6–11.

Ozmen H, Eiulas FA, Karatas F, Cukurovali A, Yakin O (2006). Comparison of the concentration of trace metals (Ni, Zn, Co, Cu and Se, Fe, vitamins A,C,E and lipid peroxidation in patients with prostate Cancer. Clin. Chem. Lab. Med., 44(2): 175-179.

Pathak SK, Sharma RA, Steward WP, Mellon JK, Griffiths TR, Gescher AJ (2005). Oxidative stress and cyclooxygenase activity in prostate carcinogenesis: targets for chemopreventive strategies. Eur. J. Cancer, 41: 61–70.

Ramanujam MD (2004). Free radicals and antioxidants Current Status. www.meclindia. Net/article/ antioxidants asp.

Ruffin MT, Rock CL (2001). Do antioxidants still have a role in the prevention of human cancer ? Curr. Oncol. Rep., 3(4): 306-313.

Sawa T, Ohshima H (2006). Nitrative DNA damage in inflammation and its possible role in carcinogenesis. Nitric oxide. 14: 91-100.

Segev Y, Native O (2006). Nutrition and pharmacological treatment for prevention of prostate Cancer. Harefuah, 145(1): 47-51, 76-7.

Simopoulos AP (2004). The traditional diet of Greece and cancer. Eur. J. Cancer Prev., 13(3): 219-230.

Szabo C, Ohshima H (1997). DNA damage induced by peroxynitrite; subsequent biological effects. Nitric oxide. 1: 373 – 385.

Thalmann GN, Sikes RA, Wu TT, Degeorges A, Chang SM, Ozen M, Pathak S, Chung LW (2000). LNCaP progression model of human prostate cancer: androgen independence and osseous metastasis. Prostate. 44(2): 91-103; 44(2).

Varshney R, Kale RK (1990). Effect of calmodulin antagonist on radiation-induced lipid peroxidation in microsomes. Int. J. Rad. Boil., 58: 733-743.

Xia J, Browing JD, O' Dell BL (1999). Decreased plasma membrane thiol concentration is associated with increased osmotic fragility of erythrocyte in Zinc-deficient rats. J. Nutr., 129: 814-819.

24

Unutilized energy reserves and mineral contents of fibroid tissues suggesting perturbed membrane transport processes

Ibegbulem C. O.[1]*, Agha N. C.[1] and Emeka-Nwabunnia I.[2]

[1]Department of Biochemistry, Federal University of Technology, Owerri, Nigeria.
[2]Department of Biotechnology, Federal University of Technology, Owerri, Nigeria.

Compositions of intramural uterine fibroid (IUF) and subserous uterine fibroid (SUF) tissues were studied. Results showed that IUF and SUF, respectively, contained (%) 77.26 ± 0.05 and 77.49 ± 0.02 moisture and 22.74 ± 0.05 and 22.52 ± 0.02 dry matter, on wet-weight basis. Their respective moisture, crude fat, crude protein, ash and total carbohydrates (%) were 4.67 ± 0.05 and 4.38 ± 0.17, 8.29 ± 0.41 and 5.84 ± 0.43, 72.66 ± 0.72 and 65.16 ± 0.91, 4.39 ± 0.08 and 4.22 ± 0.03 and 10.01 ± 1.18 and 20.40 ± 1.45 while their energy contents (kcal/ 100g) were 405.29 and 394.80, respectively, on dry-matter basis. Their crude fats contained fatty acids, triacylglycerols and cholesterol while their total carbohydrates contained glucose and glycogen. Ca^{2+}, Fe^{2+}, Zn^{2+}, K^+, Na^+ and Mg^{2+} contents (mg/ 100g) showed that Na^+ was the most abundant mineral amongst them unlike the norm in cells where K^+ is the most abundant. The tissues contained high $[Ca^{2+}]$ and $[Fe^{2+}]$, low $[Zn^{2+}]$, $[K^+]$ and $[Mg^{2+}]$ and comparable $[Na^+]$ relative to cells. These suggested that Na^+/ K^+-ATPase activity may have been decreased or inhibited by either high [oestrogen] which is a pre-disposing factor to fibroid or low $[Mg^{2+}]$ which may have decreased or inhibited the activity of the electrochemical Na^+ gradient-dependent Ca^{2+} transporter. Zn^{2+} transporter activity seemed to have been decreased or inhibited while cellular influx of Fe^{2+} seemed to have been increased. The results showed that the IUF and SUF had unutilized energy reserves and mineral contents suggestive of perturbed membrane transport activities.

Key words: Energy reserve, fibroid, minerals, membrane transport, proximate composition.

INTRODUCTION

The uterus, or womb, is a hollow organ with heavy, muscular walls. It is located between the urinary bladder in front and the rectum behind. It receives the oviduct on the right and on the left of its upper portion and opens into the vagina below through its cervix. It is where the fertilized egg is implanted and the foetus develops prior to its birth. Beyond its role in pregnancy, uterine diseases abound.

One of such uterine diseases is fibroid tumour of the uterus (leiomyomas or fibromyoma). Fibroid can grow in different parts of the uterus and are named according to what part of the uterus they are found, for instance, subserous (grow in the outer uterine wall), intramural (grow inside or within the uterine wall), submucous (grow in the inner uterine wall), pedunculated (attached by stalk to the outer or inner uterine wall), interligamentous (grow sideways between ligaments which support the uterus in the abdominal region) and parasitic (rarest form which occurs when it attaches itself to another organ) (Cotran et al., 1999; Peddadah et al., 2008; Stewart, 2011). It is hormone dependent. One of such hormones implicated in the etiology of uterine fibroid is high oestrogen concen-trations (Baird and Newbold, 2005). High progesterone concentration is also thought to play a role in fibroid

---

*Corresponding author. E-mail: ibemog@yahoo.com.

growth. Histological studies of fibroid tissues from patients treated with progesterone showed more cellular growth than those from patients that were not treated with progesterone (DeCherney and Nathan, 2003). Fibroids are benign (non cancerous) neoplasm. The tumours are found in at least 25 percent of women in active reproductive life and are more common in blacks (Cotran et al., 1999). Tumours of this sort are rarely found elsewhere in the body. Indman (2010) and Layyous (2010) reported that surgery (myomectomy) is a treatment option. They reported that fibroid calcification of the fibroid was due to the deposition of calcium.

Fibroids are thought to be caused by environmental and genetic factors or a combination of them and the exact composition of their tissues are still debatable. Shryock and Swartout (1980) and Layyous (2010) reported that fibroid tumours were composed partly of muscle tissues growing from and resembling the muscle in the walls of the organ; with this special muscle tissue being intermingled with varying amounts of fibrous connective tissue.

The objective of this study was to evaluate some compositions of intramural uterine fibroid (IUF) and subserous uterine fibroid (SUF) tissues. Results of the study may give insights into possible reasons for such compositions.

MATERIALS AND METHODS

Procurement of tissues

The fibroid tissues used were procured with informed consent from ten patients who underwent myomectomy. Patients with IUF generally had more than one fibroid growth of varying sizes than their SUF counterparts.

Histological examination of tissues

Histological study of each representative tissue was carried out using the method of Okoro (2002). The procured and fixed tissues were dehydrated through different grades of alcohol, cleared in xylene, infiltrated with melted paraffin wax and picked on albumenized slides. Staining was done using Haemotoxylin-Eosin (H and E), dried and mounted using distrene tricresyl phosphate xylene (DPX). The stained and mounted slides were examined using a light microscope and photographed at a magnification of x400.

Assay for proximate composition

The tissues were analysed for their proximate compositions using the methods of AOAC (1990). The protocol was that tissues were dried at 90°C for 24 h, their fats were exhaustively extracted with petroleum ether (crude fat contents) and the defatted samples used for crude protein and ash determinations. Total carbohydrates were estimated by difference and energy contents (kcal/ 100 g) determined by summing up the products got by multiplying the crude protein, crude fat and total carbohydrates contents by the factors of 4, 9 and 4, respectively, as described by Wardlaw and Kessel (2002).

Detection of some lipid constituents

Their crude fat extracts got above were tested for the presence of fatty acids, triacylglycerols and cholesterol using the soap formation, acrolein and Salkowski tests, respectively, as described by Plummer (1971) and Mathotra (1989).

Detection of some carbohydrate constituents

A quantity, 1.0 g of the respective tissue, was homogenized in 50 ml of ice-cold 5% trichloroacetic acid (TCA) and their filtrates tested for the presence of glucose and glycogen, respectively, using the Barfoed and iodine tests, as described by Plummer (1971).

Assay for some mineral contents

Analyses for some of their mineral contents (mg/100 g) were determined using the mixed-acid digestion and atomic absorption spectrophotometer of 1.0 g wet tissue as described by Allen et al. (1983) and AOAC (1990).

Statistical analysis

The results were compared using the students' t-test of significance at 95 percent confidence limit.

RESULTS

Figure 1A shows that the IUF had larger fibrous stroma than the SUF which appeared muscular. The epithelium was columnar because of the observed numerous long or rectangular nuclei. Tiny papillae were observed. Bundle of smooth muscle was present; appearing white on sectioning before staining.

Figure 1B shows that the SUF had numerous fibrous cysts within the stroma. No papillary projection was observed. Some of the fibrous stroma cells were large. The nuclei were oval shaped. Bundle of smooth muscle was present; appearing white on sectioning before staining.

Table 1 shows the moisture and dry matter contents of IUF and SUF (on wet-weight basis). The results showed that their compositions in the tissues did not vary significantly ($p > 0.05$).

Table 2 shows the proximate composition of the fibroid tissues (on dry-weight basis). Results showed that the IUF contained more ($p < 0.05$) crude fats and proteins than SUF, while the SUF contained more ($p < 0.05$) total carbohydrates than the IUF. Their ash and moisture contents did not vary significantly ($p > 0.05$).

Table 3 shows the qualitative detection of some the tissues' crude fats and total carbohydrate constituents. Results showed that crude fats contained fatty acids, triacylglycerols and cholesterol while the total carbohydrates contained glucose and glycogen.

Table 4 shows some types of minerals that were contained in the tissues. Results indicated that the mineral of highest concentration was Na^+ while the

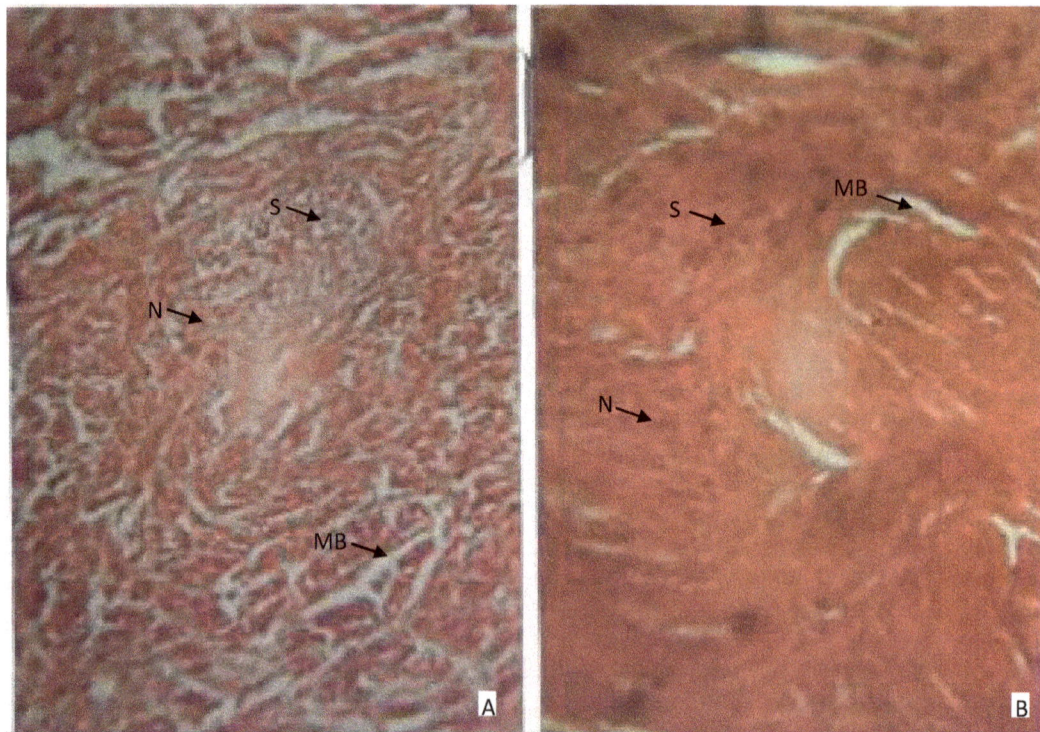

Figure 1. Histomorphology of SUF (A) and IUF (B) S = stroma; N = nucleus; MB = muscle bundle.

Table 1. Moisture and dry matter contents of the fibroid tissues (%, wet-weight)*.

Sample	Moisture	Dry matter
IUF	77.26 ± 0.05	22.74 ± 0.05
SUF	77.49 ± 0.02	22.52 ± 0.02

*Values are means ± S.D of duplicate determinations. IUF = intramural uterine fibroid. SUF = subserosal uterine fibroid.

mineral of least concentrations was Zn^{2+}. It also showed that the IUF was less ($p < 0.05$) mineralized than the SUF.

DISCUSSION

The exact compositions of fibroid tissues have been a source of debate among researchers. However, a general assumption is that they may have the same basic compositions of uterine muscle tissues.

The tissues used in the study were confirmed to be fibroids (Figures 1A and 1B). The tissues showed that there may have been changes in hormonal activities because each stroma was composed of somewhat plump fibroblasts resembling the theca.

The moisture contents of the tissues (Table 1) compared favourably with that of the average adult human body (60%) as reported by Tomlinson et al. (1997). Their proximate compositions (Table 2) indicated that high proportions of their dry weights were proteins.

Carbohydrates (Tables 2 and 3) formed lower parts of their compositions. Tomlinson et al. (1997) reported that a small amount of glucose was available to satisfy immediate energy needs and that fat formed a greater part of the body composition of a 65 kg woman. The ash contents of the tissues were the lowest. Major minerals occur in the body in molar quantities; the minor minerals occur in millimolar quantities while the trace elements occur in micromolar quantities but are essential for life (Tomlinson et al., 1997). The study also showed that the tissues contained good energy reserves which the body could have put to good use. Their energy contents were high and should have been very useful to the body if they had not been sequestered in the fibroids. However, these stores of energy can only be made available to the body after menopause when the myomas are expected to regress, diminish or disappear; if surgeries are not performed.

The presence of fatty acids (normally stored in triacylglycerols) and glucose (normally stored as glycogen) (Table 3) indicated that the tissues were

Table 2. Proximate composition (%) and energy contents (kcal/ 100g) of the fibroid tissues (dry matter basis)*.

Sample	Moisture	Crude fat	Crude protein	Ash	Total carbohydrates	Energy content (kcal/ 100g)
IUF	4.67 ± 0.05	8.29 ± 0.41	72.66 ± 0.72	4.39 ± 0.08	10.01 ± 1.19	405.29
SUF	4.38 ± 0.17	5.84 ± 0.43	65.16 ± 0.91	4.22 ± 0.03	20.40 ± 1.45	394.8

*Values are means ± S.D of duplicate determinations.

Table 3. Qualitative detection of some crude fat and total carbohydrate constituents.

Sample	Fatty acids	Triacylglycerols	Cholesterol	Glucose	Glycogen
IUF	+	+	+	+	+
SUF	+	+	+	+	+

+ = present.

Table 4. Some mineral contents of the fibroid tissues, wet weight basis (mg/ 100g tissue)*.

Mineral	IUF	SUF
Ca^{2+}	6.04 ± 0.02	9.71 ± 0.01
Fe^{2+}	2.67 ± 0.01	4.50 ± 0.01
Zn^{2+}	2.29 ± 0.00	2.91 ± 0.02
K^+	11.14 ± 0.03	36.78 ± 0.01
Na^+	30.28 ± 0.02	52.60 ± 0.02
Mg^{2+}	2.84 ± 0.01	7.91 ± 0.04

*Values are means ± S.D of duplicate determinations.

metabolically active. The oils may explain the oily feel of the tissues. Garrett and Grisham (1999) reported that sugars (a disaccharide of glucose and galactose) were found covalently attached to 5-hydroxylysine residues in the hole regions of collagen. The cholesterol may have been stored, or served as precursors for synthesis of sex hormones. Further research on these constituents is suggested as their detection was exploratory.

The most abundant mineral in the tissues was Na^+ (Table 4) unlike the norm in cells where the most abundant mineral is K^+ (Cooper, 2000; Nelson and Cox, 2000; Devlin, 2006). Cells and tissues maintain the order: $[K^+] > [Na^+] > [Mg^{2+}] > [Zn^{2+}] > [Ca^{2+}] > [Fe^{2+}]$ as confirmed in the reports of Garrett and Grisham (1999), Cooper (2000), Nelson and Cox (2000), Huerta-Leidenz et al. (2003) and Devlin (2006). The fibroid tissues rather maintained the order: $Na^+ > K^+ > Ca^{2+} > Mg^{2+} > Fe^{2+} > Zn^{2+}$. This increased the concentrations of Na^+, Ca^{2+} and Fe^{2+} and reduced the concentrations of K^+, Mg^{2+} and Zn^{2+}. These showed that the activities of their Na^+/ K^+-ATPases may have been decreased or inhibited by either high oestrogen or low Mg^{2+} concentrations. Davis et al. (1978) reported that synthetic ethinyl oestrogen decreased hepatic Na^+/ K^+-ATPase activity and bile flow to 50 percent and altered the composition and structure of surface membrane lipids in rats. Mg^{2+} on its own is required as a cofactor for the phosphorylation of a specific aspartic acid residue on the α - subunit of the Na^+/ K^+-ATPase to form a β- aspartylphosphate, during ion translocation (Garrett and Grisham, 1999; Devlin, 2006). Increasing Mg^{2+} concentrations resulted in a significant activation of Na^+/K^+ - ATPase, which related to Mg^{2+} concentration (Romanini et al., 1991). Mg^{2+} and Ca^{2+} have been reported as universal regulators of the cell and effectively influence the functional activity and conformational states of Na^+/K^+-ATPase (Kravtsov and Kratsova, 2001). Many membrane transporters depend on the electrochemical Na^+ gradient established when it (Na^+) is transported to the extracellular space by the Mg^{2+}- dependent Na^+/K^+-ATPase (Cooper, 2000; Devlin, 2006). One of such transporters is the Na^+/Ca^{2+} antiporter. If the activity of the Na^+/K^+-ATPase is decreased, it would mean that there shall be no Na^+ gradient to drive Ca^{2+} translocation to the cell's exterior milieu. This would mean an increase in cytosolic concentration of calcium, leading to calcification of the cell and tissue in general. Our results (Table 4) showed calcification of the tissues, especially the SUF. Garrett and Grisham (1999) reported that hypertensive patients that had the sodium pump of the cells lining the blood vessel wall inhibited by cardiac glycosides or cardiotonic steroids accumulated sodium and calcium in those cells. Calcification of fibroid tissue is thought to interfere with enzymes that dissolve the fibrin within, but such calcification disappeared on administration of magnesium supplements (Indman, 2010; Layyous, 2010) which may be due to the restoration of Na^+/K^+-ATPase activity (since Mg^{2+} is a cofactor) leading to the re-establishment of the electrochemical Na^+ gradient and restoration of Na^+/Ca^{2+} antiporter activity. The tissues' Zn^{2+} and Fe^{2+} contents (Table 4) indicated that cellular influx of Zn^{2+} via the Zn^{2+} transporter may also have been decreased while cellular influx of Fe^{2+} via increased synthesis of transferrin receptors and decreased synthesis of apoferritin may have been increased, especially in SUF. Murray and Stein (1968) though reported that injected oestrogen did not change the absorption of iron by mature female rats. The mineralization of the tissues by calcium, or some other minerals that were not assayed for, may be responsible for the firm texture of the fibroid tissues.

Conclusion

The study showed that the tissues sequestered high amount of energy and contained minerals whose concentrations indicated that the normal activities of their transporters may have been perturbed, either by being decreased, inhibited or increased.

ACKNOWLEDGEMENT

The contributions of the patients who chose to remain anonymous to the success of this report are acknowledged. We shall respect this onerous desire. We also acknowledge the financial support (research grant) from the Federal University of Technology, Owerri, Nigeria.

REFERENCES

Allen SE, Parinso JA, Quarmry C (1983). Chemical analysis of ecological materials. Blackwell, Oxford.

AOAC (1990). Official method of analysis, 15th edn. Association of Official Analytical Chemists, Virginia.

Baird DD, Newbold R (2005). Prenatal diethylstilbesterol (DES) exposure is associated with uterine leiomyoma development. Am. J. Reprod. Toxicol., 20: 81-84.

Cooper GM (2000). The cell: A molecular approach, 2nd edn. ASM Press, Washington, D.C.

Cotran RS, Kumar V, Collins T (1999). Robbins pathologic basis of disease, 6th edn. W.B. Saunders, Philadelphia.

Davis RA, Kern Jr. F, Showalter R, Sutherland E, Sinensky M, Simon FR (1978). Alteration of hepatic Na^+/K^+-ATPase and bile flow by oestrogen: Effects on liver surface membrane lipid structure and function. Proc. Natl. Acad. Sci., 75(9): 4130-4134.

DeCherney AH, Nathan L (2003). Current obstetric and gynecologic diagnosis and treatment, 9th edn. McGraw-Hill, New York.

Devlin TM (2006). Textbook of biochemistry with clinical correlations, 6th edn. Wiley-Liss, New Jersey.

Garrett RH, Grisham CM (1999). Biochemistry, 2nd edition. Brooks / Cole, Pacific Groove.

Huerta-Leidenz N, Arenas de Moreno L, Moron-Fuenmayor O, Uzcátequi-Bracho S (2003). Mineral composition of raw longissimus muscle derived from beef carcasses produced and grade in Venezuela. Arch. Latinoam. Nutr., 53(1): 96-101.

Kravtsov AV, Kravtsova VV (2001). Regulation of Na^+/K^+-ATPase: effects of Mg and Ca ions. Ukr. Biokhim. Zh., 73(2): 5-27.

Mathotra VK (1989). Practical biochemistry for students, 3rd edn. Jaypee Brothers Medical, New Delhi.

Murray MJ, Stein N (1968). The effect of administered oestrogens and androgens on the absorption of iron by rats. Br. J. Haematol., 14(4): 407-409.

Nelson DL, Cox MM (2000). Principles of biochemistry, 3rd edn. Worth Publishers, New York.

Okoro I (2002). Manual of practical histology, 2nd edn. Peace Publishers, Owerri.

Peddadah SD, Laughhn SK, Miner K, Guyton JP, Haneke K, Vahdat HL, Semelka R (2008). Growth of uterine leiomyomata among premenopausal black and white women. J. Endocrinol., 3(5): 19887 – 19888.

Plummer DT (1971). An introduction to practical biochemistry. McGraw-Hill, London.

Romanini C, Tranquilli AL, Valensise H, Cester N, Benedetti G, Cuqini AM, Mazzanti L (1991). In vitro effect of magnesium on the Na/K – ATPase isolated from human placenta. Magnes. Res., 4(1): 41-43.

Shryock H, Swartout HO (1980). Your health and you. The Stanborough Press, Lincolnshire, 2.

Tomlinson S, Heagerty AM, Weetman AP (1997). Mechanisms of disease: an introduction to clinical science. Cambridge University Press, Cambridge.

Wardlaw GM, Kessel MW (2002). Perspective in nutrition, 5th edn. McGraw-Hill, Boston.

Web sites

Indman PD (2010). All about myomectomy. http://www.myomectomy.net/index.htm [August 31, 2011].

Layyous N (2010). Uterine fibroid features and management options. http://www.layyous.com/root folder/fibroid_eng.htm [August 30, 2011].

Stewart EA (2011). Epidemiology, clinical manifestation, diagnosis and natural history of uterine leiomyomas (fibroids). http://www.update.com/contents/epidemiology-clinical-manifestation-diagnosis-and-natural-history-of-uterine-leiomyomas-fibroids [September 5, 2011].

25

Multipotent mesenchymal stem cells (MSCs) from human umbilical cord: Potential differentiation of germ cells

Jinlian Hua[1,2]*, Pubin Qiu[1,2], Haijing Zhu[1,2], Hui Cao[1,2], Fang Wang[1,2] and Wei Li[3]

[1]Key Laboratory for Reproductive Physiology and Embryo Biotechnology of Agriculture, Shaanxi Centre of Stem Cells Engineering and Technology, Ministry of China, Shaanxi, China.
[2]Key Lab for Agriculture Molecular Biotechnology Centre, Northwest, A and F University, Yangling, Shaanxi, 712100 China.
[3]Beike Bio-Technology Co., Ltd of Jiangsu Province, Taizhou, Jiangsu Province, 225300 China.

Previous controversy exists as to whether umbilical cord (UC) can serve as a source of multipotent mesenchymal stem cells (MSCs) with their characteristics. Different methods and reagents have been used to induce the differentiation of UC-MSCs into functional cells. We investigated the isolation of UC-MSCs and their potential differentiation into neurons, cardiomyocytes and germ cells *in vitro*. The phenotypes, proliferation potential and markers of UC-MSCs were analysed by growth curves, RT-PCR and immunofluorescence, respectively. Then the cells were induced into neurons, cardiomyocytes and germ cells besides osteoblasts and adipocytes. Here, we report to obtain single cell-derived, clonally expanded MSCs that are of multilineage differentiation potential. The immunophenotype of these cells is consistent with those reported in bone marrow MSCs and embryonic stem cells (ESCs). Surprisingly, these cells can differentiate into cardiomyocytes, neural cells, even germ cells besides osteoblasts and adipocytes under appropriate induction conditions. Thus, these cells may be multipotent MSCs as evidenced by their ability to differentiate into cell types of all three germ layers. These cells may serve as an alternative source of MSCs to bone marrow and this will provide us a model to study the mechanism of cardiomyocyte, neural cell, even germ cell differentiation and new strategies for the therapy of infertility and sterility.

Key words: Multipotent, mesenchymal stem cells (MSCs), umbilical cord (UC), cardiomyocyte, neural cells; germ cells.

INTRODUCTION

Mesenchymal stem cells (MSCs) derived from bone marrow are well-characterized population of adult stem cells, which can form a variety of cell types, including fat cells, cartilage cells, bone cells, tendon cells and ligaments cells, muscles cells, skin cells and even nerve cells (Choong et al., 2007; Pittenger et al., 1999; Jiang et al., 2002). Bone marrow transplant has been used for a long time to treat leukemia and many types of cancer, as well as various blood disorders. Bone marrow contains a promising source of multipotent stem cells-mesenchymal stem cells (MSCs) and hematopoietic stem cells (HSCs) (Pittenger et al., 1999). Recently, it has been reported that murine MSCs derived bone marrow may trans-differentiate into gametes (sperms or follicles) *in vivo* and *in vitro*. These cells share genes typical of germ cells and proposed that bone marrow stem cells can migrate and colonize the ovaries to maintain a plentiful stock for reproduction and may differentiate into sperms in mice (Johnson et al., 2005; Nayernia et al., 2006). However, aspirating bone marrow from the donor is invasive, and especially, the differentiating potential and the number of MSCs derived from bone marrow decreases gradually with age (Lee et al., 2004). Therefore, many scientists have been looking for alternative sources of MSCs and

*Corresponding author.E-mail: jinlianhua@nwsuaf.edu.cn, Jlhua2003@126.com.

found that umbilical cord blood (UCB) and UC maybe an excellent alternative source of bone marrow stem cells because these cells are younger than other adult stem cells (Barachini et al., 2009; Lee et al., 2004; Secco et al., 2008). Importantly, UC stem cell transplants are less prone to rejection issue than either bone marrow or peripheral blood stem cells and these 'waste' stem cells could be cryopreserved and stored in a stem cell bank for the donor, his families and others; What's more, these young stem cells have not yet developed the features that can be recognized and attacked by the recipient's immune system. Also, UC lacks well-developed immune cellswith less chances that the transplanted cells will attack the recipient's body, a problem called graft versus host disease (Liu et al., 2010). Both the versatility and availability of UC stem cells makes them a potent resource for transplant therapies (Kumar et al., 2006 ; Cutler et al., 2010).

UCB transplantation has been used in clinical practice for many years (Harris. 2008). Recently, scientists have shown that multipotent MSCs were obtained from UC and these MSC like cells have the ability to differentiate into multilineages including bone, adipocytes, osteoblasts and hepatocyte-like cells, and endothelial cells (Flynn et al., 2007; Lee et al., 2004; Secco et al., 2008; Zhang et al., 2009; Xu et al., 2010; Yoo et al., 2010). However, controversy exists as to whether UC contains real multipotent MSCs, which are capable of differentiating into cells of three different connective tissue lineages such as bone, cartilage, and adipose tissues, being the best candidates for tissue engineering of musculoskeletal tissues, and even differentiating into cardiomyocytes, neuron-like cells and hepatocytes (Lee et al., 2004; Kang et al., 2006; Orlandi et al., 2008; Secco et al., 2009). However, so far, no progress and evidences have been reported in the isolation and characterization of MSCs from UC to differentiate into functional germ cells. Therefore, the aim of this study is to investigate the possibility of obtaining clonally expanded MSCs that have the potential for nearly pluripotent differentiation including potential differentiation into germ cells.

MATERIALS AND METHODS

Culture of MSCs derived from UC

The MSCs derived from male term (post-natal) babies UC were provided by North Branch Bio-Technology Co., Ltd of Jiangsu (http://www.stemcellsbank.com.cn, Taizhou, Jiangsu Province, China). The cells were plated in coated tissue culture T75 flasks (Becton Dickinson) in StemPro® MSC SFM (Invitrogen) medium. The cell density was 1×10^5/ml. The cells were allowed to adhere for 2 days and non-adherent cells were washed out with medium changes. The media were used to initiate growth of the adherent UC-MSCs: DMEM/F12 (Invitrogen) with 10% fetal bovine serum (FBS, Hyclone), supplemented with 0.1 mM 2-mercaptoethanol (Invitrogen), penicillin (100 U/ml; Sigma), streptomycin (0.1 mg/ml; Sigma), and 2 mM glutamine (Invitrogen). Expansion of the cells

was performed in the same media. Cells were incubated at 37°C, 5% CO_2 in a humidified atmosphere. When cells reached 80% confluency, they were detached with 0.125% trypsin (Invitrogen), centrifuged at 1500 rpm for 5 min, and replated at the ratio of 1:3 under the same culture conditions. The medium was first changed at 24 h after plating, and then changed at every other day.

To obtain single cell-derived, clonally expanded MSCs, the isolated plate-adhering fifth passage cells, were serially diluted and plated on to 96-well plates at a final density of 10 cells per well. Colonies that grew were cultured and tested for their differentiation potential. Separated fibroblast like colonies termed CFU-Fs were identified at a mean interval of 1 to 3 weeks after initial plating. To study mesenchymal cells obtained from an individual post-natal fetus, all colonies growing in a 60 mm plate were trypsinized and obtained. The adherent stromal bone-marrow fibroblasts like (CFU-F like colonies) were fixed with methanol and stained with Wright-Giemsa staining. Also the growth curve of seventh and ninth passage UC-MSCs were evaluated respectively.

RNA isolation and RT-PCR

RNA was extracted from 30×10^5 UC-MSCs and induced differentiated cells using Trizol (Qiagen, Beijing) according to the manufacturer's instructions. The mRNA was reverse transcribed to cDNA using Advantage RT-for-PCR (Takara, Dalian) based on the manufacturer's instructions. cDNA was amplified using a ABI GeneAmp PCR System 2400 (Takara, Dalian) at 94°C for 40 s, 56°C for 50 s, and 72°C for 60 s for 35 cycles, after initial denaturation at 94°C for 5 min. Semi-quantitative RT-PCR was carried out with 0.5 µl cDNA, 30 pmol each of forward and reverse primers and 2 units Platinum Taq polymerase (Takara,Dalian) in a final volume of 15 µl. The solution was incubated at 94°C for 2 min and then subjected to 35 cycles of amplification, each consisting of 95°C for 30 s, 52 to 58°C for 30 to 45 s (annealing) and 72°C for 60 s (primer extension). At the end of the temperature cycles the solution was incubated at 72°C for 10 min. The PCR products were subjected to electrophoresis on 1.0% (w/v) agarose gels containing 1 mg/ml ethidium bromide and the products were viewed and photographed under UV light. β-actin was used as an internal control. The primers used for RT-PCR analyses are shown in Table 1. Primers were designed to span exons to distinguish cDNA from genomic DNA products.

Flow cytometry analysis

For cell surface antigen phenotyping, fifth- to seventh-passage cells were detached and stained with fluorescein- or phycoerythrin-coupled antibodies and analyzed with FACS Calibur (Becton Dickinson). Cells were treated with TrypLE (Millipore, USA), harvested, and washed twice with culture medium. Before staining, cells were allowed to recover for 20 min in suspension. Cell staining was performed using mouse monoclonal antibodies followed by fuorescein isothiocyanate (FITC)-conjugated affinity-purifed mouse fuorochrome-conjugated isotype control antibodies, or FITC or phyco-erythrin (PE)-coupled antibodies against the common leukocyte antigen CD45 (Becton Dickinson, USA), the surface-expressed 5′-ectonucleotidase CD71 (Becton Dickinson), the β1-integrin CD29 (Becton Dickinson), CD73 (Becton Dickinson); CD105 (Becton Dickinson); CD11a (Becton Dickinson), CD90 (Becton Dickinson), CD166 (Becton Dickinson), CD117 (Abcam, UK), CD34 (Becton Dickinson) and CD44 (Becton Dickinson). All antibodies were used following the manufacturers' instructions. Binding of antibodies against the markers as primary antibodies was detected by anti-mouse immunoglobulin G (IgG) conjugate (Becton Dickinson), or isotype-specifc FITC- or PE-conjugated goat

Table 1. The primer sequences.

Gene	Forward primer	Reverse primer		Product size (bp)
Sox2	GCCCAGGAGAACCCCAAGAT	GGGTGCCCTGCTGCGAGTA	58	520
Oct4	GAAGCTGGACAAGGAGAAGCT	CATGCTCTCCAGGTTGCCTC	58	379
hTERT	GTGTGCTGCAGCTCCCATTTC	GCTGCGT CTGGGCTGTCC	58	264
Nanog	GCGAATCTTCACCAATG	TTTCTGCCACCTCTTAC	54	407
B-ACTIN	GCGGCATCCACGAAACTAC	TGATCTCCTTCTGCATCCTGTC	58	138
GATA4	TCCCTCTTCCCTCCTCAAATTC	TCAGCGTGTAAAGGCATCTG	54	193
Nkx2.5	AGCACTTCTCCGCTCACTTC	CCGTGCACAGAGTGGTACTG	60	232
Dazl	ATGAAAGATAAAACCACCAACC	TGTTGACAGCCTGGTCCACTGA	58	391
Stella	TCCCTCTTCCCTCCTCAAATTC	TCAGCGTGTAAAGGCATCTG	60	238
SCP3	CTAGAATTGTTCAGAGCCAGAG	GTTCAAGTTCTTTCTTCAAAG	60	247

anti-mouse IgG F(ab')2 fragments (Becton Dickinson). Results were analysed based on the mean percentage of positive cells and standard deviation from multiple experiments.

Spontaneous differentiation in vitro

For the formation of EBs, 7th- to 15th-passage UC-MSCs were collected from the culture dish, dissociated by Tryple into single cells and resuspended at 3×10^5 cells/ml, in culture medium DMEM consisting of 10% FCS (Hyclone), 0.1 mM 2-mercaptoethanol (Sigma), 2 mM glutamine (Gibco) for 3 days. Then the cells were transferred into normal sterile culture dishes for 7 to 10 days. Cultures were observed each day and the three germ layer markers: NESTIN (ectoderm), BRACHUARY (mesoderm) and AFP (endoderm) were analysed by RT-PCR at 7 day EBs.

Generation of osteoblasts

11th- to 13th-passage cells were treated with osteogenic medium for 3 weeks with medium changes twice weekly. Osteogenesis was assessed at weekly intervals. Osteogenic medium consists of DMEM supplemented with 10^{-8} M dexamethasone (Sigma-Aldrich, St Louis, MO), 10 mM β-glycerol phosphate (Sigma-Aldrich), and 50 μg/ml ascorbic acid (Sigma-Aldrich). The potentiality of osteoblast differentiation was evaluated by Alizarin-red S staining. Briefly, cells were fixed with 4% paraformaldehyde (PFA) and stained with 1% Alizarin-red S (Sigma-Aldrich) solution in water for 10 min.

Generation of adipocytes

11th-to 13th-passage cells were treated with adipogenic medium for 1 to 3 weeks. Medium changes were carried out twice weekly and adipogenesis was assessed at weekly intervals. Adipogenic medium consists of DMEM supplemented with 0.25 mM 3-isobutyl-1-methylxanthine (IBMX; Sigma-Aldrich), 10^{-8} M dexamethasone (Sigma-Aldrich, St Louis, MO), 5 μg/ml insulin (Sigma-Aldrich), and 10% FBS (Hyclone). Adipocyte differentiation was tested by oil-red O staining. Cells were fixed with 4% PFA and stained with oil-red O (Sigma-Aldrich) for 10 min.

Generation of neural cells

5th- to 7th-passage cells, seeded at a density of 3000 cells/cm^2,

were treated with neural stem cell medium including 20 ng/ml EGF (Chemicon), 10 ng/ml bFGF (Millipore), B27 (Invitrogen) and 1% Insulin-transferrin-Selenium (ITS, Invitrogen) in DMEM/F12 (Invitrogen) for 2 weeks. The neural cells were identified by morphology and immunohistochemistry with neuron specific enolase (NSE, 1:200, Millipore) and β-tubulin III (1:1000, DSHB) antibodies respectively. Some induced samples were evaluated by aniline blue staining. Cells were fixed in 4% PFA for 30 min at room temperature, then stained with 5% w/v aniline blue in PBS (pH 3.5). Each slide was then washed with PBS, counted and examined under the inverted microscope at 200 magnification.

Generation of cardiomyocytes

After proliferating to nearly a layer, these cells were induced by 10^{-7} M RA (Sigma) and 0.75% DMSO (Sigma) in DMEM with 10% FBS, or treated with 10 μM 5-azacytidine (5-aza, Sigma, St. Louis) for 48 h, and then were induced by 10% FBS in DMEM. The induced cardiomyocytes were identified by morphology, immune-histochemistry with human cardiac a-actin (1:500, Sigma), CT3 (1:1000, DSHB), Islet1 (1:1000, DSHB) antibodies respectively. Also the induction efficiencies were evaluated by cardiac specific markers (cardiac α-actin, GATA4 and Nkx2.5) based on RT-PCR. The primers are shown in Table 1.

Potential differentiation of germ cells

The UC-MSCs were used for germ cell differentiation by using with 2×10^{-6} M RA and 10 ng/ml BMP4 in DMEM for 7-14 d (Danner et al., 2007). The induced cells were formed and identified by morphology and immunohistochemistry with VASA (1:1000, Abcam) and SCP3 (1:300, Santa Cruz) antibodies respectively, which are markers of germ cells, sperms or oocytes (Clark et al., 2004; Hua et al., 2009; Lacham-kaplan et al, 2006; Toyooka et al., 2003).The differentiation efficiencies were evaluated by expression of specific markers of germ cells (Dazl,Stella and SCP3) based on RT-PCR. The primers are shown in Table 1. Induced samples were evaluated by alkaline phosphatase (AP) described (Piedrahita et al., 1998). Briefly, culture plates were rinsed three times with PBS and fixed in 4% PFA for 10 min at room temperature. Fixed cells were washed three times with PBS and stained in naphtol AS-MX phosphate (200 μg/ml, Sigma) and Fast Red TR salt (1 mg/ml, Sigma) in 100 mM tris-buffer (pH 8.2) for 10 to 30 min at room temperature. Staining was terminated by washing cultures in PBS to evaluate the characteristics and count the number of AP positive cells or colonies.

Immunofluorescence

For staining of intracellular proteins, cells were fixed overnight with 4% PFA for 15 min at 4°C and permeabilized with 0.1% Triton X-100 (Sigma-Aldrich) for 10 min. Slides and dishes were incubated with the mouse or rabbit primary antibodies against human Oct4 (1:500, Chemicon), Sox2 (1:200, Chemicon), Klf4 (1:200, Chemicon), C-myc (1:200, Chemicon), SSEA4 (1:200, Chemicon) and hTERT (1:100, Santa Cruz) for 1 hour at room temperature or overnight at 4°C, followed by fluorescein- or phycoerythrin-coupled goat anti-mouse or rabbit IgG secondary antibodies (1:500, Invitrogen) for 1 h. Between incubations, slides and dishes were washed with PBS. The nuclei of cells were counterstained with Hoechst33342, and then observed and analysed with Leica microscope.

RESULTS

UC-MSCs were generated from 98% of UC and the mean number of MSCs was 2.11×10^5/g in North Branch Bio-Technology Co., Ltd of Jiangsu Province (Taizhou, Jiangsu Province, China). The primary cells reached 60 to 80% confluence at 4 to 9 day. The individual UC-MSCs appeared as spindle-shaped-like fibroblasts and the cells were unique in their phenotypes and assumed a monolayer configuration on reaching confluency during culture (Figure 1A). Giemsa staining of UC-MSCs at passage 5 to 8 indicated that most of the cells were of mononuclear fibroblast morphology and CFU-F colonies (Figures 1B to C). Overgrown confluent mononuclear cells in culture formed colonies (Figures 1D and E). Normally, these UC-MSCs were negative for AP staining (Figure 1F). The cell growth curve assay showed that these cells have the characteristics of rapid proliferation in vitro, and multiplied nearly 20 folds in 8 days (Figure 1G), with the mean population doubling (PD) time being 76.8±8.4 h (P0), 72.8±5.4 h (P2), 38.8 h (P7). These cells proli-ferated up to passage 20 and maintained normal MSC phenotypes. The mean clonal capacity of these cells was 15±5%. RT-PCR and immunofluoresce staining demon-strated that these cells expressed Oct4, Sox2, Klf4, Nanog, C-myc, SSEA4 and hTERT (Figures 1H and 2).

Immunophenotypic characterization of UC-derived fibroblast-like MSCs were extensively expanded, and characterization by flow cytometry revealed that the cells isolated by the described method were negative for CD71, CD34, CD45 (leukocyte common antigen), and CD117 (C-KIT), indicating these cells are not of hemato-poietic origin (Figure 3). UC-derived cells were found to be showed strong positive homogeneous staining for markers of mesenchymal progenitors at different passages (P1, P3, P4 and P6, Table 2).

These markers included being positive for integrins CD29 (β1-integrin), matrix receptors CD44 (hyaluronate receptor), CD90 (Thy-1) and CD105 (endoglin), CD73, CD166, and were negative for hematopoietic origin markers: CD34, CD14, CD71, CD45 and HLA-DR at different passages, which are consistent with the findings for bone marrow and cord blood MSCs (Yu et al., 2004; Rebelatto et al., 2008).

In vitro differentiation of osteoblasts and adipocytes from UC-derived MSCs

To investigate the osteogenic potential of the UC-derived cells, 11th-to 13th-passage cells were plated at a density of 2×10^3 cells/cm^2 and cultured under conditions appropriate for inducing differentiation for each lineage. When induced to differentiate under osteogenic conditions, the spindle shape of UC-derived cells become flattened, broadened, and aggregated with increasing time of induction and formed mineralized matrix as evidenced by Alizarin red staining (Figure 4A and B). The mean percentage of Alizarin red staining was significantly higher in induced cultures than that untreated cells at 14 days (65±8% vs 10±5%, P<0.01%).

To assess the adipogenic potential, 11th to 13th-passage cells were plated at a density of 2×10^3 cells/cm^2 and cultured in adipogenic medium. Morphologic changes in cells as well as the formation of neutral lipid vacuoles were noticeable as early as 1 week after induction and visualized by staining with oil-red O. The mean percentage of oil-red O staining was significantly higher in induced cultures than that untreated cells at 14 days (60±10% vs 10±5%, P<0.01%) (Figures 4C and D).

In vitro differentiation of neural and cardiomyocyte-like cells from UC-derived MSCs

Using the neural cell induction method, 5th- to 7 th-passage UC-derived cells were seeded at a density of 3×10^3 cells/cm^2 and tested for their neural differentiation potential. After 10 days of differentiation, 50% of cells in the plate acquired the morphology of neuroglial cells exhibiting a refractile cell body with extended neurite-like structures, 30% of cells had partially acquired the morphology of neural-like cells, and 10% of cells exhibited morphology resembling those of restricted precursors of the neuroectodermal lineage. 20% of cells stained positive for immunofluorescence assay against β-tubulin III and NSE (Figures 5A, B, C) at 6 days after differentiation and increased to approximately 50% by day 10. Also, 80% induced cells were positive for aniline blue staining (Figure 5D).

To determine whether UC-derived cells can differen-tiate into cardiomyocyte-like cells in vitro, cells were allowed to grow to 60% confluence prior to induction. The cuboidal or multinucleus morphology of cardiomyocyte-like cells were observed as early as 10 days after culturing under cardiac conditions in the presence of 5-AZA. The expression of cardiac cells associated genes: cardiac β-actin, Islet1 and CT3 were detected at the indicated time points analysed by immunofluorescent

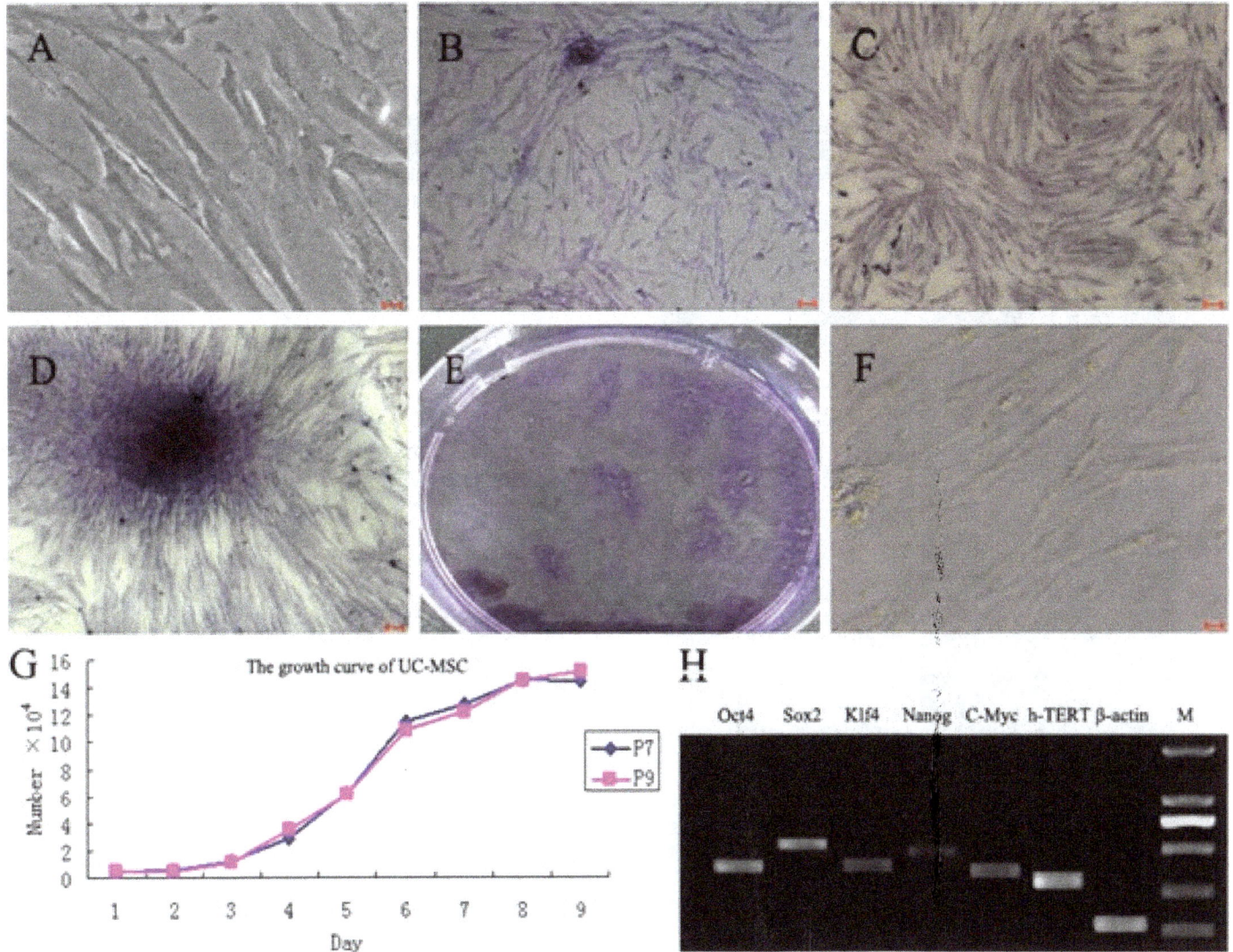

Figure 1. Characterization of UC-MSCs. The spindle fibroblast like UC-MSCs (A), Bar=50 μm; Giemsa staining showed that UC-MSCs were fibroblast and mononuclear and CFU-Fs were formed (B, C), Bar=100 μm; UC-MSCs formed compact colonies identified with Giemsa staining (D,E), Bar=100 μm. UC-MSCs were negative for AP staining (F), Bar=100 μm. G, The growth curve of UC-MSCs at Passage 7 and 9. H, Specific pluripotent markers of MSCs were analysed by RT-PCR (Oct4, Sox2, Klf4, Nanog, C-myc and hTERT,β-actin was used as internal control).

staining (Figure 6). Expressions of cardiac specific markers-Nkx2-5 and cardiac β-actin were up-regulated induced by 5-aza or RA in combination with DMSO compared with the untreated group based on semi-quantitative RT-PCR. However, GATA4 was only increased in 5-AZA induced cells. Under cardiac conditions expression of β -actin, an early developmental marker gene of cardiomyocytes, was detectable by day 7 and remained detectable up to day 15.The expression of CT3, a late marker gene of cardiomyocytes, was detected by day 14 and increased with time of differentiation. Undifferentiated cells did not express cardiac a-actin or CT3. These results demonstrated that human UC-MSCs may differentiate into cardiomyocytes in our induced cultures (Martin-Rendon et al., 2008).

In vitro differentiation of germ cells from UC-derived MSCs

To investigate the potential of UB-MSC for differentiation into germ cells, the cells were treated with 2×10^{-6} M RA. It was found that there were some round cells and a small number of germ cells formed in the treated cultures. The expression of the meiotic and germ cell markers SCP3 and VASA increased in RA-treated cultures compared with the untreated group based on semi-quantitative RT-PCR and immunofluorescent staining. Some round and spindle-shaped cells derived after the treatment of UC-MSCs with RA showed expressions of specific markers such as SCP3 and VASA by immunohistochemical analysis (Figure 7). After RA

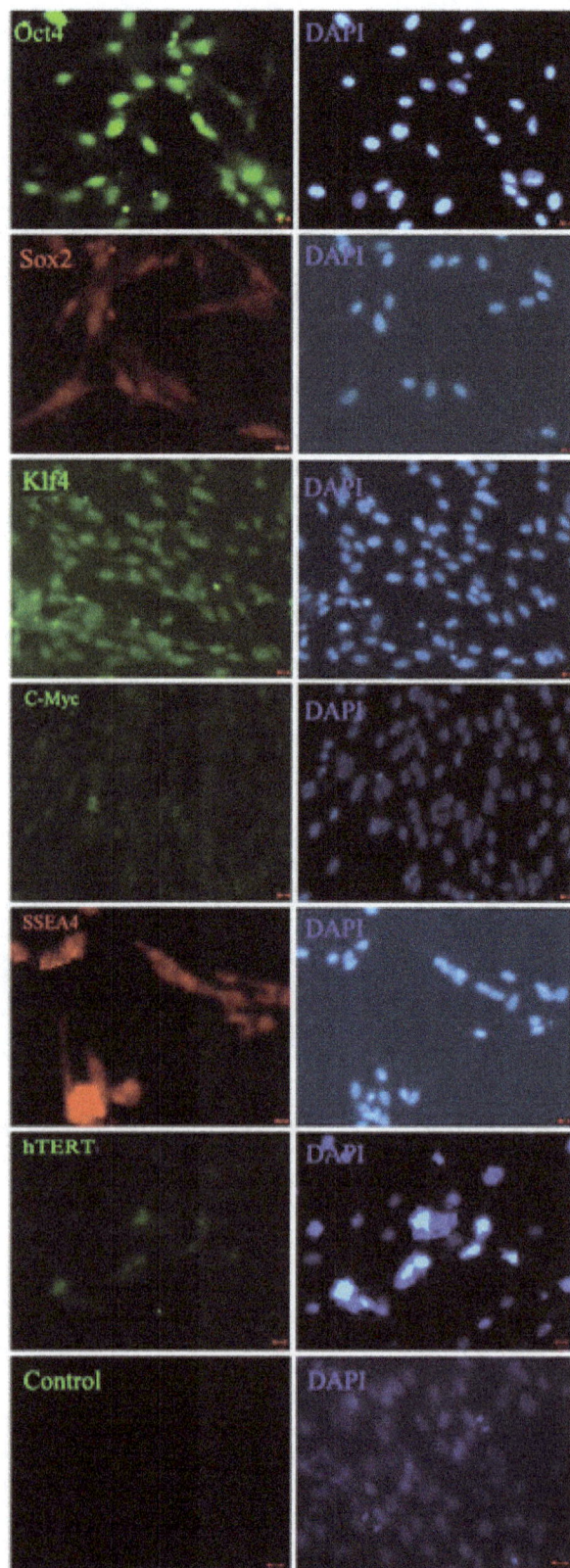

Figure 2. Immunofluorescence analysis of UC-MSCs. UC-MSCs were positive for pluripotent ESC markers:Oct4, Sox2, Klf4; C-myc; SSEA4 and hTERT; the right column were the Hoechst33342 nuclear staining for the left column, Control (Negative control). Bar=20 μm.

treatment, an increase in the number of cells which expressed VASA was detected compared with untreated cells (20±5% vs 8±5%, P<0.01%). However, no typical spermatid-like cells were observed in RA treated and it has not yet been determined whether these male germ cells can enter meiosis and form functional gametes. These results indicated that small subpopulations of UC-MSCs are able to differentiate into germ cells.

DISCUSSION

The morphology of these cells from UC resembles that of MSCs isolated from the bone marrow and other tissues (Lee et al., 2004; Pittenger et al., 1999; Gonzalez et al., 2009). MSCs have been studied extensively and many independent research groups worldwide have successfully isolated MSCs from a variety of sources, most commonly, from the bone marrow (Deet et al., 2001; Pittenger et al., 1999; Sottile et al., 2002). However, controversy exists over whether such stem cells are present in UC, and to date, little evidence in the literature substantiates the existence of such cells in UC. Erices et al. (2000) and Goodwin et al. (2001) independently reported the successful isolation of progenitor cells of mesenchymal origin from UC, whereas Mareschi et al. (2001) concluded against such findings.

MSCs are fibroblast-like in morphology, self-renewable, and capable of differentiating into, at least, three connective tissue lineages of the mesoderm including bone, cartilage, and adipocytes (Lee et al., 2004; Pittenger et al., 1999). We have shown in the present study that a plate adherent population of fibroblast-like cells isolated from UC, indeed, can be extensively clonally expanded *in vitro* while retaining the potential to differentiate, under *in vitro* conditions, into multiple lineages of the mesoderm (Rebelatto et al., 2008). Goodwin et al. (2001) showed that these cells also expressed neural markers, a heterogeneous plate-adhering population from total mononuclear cells and the presence of neural progenitors in UC has been demonstrated (Sanchez-Ramos et al., 2001). In this study, we demonstrated that our relatively more homogeneous UC-derived MSCs exhibited typical MSC phenotype analysed by morphology, RT-PCR analysis, and histochemical, cytochemical, and immune-cytochemical evaluations (Lee et al., 2004; Pittenger et al., 1999). These cells expressed Oct4, Sox2, Klf4, C-myc and hTERT, which are markers of pluripotent ESCs and induced pluripotent stem cells, and also these cells may differentiate into osteoblasts, adipocytes, neuroglial-like, cardiomyocyte-like cells and germ cells. This finding demonstrated that UC-MSCs share some features with MSCs (Baharvand et al., 2010; Panepucci et al., 2004). However, these UC-MSCs did not proliferate as pluripotent stem cells, iPSCs and cannot form teratomas in nude mice in our study and previous reports

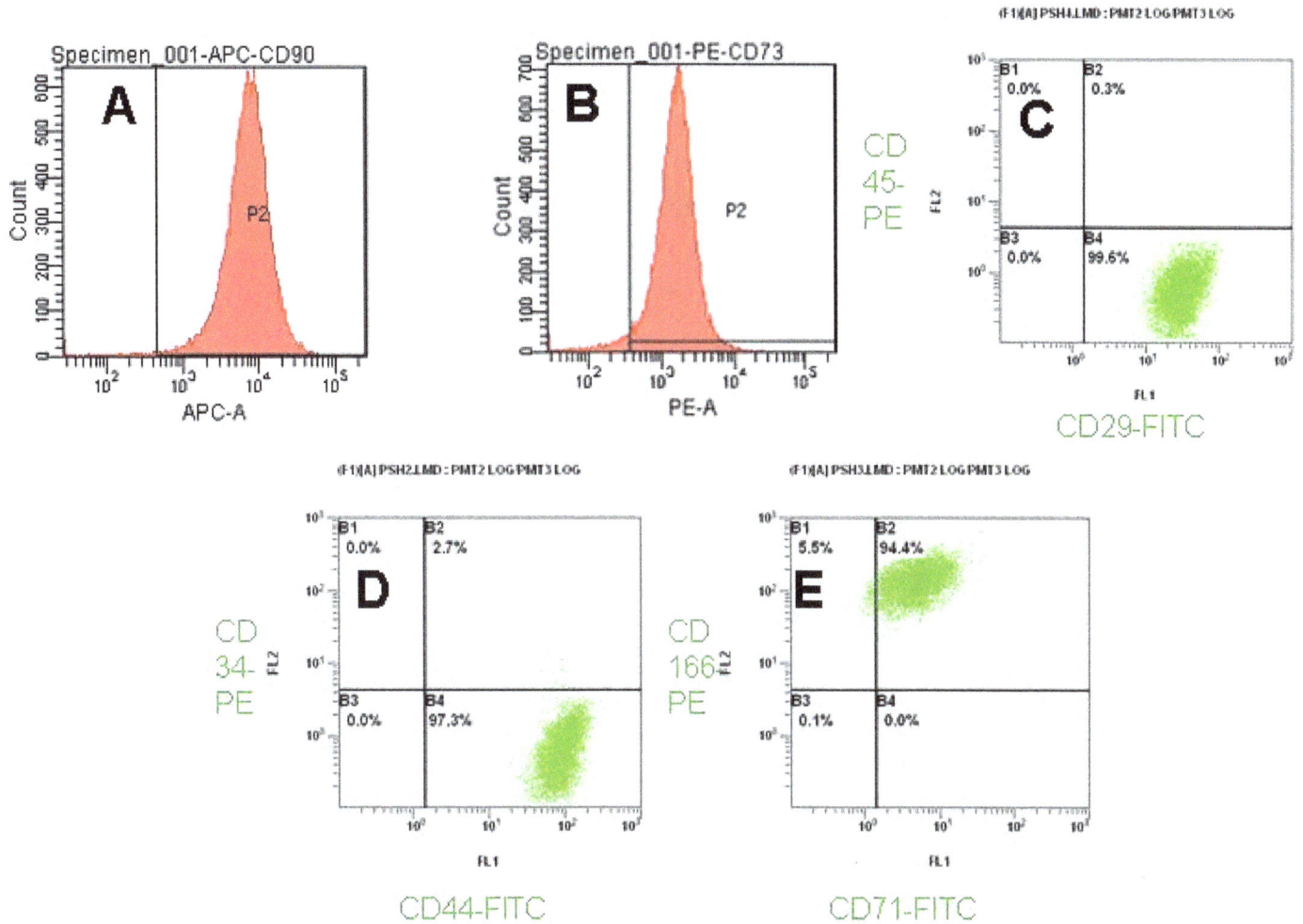

Figure 3. FACS analysis showed that UC-MSCs were positive for CD90, CD29, CD44, CD166, and negative for CD34, CD45, CD71(F-G).

Table 2. UC-MSCs were analysed with FACS at different passages.

Passage	CD73+ (%)	CD90+ (%)	CD105+ (%)	CD14+ (%)	CD34+ (%)	CD45+ (%)	HLA-DR+ (%)
P1	93.5	94.6	97.5	0.5	0.9	1.4	1.2
P3	96.6	99.8	98.6	0.4	0.9	0.3	1.2
P4	92.4	99.4	97.4	0.5	0.5	0.4	0.8
P6	92.7	99.6	98.6	0.2	0.2	0.1	0.2

(Baharvand et al., 2010; Goodwin et al., 2001; Xu et al., 2010). Therefore, we named them as multipotent UC-MSCs.

This report showed that UC does contain MSCs, even contain some pluripotent cells. With the approach reported in this study, it is possible to obtain single cell-derived, expanded MSCs from UC with remarkable potential to differentiate into multiple lineages of mesodermal and non-mesodermal origin. With this technique, the application of UC can be further extended and used as a new alternative source of MSCs to bone marrow (Lee et al., 2004). The self-renewal capacity of

these cells is remarkable and is expected of multipotent stem cells. Flow cytometric analysis showed that these cells were negative for various hematopoietic lineage markers but positive for human MSC markers, as well as other various other integrins and matrix receptors. They were consistent with that reported in the literature for the bone marrow counterpart, indicating the MSC nature of these UC-derived cells (Lee et al., 2004). In addition to the potential of MSCs to differentiate into multiple lineages of the mesoderm, recent reports in the literature have indicated that bone marrow MSCs are capable of transdifferentiating into germ layer boundaries (Kopen et

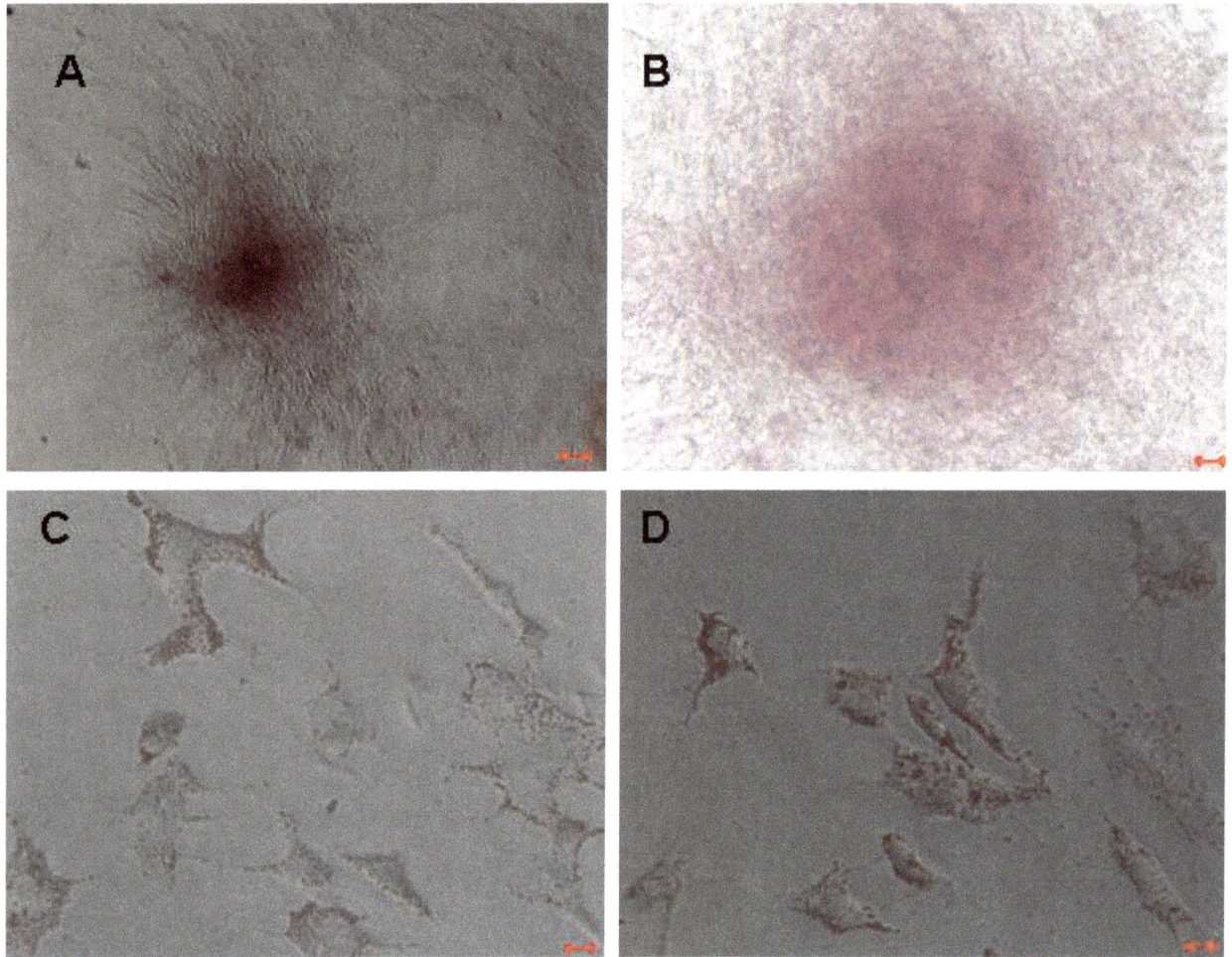

Figure 4. Osteoblast and adipocyte like cells were formed after induction. Aggregates were positive for Alizarin red staining (A, B); Lipid vacuoles were noticeable and visualized by staining with oil-red O in treated cells (C, D).

al., 1999; Rebelatto et al., 2008).

Most reports showed that the UC-derived cells may differentiate into mesoderm cell types including osteoblasts, chondrocytes and adipocytes under *in vitro* conditions (Lee et al., 2004; Pittenger et al., 1999). However, little progress and controvercial evidences have been reported in the isolation, characterization, and differentiation of MSCs from UC into functional neural, cardiomyocyte and germ cells (Lee et al., 2004; Rebelatto et al., 2008). Under neurogenic conditions, UC-derived cells exhibited the morphology of neural cells and expressed specific markers of neuroglial cells, which was confirmed by immunofluorescence assays for neural-specific proteins: β-tubulin II and NSE.

When UC-derived cells were cultured under cardiac conditions, most cells acquired a cuboidal morphology as opposed to the fibroblast-like morphology of undifferentiated cells. Cardiac β-actin, CT3 and Islet1 were detectable treated by 5-AZA or RA and DMSO. Previous studies showed that cardiac β-actin and CT3 were specific markers of cardiomyocytes (Rangappa et

al., 2003; Orlandi et al., 2009). Nkx2-5 and/or Islet1 positive cardiac progenitors contribute to proepicardium during heart development (Zhou et al., 2008). These results demonstrated that human UC-MSCs may differentiate into cardiomyocytes in our induced cultures (Martin-Rendon et al., 2008).

Previous reports have shown that ESCs, fetal porcine skin stem cells and bone marrow MSCs can differentiate into germ cells *in vitro* (Clark et al., 2004; Dyce et al., 2006; Hua et al., 2009; Lacham-kaplan et al., 2006; Nayernia et al., 2006). Stem cells may provide a new potential source of male and female germ cells that could be used for infertility and sterility. However, the mechanisms are unclear and the efficiency of germ like cells derivation of stem cells is still low, with few could go through or complete the meiosis phase (Nayernia et al., 2006; Hua and Sidhu, 2008; Hua et al., 2009). In this study, small number of cells were positive for specific germ cell markers-SCP3 and VASA, and levels of expression of Stella, Dazl and SCP3 were increased inRA treated compared to control. These genes are

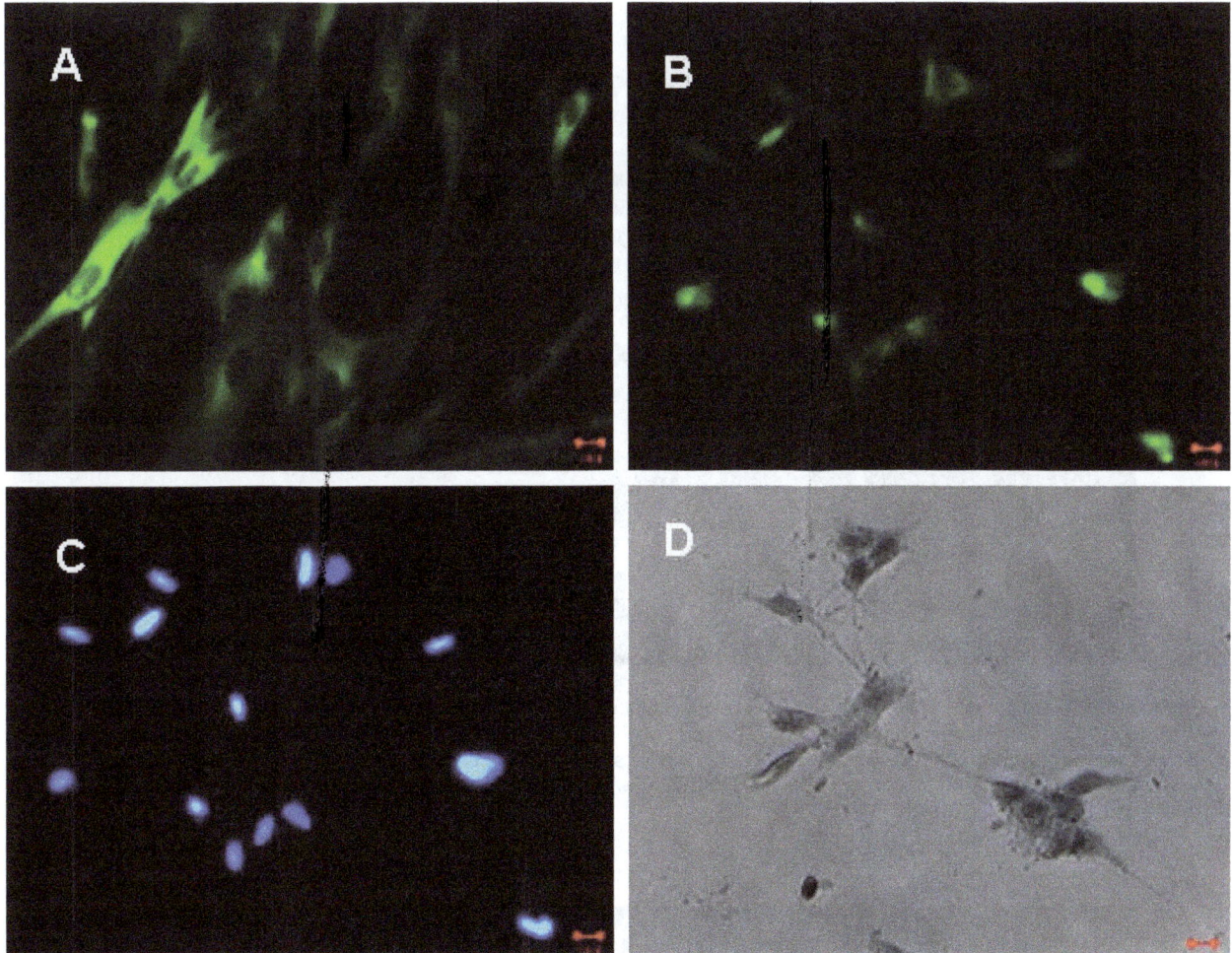

Figure 5. Neural like cells were formed after induction. A, β-tubulin III positive cells; B, NSE positive cells; C, The nuclei of induced cells were stained with Hoechst33342; D, Neural like cells were positive for aniline blue staining.

expressed specifcally in PGCs, spermatogonial stem cells and male germ cells (Clark et al., 2004). The expression of germ cell markers in treated UC-MSCs was consistent with previous observations in ESCs and bone marrow MSCs (Nayernia et al., 2006). They indicated spontaneous differentiation of part or all of the population of MSCs and ESCs into germ cells *in vitro* induced by RA. These results demonstrated that UC-MSCs are pluripotent similar as VESL pluripotent cells derived from bone marrow and with the capacity to differentiate into germ cells (Lu et al., 2008).

Conclusion

This study demonstrated the isolation and characterization of a non-hematopoietic MSC population from the human UC and provides evidences that UC-MSCs have the ability to give rise to non-hematopoietic cells with characteristics of osteoblast, adiopcytes, neural cells, cardiomyocytes and even germ cells. Therefore, these cells may serve as a novel alternative source of multipotent autologous stem cells for cell replacement therapies in various diseases and provide a model to study the mechanism of germ cells differentiation and new strategies for the therapy of infertility and sterility.

ACKNOWLEDGEMENTS

This work was supported by grants from the Program (30972097) from National Natural Science Foundation of China, Key Program of State Education Ministry (109148), Program for New Century Excellent Talents in University (NCET-09-0654), Key Program of State Education Ministry (109148), China Postdoctoral Science Foundation funded project (20080431253, 200801438), the Basic Technological and Research Programme of Jiangsu Province (BM2008146). The authors are grateful to Ms Long Wang's careful editing and revision. We appreciate the editor and reviewer's excellent work and suggestions.

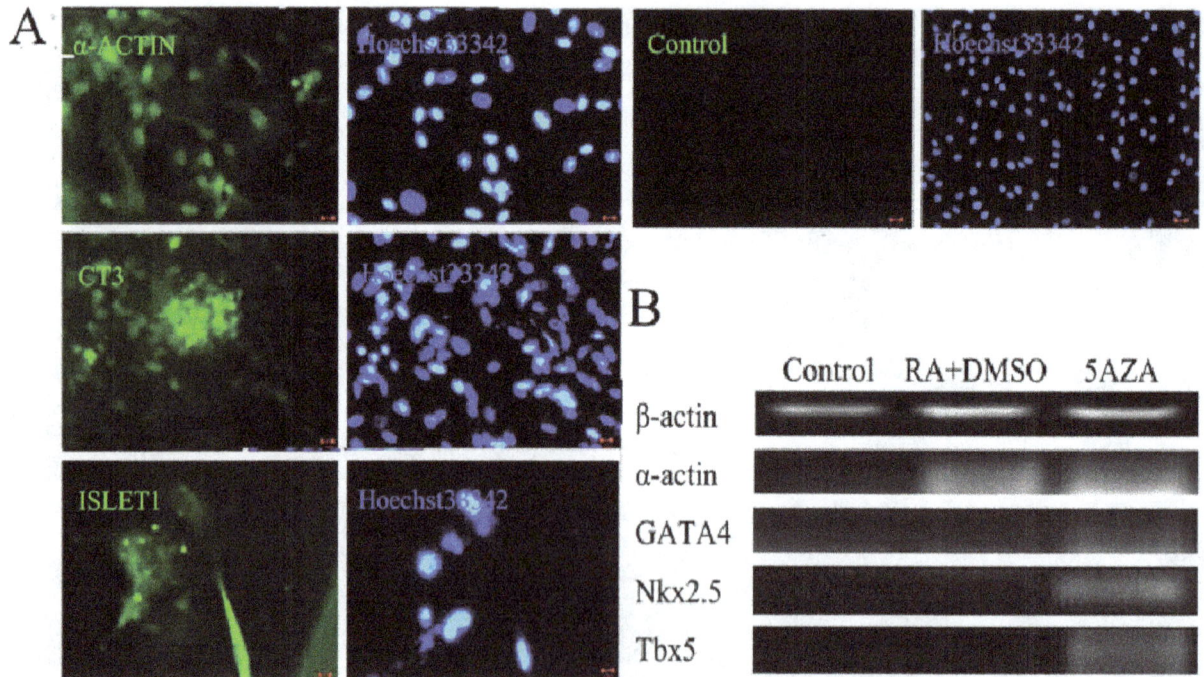

Figure 6. A, Cardiomyocyte-like cells were formed in induced cultures. Cardiac α-ACTIN, CT3 and ISLET1 positive cells were formed in the treated cultures (left lane); The nuclei of cells were stained with Hoechst33342 respectively in the right lane. Control (Negative control). B, Cardiac specific markers analysed by semi- RT-PCR showed that cardiac α-actin was up-regulated induced by RA in combination with DMSO or 5-AZA; GATA4, Nkx2 and Tbx5 were increased in 5-AZA induced cells (G).

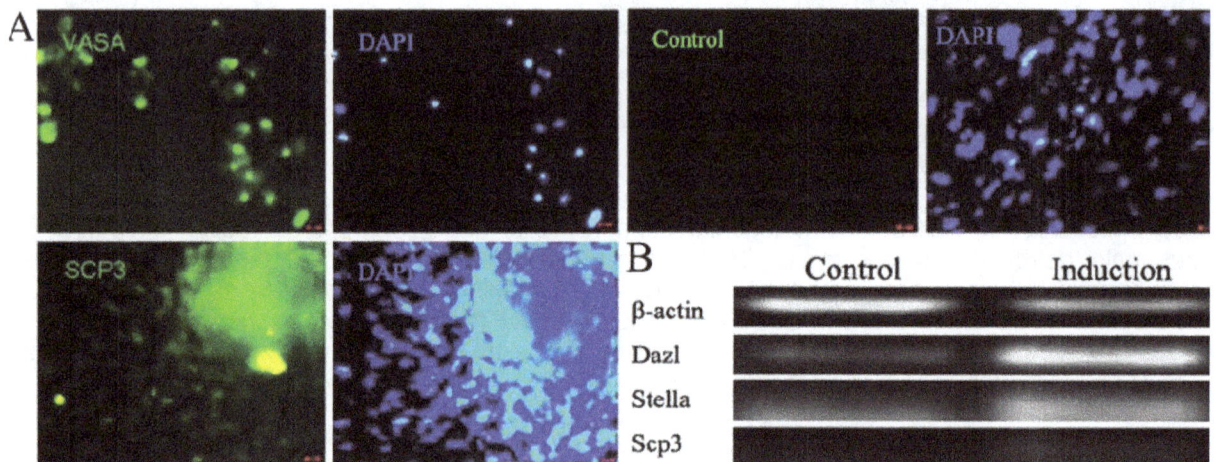

Figure 7. Immunofluorescence analysis of the germ cell differentiation after induction. The percentage of VASA and SCP3 positive cells were increased compared to control. The nuclei of cells were stained with Hoechst33342 in the right lane. Germ cell specific markers analysed by semi- RT-PCR showed that Dazl, Stella and Scp3 were up-regulated induced by RA(B).

REFERENCES

Baharvand H, Totonchi M, Taei A (2010). Human-induced pluripotent stem cells: derivation, propagation, and freezing in serum- and feeder layer-free culture conditions. Methods Mol. Biol., 584: 425-443.

Barachini S, Trombi L, Danti S (2009). Morpho-functional characterization of human mesenchymal stem cells from umbilical cord blood for potential uses in regenerative medicine. Stem Cells

Dev., 18: 293-305.

Choong P-F, Mok P-L, Cheong S-K (2007). Generating neuron-like cells from BM-derived mesenchymal stromal cells *in vitro*. Cytotherapy, 9: 170-183.

Clark AT, Bodnar MS, Fox M (2004). Spontaneous differentiation of germ cells from human embryonic stem cells *in vitro*. Hum. Mol. Genet., 13: 727-739.

Cutler AJ, Limbani V, Girdlestone J, Navarrete CV (2010). Umbilical cord-derived mesenchymal stromal cells modulate monocyte function to suppress T cell proliferation. J. Immunol., 185: 6617-6623.

Danner S, Kajahn J, Geismann C (2007). Derivation of oocyte-like cells from a clonal pancreatic stem cell line. Mol. Hum. Reprod., 13: 11-20.

Dyce PW, Wen L, Li J (2006). *In vitro* germline potential of stem cells derived from fetal porcine skin. Nat. Cell Biol., 8: 384-390.

Erices A, Conget P, Minguell J (2000). Mesenchymal progenitor cells in human umbilical cord blood. Br. J. Haematol., 109: 235-242.

Flynn A, Barry F, O'Brien T (2007). UC blood-derived mesenchymal stromal cells: an overview. Cytotherapy, 9: 717-726.

Goodwin HS, Bicknese AR, Chien S-N (2001). Multilineage differentiation activity by cells isolated from umbilical cord blood: expression of bone, fat, and neural markers. Biol. Blood Marrow Transplant., 7: 581-588.

Gonzalez R, Griparic L, Vargas V (2009). A putative mesenchymal stem cells population isolated from adult human testes. Biochem. Biophys. Res. Commun., 385: 570-575.

Hao L, Gao L, Chen XH (2010). Human umbilical cord blood-derived stromal cells prevent graft-versus-host disease in mice following haplo-identical stem cell transplantation. Cytotherapy. Jul 22. [Epub ahead of print]

Harris DT (2008). Collection, processing, and banking of umbilical cord blood stem cells for clinical use in transplantation and regenerative medicine. Labmed. 39: 173-178.

Hua J, Pan S, Yang C (2009). Derivation of male germ cell-like lineage from human bone marrow stem cells. Biomed. Reprod. Online.19: 99-105.

Hua J, Sidhu KS (2008). Recent advances in the derivation of germ cells from the embryonic stem cells. Stem Cells Dev., 17: 399-411.

Jiang Y, Jahagirdar BN, Reinhardt RL (2002). Pluripotency of mesenchymal stem cells derived from adult marrow. Nature, 418:41-49.

Johnson J, Bagley J, Skaznik-Wikiel M (2005). Oocyte generation in adult mammalian ovaries by putative germ cells in bone marrow and peripheral blood. Cell, 122: 303-315.

Lee OK, Kuo TK, Chen WM (2004). Isolation of multipotent mesenchymal stem cells from umbilical cord blood. Blood, 103: 1669-1675.

Liu Y, Mu R, Wang S, Long L, Liu X, Li R, Sun J, Guo J, Zhang X, Guo J, Yu P, Li C, Liu X, Huang Z, Wang D, Li H, Gu Z, Liu B, Li Z (2010). Therapeutic potential of human umbilical cord mesenchymal stem cells in the treatment of rheumatoid arthritis.Arthritis Res. Ther., 12: R210.

Kang XQ, Zang WJ, Bao LJ (2006). Differentiating characterization of human umbilical cord blood-derived mesenchymal stem cells *in vitro*.Cell Biol. Int., 30: 569-75.

Kopen GC, Prockop DJ, Phinney DG (1999). Marrow stromal cells migrate throughout forebrain and cerebellum, and they differentiate into astrocytes after injection into neonatal mouse brains. Proc. Natl. Acad. Sci. U S A., 96: 10711-10716.

Kumar S, Singh NP (2006). Stem cells: A new paradigm. Indian J. Hum. Genet., 12: 4-10.

Lacham-kaplan O, Chy H, Trounson A (2006). Testicular cell conditioned medium supports differentiation of embryonic stemcells into ovarian structures containing oocytes. Stem Cells, 24: 266–273.

Lu R, Miao D (2008). Very small embryonic like(VSEL) stem cells. J. Nanjing Med. University. 22: 265-268.

Mareschi K, Biasin E, Piacibello W (2001). Isolation of human mesenchymal stem cells: bone marrow versus umbilical cord blood. Haematologica, 86: 1099-1100.

Martin-Rendon E, Sweeney D, Lu F (2008). 5-Azacytidine-treated human mesenchymal stem/progenitor cells derived from umbilical cord, cord blood and bone marrow do not generate cardiomyocytes *in vitro* at high frequencies. Vox Sang, 95(2): 137-148.

Nayernia K, Nolte J, Michelmann HW (2006). *In vitro* differentiated embryonic stem cells give rise to male gametes that can generate offspring mice. Dev. Cell., 11: 125–132.

Orlandi A, Pagani F, Avitabile D (2008). Functional properties of cells obtained from human cord blood CD34 stem cells and mouse cardiac myocytes in coculture. Am. J. Physiol. Heart Circ. Physiol., e294: H1541–H1549.

Orlandi A, Hao H, Ferlosio A (2009). Alpha actin isoforms expression in human and rat adult cardiac conduction system. Differentiation, 77: 360-368.

Panepucci RA, Siufi JL, Silva WA Jr, Proto-Siquiera R, Neder L, Orellana M, Rocha V, Covas DT, Zago MA (2004). Comparison of gene expression of umbilical cord vein and bone marrow-derived mesenchymal stem cells. Stem Cells, 22(7): 1263-78.

Piedrahita JA, Moore K, Oetama BJ (1998). Generation of transgenic porcine chimeras using primordial germ cell-derived colonies. Biol. Reprod., 58: 3121–3129.

Pittenger MF, Mackay AM, Beck SC, Jaiswal RK, Douglas R, Mosca JD, Moorman MA, Simonetti DW, Craig S, Marshak DR (1999). Multilineage potential of adult human mesenchymal stem cells. Sci., 284: 143-147.

Rangappa S, Entwistle JWC, Wechsler AS (2003). Cardiomyocytemediated contact programs human mesenchymal stem cells to express cardiogenic phenotype. J. Thorac. Cardiovasc. Surg., 126: 124-132

Rebelatto CK, Aguiar AM, Moretao M (2008). Dissimilar differentiation of mesenchymal stem cells from bone marrow, umbilical cord blood, and adipose tissue. Exp. Biol. Med., 233: 901–913.

Sanchez-Ramos JR, Song S, Kamath SG, Zigova T, Willing A, Cardozo-Pelaez F, Stedeford T, Chopp M, Sanberg PR (2001). Expression of neural markers in human umbilical cord blood. Exp. Neurol., 171: 109-115.

Secco M, Moreira YB, Zucconi E (2009). Gene expression profile of mesenchymal stem cells from paired umbilical cord units: cord is different from blood. Stem Cell Rev. Rep. DOI 10.1007/s12015-009-9098-5: 387-401.

Secco M, Zucconi E, Vieira NM (2008). Multipotent stem cells from umbilical cord: cord is richer than blood. Stem Cells, 26: 146-150.

Sottile V, Halleux C, Bassilana F (2002). Stem cell characteristics of human trabecular bone-derived cells. Bone. 30:699-704.

Toyooka Y, Tsunekawa N, Akasu R (2003). Embryonic stem cells can form germ cells *in vitro*. Proc. Natl. Acad. Sci. USA., 100: 11457-11462.

Xu Y, Meng H, Li C, Hao M, Wang Y, Yu Z, Li Q, Han J, Zhai Q, Qiu L (2010). Umbilical cord-derived mesenchymal stem cells isolated by a novel explantation technique can differentiate into functional endothelial cells and promote revascularization.Stem Cells Dev., 19: 1511-1522.

Yoo BY, Shin YH, Yoon HH, Seo YK, Song KY, Park JK (2010). Lication Of Mesenchymal Stem Cells Derived From Bone Marrow And Umbilical Cord In Human Hair Multiplication. J. Dermatol. Sci., 60: 74-83.

Yu M, Xiao Z, Shen L (2004). Mid-trimester fetal blood-derived adherent cells share characteristics similar to mesenchymal stem cells but fullterm umbilical cord blood does not. Br. J. Haematol. 124:666-675.

Zhang YN, Lie PC, Wei X (2009). Differentiation of mesenchymal stromal cells derived from umbilical cord Wharton's jelly into hepatocyte-like cells. 11: 548-558.

Zhou B, von Gise A, Ma Q (2008). Nkx2-5- and Isl1-expressing cardiac progenitors contribute to proepicardium. Biochem. Biophys. Res. Commun., 375: 450-453.

26

Biochemical markers in semen and their correlation with fertility hormones and semen quality among Sudanese infertile patients

Abdelmula M. Abdella[1]*, Al-Fadhil E. Omer[1] and Badruldeen H. Al-Aabed[2]

[1]Faculty of Medical Laboratory Science, Al-Neelain University, Khartoum, Sudan.
[2]Faculty of Medicine, University of Science and Technology, Khartoum, Sudan.

Several biochemicals in semen are secreted by the accessory glands in the reproductive tract. These biochemicals can be used as diagnostic predictors for the disorders in the male reproductive system. To assess the level of biochemical markers in semen, their relation to fertility hormones and spermogram among Sudanese infertile patients were studied. The biochemical markers studied were fructose, citric acid, zinc and neutral α-glucosidase. Their levels in semen were estimated using analytical photometry, spectrophotometry and atomic absorption spectrometry. Estimation covered 500 infertile males (150 azoospermic, 150 oligospermic, 100 asthenozoospermic and 100 with abnormal sperm morphology), as well as 100 normospermic control males. Fertility hormones were assayed in patients and controls by ELISA. Seminal neutral α-glucosidase and citric acid levels were found significantly reduced in azoospermic and oligospermic patients, while zinc levels was reduced in all infertile patients ($p < 0.05$). Semen fructose level was found within the normal range. Significant negative correlation was noticed between neutral α-glucosidase and both follicle stimulating and luteinising hormones (in azoospermic patients), and prolactin hormone in oligospermic patients ($r < 0.05$). 2.7% of azoospermic patients had Sertoli cell syndrome only. 13% of the infertile patients had varicocele, and this was associated with a significant increase in FSH and LH and a decrease in seminal neutral α-glucosidase, citric acid and zinc ($p < 0.05$). 9.6% of the patients studied had dysfunctional sexual problems and was associated with a significant increase in prolactin. On the other hand, 7.2% of these patients were smokers and this was associated with a significant reduction in semen volume and levels of neutral α-glucosidase and zinc ($p < 0.05$). There was conflicting association between biochemical markers in semen with both reproductive hormones level and semen quality in the infertile patients, but neutral α-glucosidase level was the only biochemical markers in semen that correlated well with both gonadotropins hormones (negatively/inversely) and the semen quality.

Key words: Biochemical markers, reproductive hormones, male infertility, seminal plasma, α-glucosidase.

INTRODUCTION

Male fertility depends on the proper function of a complex system of organs and hormones The testis produce the spermatozoa, which begin partially embedded in nurturing amoebae-like cells known as Sertoli cells, which are located in the lower parts of the seminiferous tubules. As they mature and move along, they are stored in the upper part of the tubules (Vaclav et al., 1993). When the sperm has completed the development of its head and tail, they are released from the cell into the epididymis. The epididymis is a convoluted canal in which final steps of sperm maturation and development take place. The spermatozoa are stored in the epididymis for a long time, and pass through it during ejaculation. The epididymis sustains maturation process that results in the acquisition of progressive motility and fertilizing ability by spermatozoa. The epididymis synthesizes certain compounds that are secreted in the semen. These include protein,

*Corresponding author. E-mail: mula200099@yahoo.co.uk.

carnitine, lipid, glycerylphosphorylcholine, neutral α-glucosidase, carbohydrates, steroids and other small molecules (Vaclav et al., 1993).

Semen or ejaculate is the fluid discharged from the penis during the time of orgasm. Like blood, semen consists of two compartments, cellular compartment (spermatozoa), and noncellular compartment (seminal plasma). Thus, it contains the sperm, which sometimes results in pregnancy following vaginal sex with a female. Semen is a whitish, milky fluid, slightly viscous, containing water and small amounts of salt, protein, fructose, citric acid and other substances (Setchell and Waites, 1970). Spermatozoa make up only about 5 to 10% of the volume of semen. The bulk of the seminal plasma, the fluid portion of semen, is contributed by the male accessory organs of reproduction. 40 to 80% of semen is produced by the seminal vesicles. The secretions include fructose for sperm nutrition, prosta-glandins, coagulating substances and bicarbonate to buffer the acidic vaginal vault. Most of the remainder is generated by the prostate (10 to 30%). Prostatic secretions in humans contain enzymes and proteases to liquefy the seminal coagulum, high levels of citric acid, acid phosphatase, phospholipids and spermine. Small amount of seminal fluid comes from the bulbourethral (Cowper) glands (Orth, 1982). The seminal plasma contains a very complex range of organic and inorganic constituents. Fructose is the main energy source of sperm cells. A reduction of fructose level below normal value is often due to an obstruction of the ejaculatory ducts or absence of the seminal vesicles, especially when associated with a low ejaculate volume and a thin, watery consistency. Some of the other components of semen serve to increase the mobility of sperm cells; sperm function best in a slightly alkaline environment (Setchell and Waites, 1970).

Although, the seminal plasma may not contain factor that are absolutely essential for fertilization, the secretion may nevertheless optimize condition for sperm motility, survival and transport in both the semen and the female reproductive tract (Vaclav et al., 1993). The increased or decreased level of these factors visualizes the abnorma-lities that occurred along the male reproductive tract. Fructose secretion is under androgenic regulation, but many factors such as frequency of ejaculation, blood glucose levels and nutritional status can also affect the seminal plasma fructose concentration (Vaclav et al., 1993). Zinc, is accumulated and released in the prostatic fluid; the anatomical localization of Zn^{2+} in the human prostate, is in the epithelium of the glands. Dihydrotes-tosterone may initiate the synthesis of zinc binding protein required for the accumulation of zinc within the prostatic cells and lumen. As the epithelial cells become saturated with zinc, this cation may switch off testosterone reduction by binding to thiol group at or near the cofactor binding site of the 5-reductase. Following ejaculation, 5-reductase activity may be restored, allowing the renewed accumulation of zinc within the prostate (Carol, 1998).

In the early stages, the sperm cannot swim in a forward direction and can only vibrate its tail weakly. By the time the sperm reaches the end of the epididymis, however, it is matured and looks like a microscopic squirming tadpole. The ability of a sperm to move forward rapidly and straight is probably the most significant determinant of male fertility (MacSween and Keith 1995).

In the penis, the sperm first pass into one of two rigid and wire-like muscular channels, called the vas deferentia. Muscle contractions in the vas deferens from sexual activity propel the sperm along past the seminal vesicles. Each vas deferens then joins to form the ejaculatory duct. This duct, which now contains the semen containing the sperms, passes down through the urethra, but during orgasm, the prostate closes off the bladder so urine cannot enter the urethra. The semen is forced through the urethra during ejaculation, the final stage of orgasm when the sperm is literally shot out of the penis (Vaclav et al., 1993). It is important to emphasize the relationship between the germ cells and the Sertoli cells, with the reproductive hormones that are essential for the successful development of normal spermatogenesis, FSH (follicle-stimulating hormone) and testosterone, act through the Sertoli cells since the receptors for these hormones are located on these cells and are not located on germ cells (Cyril et al., 1994; Plant and Marshall, 2001; Walker and Cheng, 2005).

According to our knowledge, this study was the first work in Sudan in this field. It aimed to correlate the concentrations of the biochemical markers (semen fructose, citric acid, zinc, and neutral α-glucosidase) to fertile hormones and spermogram in infertile Sudanese males, and highlight their importance in male infertility.

MATERIALS AND METHODS

Study subjects

500 infertile males were enrolled for this study in Khartoum State (Sudan). It was carried out between July 2005 and October 2007. The subjects were males attending the fertility clinics that complained of inability to achieve pregnancy for at least one year after marriage (with no apparent chronic or acute disease), and whom their wives had shown no diagnosed causes of infertility (hormone test, laparoscopy).

The infertile patients studied were divided into four groups according to World Health Organization (WHO) standards for semen quality: azoospermic patients (n = 150), oligospermic patients (n = 150), asthenozoospermic patients (n = 100) and patients with abnormal sperm morphology (n = 100). The control group included 100 fertile males who had fathered a child during the last two years, and with normal spermogram. Consent for the study was obtained from all enrolled subjects.

Demographical data were collected via a structural interview that was conducted during the first visit. A basic medical, surgical, reproductive and family history was recorded. A complete physical and genital examination was carried out. All subjects submitted a semen specimen and a blood sample.

Hormones analysis

Venous blood was collected in a plain tube for the estimation of the fertility hormones: follicle stimulating hormone (FSH), luteinising hormone (LH), prolactin and total testerone. The separated serum was assayed by the enzyme-immunolinked assay (ELISA) (Tietz, 1988; Wisdom 1976; Shome and Parlow, 1974; Baker, 1974; Chopra et al., 1971). Kits were provided from DIMA GmbH Robert-Bosch-Breite 23, 37079 Goettingen Germany.

Semen analysis

Semen samples were obtained by masturbation and collected in sterile polystyrene containers after 3-5 days of abstinence. Samples were analyzed according to the WHO criteria (WHO 1999). Seminal plasma was separated from the spermatozoa by centrifugation and divided into two containers: one metal-free polypropylene container for the estimation of zinc by atomic absorption spectrometry (AAS) with the Zeeman background correction method, and using an analytical quality control for accuracy (Michael et al., 1996) and the second container was for the estimation of: Neutral α- glucosidase (NAG) according to Cooper et al. (1990) and Guerin et al. (1986). Kits were produced by FertiPro N. V., Industriepark Noord 32, 8730 Beemem and Belgium. Citric acid was estimated according to Tietz et al. (1999) and Young et al. (1975) and fructose according to WHO Manual (1999). Analytical photometers and Jenway 6305 spectrophotometer were used for the estimation.

Quality control of assays

Samples representing the normal and pathological level for all the measured analysts in serum/or semen were used for quality control. Results ± 2SD of the target values were considered acceptable. The batches with all controls been within permissible were accepted.

Statistical analysis

Data were expressed in mean, standard error, and standard deviation. They were analyzed by student t-test, one-way analysis of variance (ANOVA) and a coefficient of correlation (r) with a significance fixed at p = 0.05.

Ethical consideration

Consent for this study was taken from the infertile patients and the healthy volunteers.

RESULTS

500 Sudanese infertile males were studied. Duration of infertility was 2-15 years in azoospermic patients, 4 to 18 years in oligospermic patients, 3-14 years in asthenozoo-spermic patients, and 2 to 11 years in patients with abnormal sperm morphology. The control group constituted 100 fertile males with normal spermo-gram.

Clinical examination and laboratory investigations revealed that 2.7% of azoospermic patients were suffering from Sertoli-cell syndrome, 13% of the infertile patients had varicocele and 9.6% of the infertile patients

had sexual dysfunction.

Seminal NAG (neutral-acetyl-β-D-glucosaminidase) and citric acid levels were significantly reduced in azoospermic and oligospermic patients, whereas zinc was reduced in all infertile subjects. Seminal fructose was slightly increased in oligospermic patients and subjects with abnormal sperm morphology. However, it was decreased in azoospermic and asthenozoospermic patients as shown in Tables 1 and 2.

Tables 3a, 3b, 4a and 4b show the correlation of seminal biochemical markers with fertility hormones in the infertile patients studied. In azoospermic patients, a slight correlation was evident between NAG and testosterone and between fructose and citric acid with prolactin. Furthermore, a significant inverse negative correlation was recorded between NAG and FSH and LH in this group of patients. In oligospermic patients correlation was clear between citric acid with LH, prolactin and NAG. In addition, a significant negative correlation was also noticed between seminal zinc, LH and prolactin in such patients. On the other hand, fructose correlated well with prolactin in asthenozoospermic patients.

DISCUSSION

Jeyendran et al. (1995) measured the neutral α-glucosidase (NAG) activity in seminal plasma of infertile patients. They found an inverse correlation of NAD activity with the spermatogram abnormalities. Semen NAG level in this study was significantly reduced among infertile patients which is in agreement with the findings of other workers (Jeyendran et al., 1995). Since NAG is secreted by the epididymis that is under the control of the testosterone hormone, the high significant reduction of NAG level recorded in azoospermic patients may be due to obstruction of the first part of the ejaculatory duct next to the epididymis (Vaclav, 1993).

Seminal fructose level was found to be increased in oligospermic and abnormal sperm morphology patients. This may have resulted from the reduced sperm count and activity leading to low consumption of the synthesized fructose. On the other hand, seminal fructose level was decreased in azoospermic and asthenozoospermic patients. This is explained by partial or complete obstruction of the seminal ducts or ageing of the accessory glands that secrete fructose (Buckett et al., 2002).

Seminal citric acid level was slightly reduced in asthenozoospermic patients and significantly reduced in azoospermic and oligospermic patients. Since seminal citric acid is secreted mainly from the prostate gland, any partial or complete obstruction of the ejaculatory ducts may reduce its level in semen. Abnormal testosterone secretion and accessory glands infection may also lead to reduction of seminal citric acid level.

Generally, zinc and citric acid concentrations as well as NAG activity were found to decrease significantly with

Table 1. Descriptive analysis of seminal biochemical markers in infertile patients and control group.

Seminal biochemical marker	Azoospermic patients (n = 150)	Oligospermic patients (n = 150)	Asthenozoospermic patients (n = 100)	Abnormal sperm morphology patients (n = 100)	Control group (n = 100)
NAG (mIU/ml)	11.2 ± 0.3	18.2 ± 0.5	23.1 ± 0.48	22.8 ± 0.5	24.6 ± 0.9
Fructose (mmol/l)	12.9 ± 0.73	13.2 ± 0.69	12.1 ± 0.81	13.4 ± 0.87	13.0 ± 0.39
Citric acid (mmol/l)	1.9 ± 0.09	2.2 ± 0.15	2.6 ± 0.19	2.8 ± 0.20	3.3 ± 0.16
Zn (mg/dl)	11.3 ± 0.27	11.5 ± 0.25	12.8 ± 0.4	12.6 ± 0.45	19.3 ± 0.6

n = the number of subjects. Data are presented as M ± SE.

Table 2. Dunnetts T3 multiple analysis of seminal biochemical markers in infertile patients and control group.

Seminal biochemical marker	Statistic	Azoospermic patient *versus* control	Oligospermic patient *versus* control	Asthenozoospermic patient *versus* control	Abnormal sperm morphology patient *versus* control
NAG	Mean difference	13.4	6.5	1.5	1.9
	Significance	0.000[‡]	0.000[‡]	0.78	0.51
Fructose	Mean difference	0.07	- 0.2	0.83	- 0.46
	Significance	1.0	1.0	0.98	1.0
Citric acid	Mean difference	1.4	1.1	0.7	0.5
	Significance	0.00[‡]	0.00[‡]	0.07	0.5
Zinc	Mean difference	8.0	7.6	6.7	6.6
	Significance	0.00[‡]	0.00[‡]	0.00[‡]	0.00[‡]

[‡]Significant at the level $p < 0.001$.

Table 3a. Statistical correlation of seminal biochemical markers with fertility hormones in azoospermic and oligospermic patients (Azoospermic patients).

Seminal biochemical marker	Statistic	Testosterone	FSH	Prolactin	LH	Zinc
NAG	Person correlation	0.14	- 0.25	0.09	- 0.20	0.08
	Significance (2-tails)	0.1	0.02[‡]	0.3	0.01[‡]	0.32
Fructose	Person correlation	-0.01	0.02	0.15	- 0.02	-0.11
	Significance (2-tails)	0.9	0.8	0.08	0.8	0.19
Citric acid	Person correlation	- 0.03	- 0.1	- 0.11	0.03	0.01
	Significance (2-tails)	0.7	0.2	0.19	0.76	0.95
Zinc	Person correlation	0.07	0.05	- 0.1	0.04	-
	Significance (2- tails)	0.4	0.5	0.2	0.7	-

FSH = Follicle stimulating hormone; LH = luteinizing hormone; [‡]correlation was significant at $p < 0.05$.

increase in seminal abnormalities. This is quite obvious in infertile azoospermic patients. NAG level is considered the most important marker because it affects the maturation and the acquisition of spermatozoa motility

(Vaclav, 1993).

Seminal NAG, fructose and citric acid in the test groups recorded relation with seminal zinc. There was weak positive relation between NAG, Zn and fructose in infertile

Table 3b. Statistical correlation of seminal biochemical markers with fertility hormones in azoospermic and oligospermic patients (Azoospermic patients) (Oligospermic patients).

Seminal biochemical marker	Statistics	Testosterone	FSH	Prolactin	LH	Zinc
NAG	Person correlation	0.11	0.05	0.16	0.05	-0.15
	Significant (2-tails)	0.2	0.6	0.05*	0.5	0.07
Fructose	Person correlation	0.03	0.03	0.08	0.07	0.07
	Significant (2-tails)	0.7	0.7	0.3	0.4	0.4
Citric acid	Person correlation	0.07	- 0.11	0.07	- 0.25	-0.12
	Significant (2-tails)	0.3	0.17	0.4	0.02‡	0.13
Zinc	Person correlation	- 0.03	- 0.08	0.2	0.20	- -
	Significant (2-tails)	0.7	0.3	0.02‡	0.03‡	

*Correlation was significant at $p < 0.001$; FSH = follicle stimulating hormone; LH = luteinizing hormone; ‡correlation was significant at $p < 0.05$.

Table 4a. Statistical correlation of seminal biochemical markers with fertility hormones in asthenozoospermic patients and abnormal sperm morphology patients (Asthenozoospermic patients).

Seminal biochemical marker	Statistics	Testosterone	FSH	Prolactin	LH	Zinc
NAG	Person correlation	- 0.04	0.06	0.08	- 0.09	-0.07
	Significant (2-tails)	0.7	0.5	0.4	0.4	0.5
Fructose	Person correlation	0.1	0.03	0.17	0.09	-0.08
	Significant (2-tails)	0.3	1.0	0.08	0.4	0.4
Citric acid	Person correlation	0.01	0.01	- 0.05	0.1	-0.16
	Significant (2-tails)	1.0	1.0	0.6	0.3	0.12
Zinc	Person correlation	0.14	0.15	0.07	- 0.13	-
	Significant (2- tails)	0.5	0.6	0.5	0.2	-

Table 4b. Statistical correlation of seminal biochemical markers with fertility hormones in asthenozoospermic patients and abnormal sperm morphology patients (Asthenozoospermic patients; abnormal sperm morphology patients).

Seminal biochemical markers	Statistics	Testosterone	FSH	Prolactin	LH	Zinc
NAG	Person correlation	0.01	0.01	- 0.1	- 0.08	0.12
	Significance (2-tails)	0.9	1.0	0.3	0.4	0.3
Fructose	Person correlation	0.16	0.13	0.16	0.1	0.1
	Significance (2-tails)	0.12	0.2	0.12	0.2	0.3
Citric acid	Person correlation	- 0.07	0.16	0.08	0.06	-0.14
	Significance (2-tails)	0.5	0.12	0.4	0.5	0.17
Zinc	Person correlation	0.14	0.02 0.8	0.07	0.1	-
	Significance (2-tails)	0.2		0.5	0.3	-

patients with azoospermia since these parameters are affected by testosterone level, and accessory glands secreting activity. This appeared clearly in azoospermic patients. The secretory activity of the accessory glands is affected by the level of seminal zinc. This phenomenon was demonstrated by the strong relation between the zinc level and seminal volume and liquefaction in azoospermic and oligospermic patients.

Testosterone is essential for all steps of spermatogenesis hence any reduction in testosterone concentration may affect sperm quality. Hunt et al. (1992) deduced the important role of zinc in testosterone production and its need in the spermatogenesis process. They also observed the positive relation between testosterone and zinc in healthy male volunteers fed on zinc-restricted diet. As a consequence of zinc deficiency, serum testosterone concentration and seminal volume per ejaculate were reduced in this study.

The fertility hormones levels in infertile patients reflected a relationship with seminal zinc level. The positive correlation of prolactin hormone with seminal zinc indicates the stimulating effect of prolactin on the prostate gland that secretes zinc. The inverse correlation of LH with seminal zinc in oligospermic patients suggests the feed-back stimulation of zinc deficiency on LH secretion by the pituitary gland.

Conclusion

Seminal NAG is the best epididymal marker in semen that showed significant correlation with both gonado-tropins hormones (negatively/inversely) and the semen quality, which can aid in the differential diagnosis of obstructive and nonobstructive azoospermia in men with infertility and replace the semen aspiration technique.

ACKNOWLEDGEMENTS

We gratefully acknowledge the assistance of the technologists in the Sudanese Saudi Laboratory, the Central Medical Laboratory Services, and the Military Hospital Laboratory (Khartoum) for their valuable cooperation. Our thanks also go to Al-Neelain University for its financial support.

REFERENCES

Baker SB (1974). Determination of protein bound Iodine. J. Biol. Chem. pp. 173-175.

Buckett WM, Lewis-Jones DI (2002). Fructose concentrations in seminal plasma from men with non-obstructive azoospermia. Arch. Androl. 48: 23–27.

Carol MP (1998). Pathophysiology 5th ed, UK .London ,Lippincott. pp. 1149-1155.

Chopra IJ, Solomon DH, Ho RS (1971). A radioimmunoassay of thyroxine. J. Clin. Endocrinol. Metab., 33(5): 865-868.

Cooper TG, Yeung CH, Nasnan C, Jcckenhovel F (1990). Improvement in the assessment of human apididymel function by the use of inhibitors in the assay of α-glucosidase in seminal plasma. Int. J. Androl., 13: 297-305.

Cyril AK, Eric N, Norman J, Samson W (1994). Applied physiology, 13th ed, USA, New York, Oxford university press. pp. 497-566.

Guerin JF, Ali HB, Rollet J, Souchier C, Czyba JC (1986). Alpha-glucosidase as a specific epididymal enzyme marker. Its validity for the etiologic diagnosis of azoospermia. J. Androl., 7: 156-162.

Hunt CD, Johrison PE, Herbel J, Mullen LK (1992). Effect of dietery zinc deplation on seminal volume and zinc loss,serum testosterone concentration and sperm morphology in young men. Am. J. Clin. Nutr., 56: 148-157.

Jeyendran RS, Milad M, Kret B (1995). New discriminatory level for glucosidase activity to diagnose epididymal obstruction or dysfunction. Arch. Androl., 35(1): 29-33.

Michael LB, Duben-Engelkirk JL, Edward PF (1996). Clinical Chemistry Principle, Procedure, Correlation, third edition, Philadelphia. New York. USA, Lippincott. pp. 279-288.

Orth JM (1982). Proliferation of Sertoli cells in fetal and postnatal rats: a quantitative autoradiographic study. Anat. Rec., 203: 485-492.

Plant TM, Marshall GR (2001). The functional significance of follicle-stimulating hormone in spermatogenesis and the control of its secretion in male primates. Endocr. Rev., 22: 764-786

MacSween RNM, Keith W (1995). Muirs Text book of pathology, 3th ed, USA, Oxford. pp. 1022-1057.

Setchell BP, Waites GMH (1970). Changes in the permeability of testicular capillaries and of "blood-testis barrier" after the injection of cadmium chloride in rat. J. Endocrinol., 47: 81-86.

Shome B, Parlow A (1974). Human follicle stimulating hormone (hFSH): first proposal for the amino acid sequence of the alpha-subunit (hFSHa) and first demonstration of its identity with the alpha-subunit of human luteinizing hormone (hLHa). J. Clin. Endocrinol. Metab., 39(1): 199-202. Sited by the kit from DIMA GmbH Robert- Bosch-Breite 23, 37079 Goettingen Germany.

Tietz CNW (1988). Textbook of clinical chemistry, W.B. Saunders Ca, Philadelphla.

Tietz CNW (1999). Textbook of clinical chemistry, W.B. Saunders Ca, Philadelphla.

Vaclav I, Bruno L (1993). Infertility: Male and Female, 2nd ed, UK. London. Charchill Livingstone, p. 739.

Walker WH, Cheng J (2005). FSH and testosterone signaling in Sertoli cells. Reproduction, 130: 15-28.

Wisdom GB (1976). Enzyme-immunoassay. Clin. Chem., 22: 1243-1255. Sited by the kit from DIMA GmbH Robert- Bosch-Breite 23, 37079 Goettingen Germany.

World Health Organization (WHO) (1999). Laboratory Manual for the Examination of Human Semen and Sperm-Cervical Mucus Interaction 4th ed. Cambridge, United Kingdom: Cambridge University Press.

Young DS, Pestaner LC, Gibberman V. (1975). Effects of drugs on clinical laboratory tcctc. Clin Chcm; 21: 5GD

27

Peripheral blood and C-reactive protein levels (CRP) in chronic periodontitis

Balwant Rai[1], Jasdeep Kaur [2], Simmi Kharb[2], Rajnish Jain[2], S. C. Anand[2] and Jaipaul Singh[2]

[1]Prabhu Dayal Memorial (PDM) Dental College and Research Institute - Haryana - Bahadurgarh-India.
[2]BJS dental collage, India
[3]Post Graduate Institute of Medical Science, Rohtak, Haryana, India.
[4]UClan, UK

Evidence for a potential link between periodontal disease and coronary heart disease (CHD) has accumulated in recent years. C-reactive protein is potential marker of cardiovascular risk and associated with periodontal disease. CRP levels were analyzed in 26 periodontitis patients and 23 healthy controls along with hemoglobin, red blood cell count (RBC), leukocyte cell count (total and differential WBC), platelet count. Total WBC, neutrophil and platelet counts and CRP levels were raised significantly in periodontitis patients as compared to controls (p<0.01) while RBC count and hemoglobin were significantly lowered in periodontitis as compared to controls (p<0.01). These findings suggest an important role of CRP in development of periodontitis. Routinely screening for CRP in periodontitis patients might be an important tool to prevent heart disease.

Key words: C-reactive protein, WBC, RBC, thrombocyte, periodontitis, coronary heart disease.

INTRODUCTION

Periodontal disease, a common chronic oral inflammatory disease, is characterized by destruction of soft tissue and bone. Atherosclorosis starts early in life, however, since disease progression is usually slow, clinical symptoms or hospitalization are rare before 40 years of age. Epidemiological associations between periodontitis and cardiovascular disease have been reported (Morrison et al., 1999; Slade et al., 2003). Periodontitis and atherosclerosis have complex aetiologies, genetic and gender predispositions and might share pathogenic mechanisms as well as common risk factors. It is becoming increasingly clear that infections and chronic inflammatory conditions such as periodontitis may influence the atherosclerotic process. They may increase haemostatic variable which pronate haemostatic plugs and thrombi and rheological variables, both of which play important roles in pathogennesis of vascular disease (Fredriksson et al., 1999; Herzberg et al., 1998). The crucial casual relation might be established by prospective treatment studies, which elucidate the connection between treatment of poor dental

health and systemic inflammatory markers (Iwamoto et al., 2003; D'Aiuto et al., 2004). Several short-term intervention studies have been reported that treatment of periodontitis reduces the serum concentrations of inflammatory markers, such as C-reactive protein (CRP), TNF-a, IL-6, which are thought to be initiating factor cardiovascular disease. The aim of this study was to determine whether patients with severe periodontitis have higher plasma concentrations of established risk markers of atherosclerosis such as CRP and peripheral blood markers.

MATERIALS AND METHODS

Twenty six subjects ranging 30 - 65 years (Table 1) who had at least 6 mm loss of clinical attachment and had been referred to the Department of Periodontology, Government Dental College, PGIMS, Rohtak were selected for the study due to their severe periodontitis and were undergoing treatment for this condition. On clinical examination there is a horizontal loss of supporting tissue by more than 1/3 of root length with bleeding on probing, furcation (the area between the multiroot tooth in relation to bone height) involvements of the multi-rooted teeth and/or angular bony defects. Probing depth and attachment level measurements were performed at six sites on tooth. Patients were excluded from the study if they had alcoholic or chronic smoker. The healthy, non periodontal con-

*Corresponding author. E-mail: drbalwantraissct@rediffmail.com.

Table 1. Clinical loss of attachment and numbers of sites involved in periodontitis assess of periodontal disease of the control and periodontal patients.

Age (in years)	No. of sites				Clinical loss of attachment (in mm)			
	Male		Female		Male		Female	
	P (n=14)	N (n=11)	P (n=12)	N (n=13)	P (n=14)	N (n=11)	P (n=12)	N (n=13)
30-47.5	9.0±1.2 (n=7)	1.3±1.2* (n=5)	9.2±1.3 (n=7)	1.4±1.2* (n=5)	8.2±1.4 (n=7)	1.4±1.3* (n=5)	8.7±1.3 (n=6)	1.7±1.4* (n=7)
47.6-65	9.2±1.3 (n=7)	1.6±1.2* (n=6)	10.4±1.4 (n=5)	1.8±1.3* (n=8)	8.4±1.2 (n=7)	1.7±1.3* (n=6)	9.3±1.2 (n=6)	2.1±1.7* (n=6)

Value = Mean ± S.D.
P = periodontal patients
N = normal control
*Significant difference (p<0.05) between normal and periodontal patients.

Table 2. Total number of peripheral blood; leukocytes, neutrophil counts, total RBC counts, Hb level and C-reactive protein plasma levels in periodontitis patients and healthy controls.

Total No. of patients	Sex	Total WBC counts (x10^9/L)	Total thrombocytes counts (x10^9/L)	Neutrophil counts (x10^9/L)	Total RBC counts (x10^12/L)	Hb level (g/dl)	C-reactive plasma level (mg/L)
Periodontitis group (n=26)	M (n=14)	8.3 ± 1.2a 8.5 – 8.2	2.5 ± 1.7a 2.7 – 2.3	6.5 ± 1.3a 6.8 – 6.4	4.4 ± 1.3a 4.7 – 4.2	10 ± 1.7a 12 – 9	4.28 ± 1.72a 4.93 – 4.12
	F (n=12)	8.1 ± 1.4b 8.3 – 7.9	2.4 ± 1.5b 2.6 – 2.1	6.4 ± 1.7b 6.6 – 6.3	4.2 ± 1.7b 4.5 – 4.1	8.5 ± 1.2b 8.7 – 8.1	4.22 ± 1.74b 4.53 – 4.12
	MF(n=26)	8.2 ± 1.5c 7.9 – 8.5	2.3 ± 1.5c 2.1 – 2.7	6.4 ± 1.8c 6.3 – 6.8	4.3 ± 1.6c 4.1 – 4.7	9.5 ± 1.7c 8.1 – 12	4.25 ± 1.73c 4.12 – 4.93
Healthy group (n=23)	M (n=11)	6.1 ± 1.3d 6.7 – 6.2	2.1 ± 1.2d 2.5 – 1.9	6.1 ± 1.3d 4.7 – 4.1	4.8 ± 1.3d 4.9 – 4.3	11.5 ± 1.2d 12.3 – 11.6	1.70 ± 1.82d 1.92 – 1.32
	F (n=13)	5.8 ± 1.2e 6.3 – 5.4	1.9 ± 1.3e 2.3 – 1.7	5.8 ± 1.2e 4.3 – 3.9	4.7 ± 1.2e 4.9 – 4.2	9.8 ± 1.3e 10.2 – 8.9	1.68 ± 1.83e 1.93 – 1.52
	MF(n=24)	5.9 ± 1.3f 5.4 – 6.7	2.0 ± 1.2f 1.7 – 2.5	4.2 ± 1.3f 3.9 – 4.7	4.7 ± 1.7f 4.2 – 4.9	10.7 ± 1.3f 8.3 – 12.3	1.69 ± 1.82f 1.32 – 1.93

Value = Mean ± S.D.
a, b, c: significant difference (p<0.05) as compared to the healthy group (d, e, f) respectively.

trol group comprised 23 subjects (Table 1; range 20 - 63 years). None of the control subjects exhibited clinical signs over 5 mm or any clinical attachment loss. None of the participants was cardiovascular disease or any other ongoing general disease or infections diagnosed. Informed consent was taken from each subject. The study was proved by PDM ethical society of our collage.

In all these cases, the peripheral blood were taken drawn for routine investigation, that is, total white blood cells count, red blood cell counts, thrombocytes count and Hb level. Plasma was obtained after centrifugation at 1500x g for 10 min and stored at -4°C until analysis of C-reactive protein (Eckersall et al., 2004). Student t-test was performed using SPSS software package (version 7.0) and the significance level was set at 95 percentage confidence interval.

RESULTS AND DISCUSSION

The clinical loss of attachment and number of sites involvement were significantly higher in periodontitis as compared to normal patients and also with increasing the age clinical loss of attachment and numbers of sites involvement were increased (Table 1) as reported in previous studies (Rai, unpublished data; Ushida et al., 2008). Total WBC, neutrophil, total thrombocytes counts and C-reactive protein (CRP) levels were increase significantly in periodontitis patients in comparison to healthy controls while RBC counts and Hb level were decreased in periodontitis in comparison to healthy controls (Table 2 and Figure 1, p<0.01). The chronic infections, such as periodontitis, are associated with increased risk for cardiovascular disease. It has been reported that higher number of leukocytes in periodontitis (Kweider et al., 1993). Subsequent studies have reported slight elevations of leukocytes in periodontitis in comparison to controls, although not always statistically significant (Fredriksson, 1999). In present study the total leukocytes counts significantly increased in periodontitis as compared to

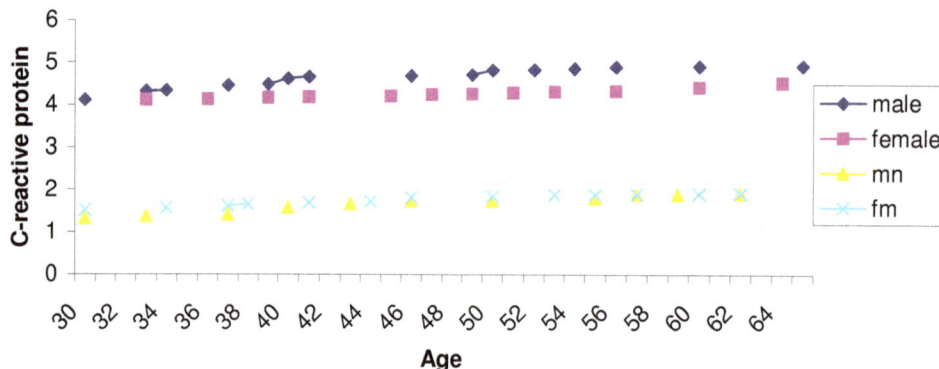

Figure 1. The level of C-reactive protein in normal and periodontitis patients in both genders in different age.

Male = periodontal patients (male)
Female = periodontal patients (female)
mn = normal males
fn = normal female

healthy controls (Table 1, $p<0.01$). A significant increase in polymorphonuclear leukocytes were also observed ($p<0.01$, Table 1). The increase in number of leukocytes in periodontitis has been suggested to be mainly due to to an increase in the number of polymorphonuclear leukocytes (Kowalik et al., 2001). Moderately elevated numbers of leukocytes have been associated with an increased risk for cardiovascular diseases (Albert and Ridker, 2004). Also, it has been suggested that the higher numbers of leukocytes increase the blood rheology. More cell numbers per unit cube made blood more viscous, and more cells may adhere to endothelial cells lining the blood vessels, therefore decrease the blood flow. Reduced blood flow could play a role in relation to cardiovascular diseases, especially in narrow or partly blocked arteries due to atheroscleratic plaque formation (Sweetnam et al., 1997).

Low hematocrit values have been reported in the periodontits patients (Hultter et al., 2001). Lowered hemoglobin was observed in our study. Our results are consistent with those reported in literature (Fredriksson et al., 1999). Few reports are available concerning red blood cell parameters in relation to periodontitis (Salvi et al., 1997). In our study, the RBC counts were decreased significantly in periodontitis as compared to healthy ($p<0.01$, Table 2).

Also, the platelets count were significantly increased in periodontitis as compared to healthy control (Table 2, $p<0.01$). This increase in thrombocyte count could be due to fact that they play an integral role in innate immunity against micro-organism (Klinger et al., 2002). Certain proteins from the periodontal pathogen porphyremonos gingivalis can stimulate thrombocytes to aggregate in a similar fashion as the clotting factor thrombin (Herzberg and Weyer, 1998). It is known that inflammatory and infectious processes can result in an increase in the number of active thrombocytes; that is "reactive thrombocy-

tosis". Therefore, increase in circulating thrombocytes could occur in periodontitis patients.

The C-reactive protein (CRP) is part of the body's normal response to infection and inflammation; chronically elevated levels are associated with a heightened risk for cardiovascular disease and mortality in both, the middle aged and the elderly (Harris et al., 1999; Lower et al., 2001). CRP might also be indicator of chronic infective processes possibly correlated with risk of coronary heart disease, such as infection by chlamydia pneumonia or chronic gastric infection with helicobactor pylori (Zhu et al., 2000). The higher plasma levels of CRP were observed in the present study in case of severe periodontitis patients in comparison to healthy control (Table 2, $p<0.01$). Recent studies (Rai, unpublished data; Ushida et al., 2008) also have reported elevated CRP levels among these with periodontitis. Conventional risk factors or conditions associated with progression of atherosclerosis include diabetes mellitus, cigarette smoking, hypertension, family history of ischaemic heart disease, hyperlipidemia, obesity, elevated plasma homocysteine, elevated CRP, menopause, male gender, nephritic syndrome and hyperthyroidism (Zhu et al., 2001). It has been well documented that CRP is a valuable marker in assessment of cardiovascular risk. And strong association of CRP with periodontal disease has been reported in literature (Zhu et al., 2000). Periodontal disease is often a chronic bacterial infection which involves inflammatory processes that also affect cardiovascular systems. Similar pathology is evident in both diseases and it is reasonable to expect some form of interaction between their pathogenic processes. Currently, American Heart Association is developing a summary on the inclusion of periodontal screening with previously established measured of risk assessment of CHD (Noack et al., 2001). There were lot of limitations of this study such important marks as CRP, alpha-TNF and interleukins (IL 6), clotting factor e.g.

prothrombin were not included in this study. Also we did not include patients with periodontitis without atherosclerosis and patients with periodontitis and atherosclerosis to prove the periodontitis is a risk factor for cardiovascular diseases. Further study is required on large scale and inflammatory factors e.g. hs-CRP, alpha-TNF and interleukins (IL 6), clotting factor e.g. prothrombin, group with periodontitis without atherosclerosis and patients with periodontitis and atherosclerosis include in study.

REFERENCES

Albert MA, Ridker PM (2004). Inflammatory biomarkers in Africans-American: A potential link to accelerated atherosclerosis. Rev. Cardiovasc. Med. 5 (Suppl. 3): S22-S27.

D'Aiuto F, Ready D, Tonetti MS (2004). Periodontal disease and C-reactive protein-associated cardiovascular risk. J. Periodontal. Res. 39: 236-241.

Eckersall P D , Conner J G , Harvie J (2004). An immunoturbidimetric assay for canine C-reactive protein. Vet. Res. Commun. 7:17-24

Fredriksson M, Figueredo C, Gustafsson A, Bergstrom K, Asman B (1999). Effect of periodontitis and smoking on blood leukocytes and acute phase protein. J. Periodontol. 70: 1355-60.

Harris TB, Ferrucci L, Tracy RP, Corti MC, Wacholder S (1999). Associations of elevated interleukin-6 and C-reactive protein levels with mortality in the elderly. Am. J. Med. 106 (5): 505-512.

Herzberg MC, Weyer MW (1998). Dental Plaque, platelets and cardiovascular diseases. Ann. Periodontol. 3: 151-60.

Hultter JM, Vander Velden U, Varoufaki A, Huffels RA, Hoek FJ, Loos BG (2001). Lower numbers of erythrocytes and lower levels of hemoglobin in periodontitis patients compared to control subjects. J. Clin. Periodontol. 28: 930-36.

Iwamoto Y, Nishimura F, Soga Y, Takeuchi K, Kurihara M, Takashiba S (2003). Antimicrobial periodontal treatment decreases serum C-reactive protein, tumor necrosis factor-alpha, but not adiponectin levels in patients with chronic periodontitis. J. Periodontol. 74: 1231-1236.

Klinger MH, Jelkmann W (2002). Role of blood platelets in infection and inflammation. J. Interferon Cytokine Res. 22: 913-22.

Kowalik MJ, Dowsett SA, Rodriguez J. Dela Rosa RM, Eckert GJ (2001). Systemic neutrophil response resulting from dental plaque accumulation. J Periodontol. 72: 146-51.

Kweider M, Lowe GD, Murray GD, Kinane DF, Mc Gowan DA (1993). Dental disease, fibrinogen and white cell count; links with myocardial infaraction? Scott Med. J. 38: 73-74.

Lower GD, Varnell JW, Rumley A, Bainton D, Sweetnam PM (2001). C-reactive protein, fibrin D-dimer, and incident ischemic heart disease in the speedwell study are inflammation and fibrin turnover linked in pathogenesis? Arterioscler Thromb Vasc. Biol. 21 (4): 403-10.

Morrison HI, Ellison LF, Taylor GW (1999). Periodontal disease and the risk of fatal coronary heart and cerebrovascular diseases. J. Cardiovascular Risk. 6(1): 7-11.

Noack B, Genco RJ, Trevisan M, Grossi S, Zambon JJ (2001). Periodontal infections contribute to elevated systemic C-reactive protein level. J. Periodontol. 72 (9): 1296-27.

Salvi GE, Lowrence HP, Offenbacher S, Beck JD (1997). Influence of risk factors on the pathogenesis of periodontitis. Periodontology 2000. 14: 173-201.

Slade GD, Ghezzi EM, Heiss G (2003). Relationship between periodontal disease and C-reactive protein among adults in the atherosclerosis risk in community study. Arch. Int. Med. 163 (10): 1172-9.

Sweetnam PM, Tomas HF, Yarnell JM, Baker IA, Elwood PC (1997). Total and differential leukocyte counts as predictors of ischemic heart disease: The caerphilly and speed well studies. Am. J. Epidemiol. 145: 416-21.

Tüter G, Kurtis B, Serdar M (2007). Evaluation of gingival crevicular fluid and serum levels of high-sensitivity C-reactive protein in chronic periodontitis patients with or without coronary artery disease. J. Periodontol. 78(12): 2319-24

Ushida Y, Koshy G, Kawashima Y, Kiji M, Umeda M, Nitta H, Nagasawa T, Ishikawa I, Izumi Y (2008). Changes in serum interleukin-6, C-reactive protein and thrombomodulin levels under periodontal ultrasonic debridement. J. Clin Periodontol. 35(11): 969-75.

Zhu J, Quyyumi AA, Norman JE (2000). Effects of total pathogen burden on coronary artery disease risk and C-reactive protein levels. Am. J .Cardiol. 85: 140-46.

28

Polymerization of human sickle cell haemoglobin (HbS) in the presence of three antimalarial drugs

Chikezie P. C.*, Chikezie C. M. and Amaragbulem P. I.

Department of Biochemistry, Imo State University, P. M. B. 2000, Owerri, Imo State, Nigeria.

To study the capacity of antimalarial drugs to interfere with polymerization of sickle cell haemoglobin (HbS), *in vitro* polymerization was induced by incubating erythrocyte haemolysate of HbS with 2 g% sodium metabisulphite ($Na_2S_2O_5$) in the presence of 2 mg% concentration of three (3) separate antimalarial drugs (Coartem™, Quinine, and Chloroquine phosphate). Spectrophotometric method was used to ascertain the level of polymerization of HbS at a regular interval of 30 seconds for a period of 180 seconds at extinction coefficient (λmax) of 700nm. The level of polymerization was expressed in percentage of the absorbance of control sample at the 180th second. The three antimalarial drugs caused significant ($p < 0.05$) reduction in HbS polymerization in the following ranges: Coartem™ (17.05-31.07%), Quinine (13.95-28.85%) and Chloroquine phosphate (10.85-33.01%). These results indicated the feasibility of the three drugs to retard HbS polymerization and as apparent potential candidates for therapy and management of sickle cell disease.

Key words: Anti-malarial drugs, polymerization, sickle cell, haemoglobin, sodium metabisulphite.

INTRODUCTION

The sickling disorder was the first description of a molecular disease and accounts for the vast majority of clinical important structural disorder to be studied. The sickle cell gene (β^S) occurs widely throughout Africa, part of Asia, the Arabian Peninsula and part of Southern Europe. In Africa, there are two areas with very high frequencies of β^S. One includes Nigeria, Ghana, and the other, Gabon and Zaire (Wainscoat, 1987; Uzoegwu and Onwurah, 2003). The carrier rate in this area is 10-30% of the population and the disease accounts for approximately 80,000 infant deaths per year (Wainscoat 1987). Sickle cell disease occurs with a much lower incidence in part of Italy, Greece, the Middle East and India. Genetic studies suggests that β^S mutation may

have arisen independently in Africa and Asia and that subsequent selection pressure by malaria has resulted in the observed high frequencies because heterogeneous carriers are more resistant to *P. falciparum* malaria during childhood (Weatheral et al., 1983).

Sickle cell anemia is caused by a single base mutation of adenine to thymine which results in a substitution of valine for glutamic acid at the sixth codon of the β-globin chain (Koch et al., 2002). This substitution has a profound structural consequence on haemoglobin and its biologic function. The reason for this phenomenon is the consequence of the substitution in HbS of polar glutamic acid residue with a non-polar valine molecule and thereby generates a sticky patch on the beta chain (Martin, 1983). Importantly, only this one β6 Val---Glu residue of each tetramer is in an intermolecular contact region. Specifically, the hydrophobic valine side chain appears to fit into a hydrophobic pocket formed by a β88 leucine residue and a β85 phenylalanine residue on an adjacent molecule. The normal glutamic acid would not easily fit into this pocket explaining at least part of why deoxyHbA does not polymerize. In conditions of reduced oxygen tension, HbS molecules form relatively insoluble polymer. The model described in sequential steps includes –

*Corresponding author. E-mail: p-chikezie@yahoo.com.

Abbreviations: 5HMF, 5-Hydroxymethyl-2-furfural; **FMC,** Federal Medical Center; **IMSUTH,** Imo State University Teaching Hospital; **PBS,** phosphate buffered saline; **LSD,** least Significant difference; **SAS,** statistical analysis system; **EDTA,** ethylene diamine tetraacetic acid; **HbS,** sickle cell haemoglobin.

nucleation, growth and subsequent alignment of the molecule into microfibrils parallel to each other with the resultant membrane deformity and damage (Bindon, 2003).

In spite of the full understanding of the pathology, physiology, and the molecular nature of the disease, a cure for sickle cell anemia still is unavailable. Strategies to treat sickle cell anemia since the early times of the disease state discovery in 1910, has focused mainly on prophylactic measures to alleviate the painful crises. While many researchers are still searching for clues to a cure, individuals can only alleviate the symptoms and prevent complications. Some common ways that sickle cell anemia can be managed are: blood transfusions, administration of painkilling drugs, intravenous fluids, oral antibiotics such as penicillin and by using the anticancer drug hydroxyurea (Bownas, 2002). Several attempts are in progress to finding new types of anti-sickling agents that specifically binds to intracellular sickle cell haemoglobin (HbS). Some anti-sickling agents that fulfill this criterion are 5-hydroxylmethyl-2-furfural (5HMF) (Abdulmalik et al., 2005), certain amino acids such as phenylalanine, lysine, and arginine (Anosike et al., 1991) and 2-imidazolines (Chang et al., 1983). Interestingly, in clinical practice, hydroxyurea is commonly used as anti-sickling agent to manage the disease, and it has been recently approved by the United States Food and Drug Administration as a drug for that purpose (Mehanna, 2001).

There are reports on the capability of some anti-malarial drugs to distort/alter certain red blood cell elements/physicochemical properties. Notable are red blood cell glutathione-S transferase activity, osmotic fragility index and level of methaemoglobin (Chikezie et al., 2009a,b; Chikezie, 2009; Chikezie et al., 2008). Therefore, the present *in vitro* study intends to ascertain the capacity of three anti-malarial drugs to interfere with HbS polymerization. The findings would provide preliminary insights into the role of these anti malarials in the sickling process, for more informed prescription and therapeutic strategies for individuals who express this genotype.

MATERIALS AND METHODS

Collection of blood samples/preparation of erythrocyte haemolysate

Five milliliters (5.0 ml) of venous blood obtained from the volunteers by venipuncture was stored in EDTA anticoagulant tubes. Blood samples were from patients attending clinics at the Federal Medical Center (FMC), Imo State University Teaching Hospital (IMSUTH), Orlu, St. John Clinic / Medical Diagnostic Laboratories, Avigram Medical Diagnostic Laboratories, and Qualitech Medical Diagnostic Laboratories. These centers are located in Owerri, Imo State, Nigeria.

The erythrocytes were washed by methods as described by Tsakiris et al. (2005). Within 2 h of collection of blood samples, portions of 1.0 ml of the samples were introduced into centrifuge

test tubes containing 3.0 ml of buffer solution pH = 7.4: 250 mM tris (hydroxyl methyl) amino ethane–HCl (Tris-HCl)/140 mM NaCl/I.0 mM $MgCl_2$/10 mM glucose). The erythrocytes were separated from plasma by centrifugation at 1200xg for 10 min, washed three times by three similar centrifugations with the buffer solution. The erythrocytes re-suspended in 1.0 ml of this buffer were stored at 4°C. The washed erythrocytes were lysed by freezing/thawing as described by Galbraith and Watts (1980) and Kamber et al. (1984). The HbS erythrocyte haemolysate was used for polymerization studies.

Anti-malarial drugs

Three (3) anti malarial drugs were used in this study: CoartemTM (Beijing NORVATIS Pharmaceutical Company, Beijing, China), Chloroquine phosphate (MAY and BAKER, Pharmaceutical Company Nigeria, Plc), and Quinine (BDH, UK). Two percent (2 mg %) (w/v) solution of the three anti malarial drugs was prepared by dissolving 2 mg of each drug in 100 ml of distilled water.

Polymerization studies of erythrocyte haemolysate

HbS polymerization was assessed by the turbidity of the solution (polymerizing mixture) by the method described by Iwu et al. (1988). The level of polymerization was ascertained by increasing absorbance of the assay mixture. A portion of 0.1 ml of the erythrocyte haemolysate containing HbS was introduced into a test tube followed by addition of 0.5 ml of Phosphate buffered saline (PBS) solution, pH = 7.4 {NaCl (9.0 g)/$Na_2HPO_4.2H_2O$ (1.71 g)/ $NaH_2PO_4.2H_2PO_4.2H_2O$ (2.43 g) per 1litre of distilled water} and 1.0 ml of distilled water. The sample was transferred into a cuvette and HbS polymerization was induced by adding 3.4 ml of 2% sodium metabisulphite solution. The change in absorbance of the assay mixture was recorded at a regular interval of 30 s for 180 s at λmax = 700 nm.

To determine the anti-sickling property of the three antimalarial drugs, the procedure above was repeated by substituting 1.0 ml of distilled water with separate 2 mg% of the three antimalarial drugs.

Calculation

Percentage polymerization = [At x 100] / [Ac_{180}^{th} second]

Where, At = Absorbance of test/control sample at time = t second; Ac_{180}^{th} second = Absorbance of control sample at 180th s.

Statistical/data analyses

The experiments were designed in a completely randomized method and data collected were analyzed by the analysis of variance procedure while treatment means were separated by the Least Significance Difference (LSD) incorporated in the Statistical Analysis System (SAS) package of 9.1 versions (2006).

RESULTS AND DISCUSSION

The change in absorbance of the control and test the samples and the corresponding percentage polymerize-tion are presented in presented in Table 1 and Figure 1, respectively. The results presented in Table 1 showed increasing absorbance of the assay mixture in the control

Table 1. Changes in absorbance of the control and test samples with time.

Time (s)	Absorbance (λ=700 nm)			
	Control	CoartemTM	Quinine	Chloroquine phosphate
0	0.00	0.00	0.00	0.00
30	0.052 ± 0.02a	0.043 ± 0.004a	0.037 ± 0.04a	0.042 ± 0.02a
60	0.103 ± 0.01b	0.071 ± 0.005a	0.074 ± 0.04a	0.069 ± 0.02a
90	0.113 ±0.02b	0.087 ± 0.008a	0.091 ± 0.04a	0.086 ± 0.02a
120	0.122 ± 0.01b	0.097 ± 0.012a	0.100 ± 0.05a	0.098 ± 0.02a
150	0.126 ± 0.02b	0.104 ± 0.015a	0.106 ± 0.05a	0.107 ± 0.02a
180	0.129 ± 0.01b	0.107 ± 0.015a	0.111 ± 0.05a	0.115 ± 0.02a

Means in the row with the same letter are not significantly different at $p < 0.05$ according to LSD.

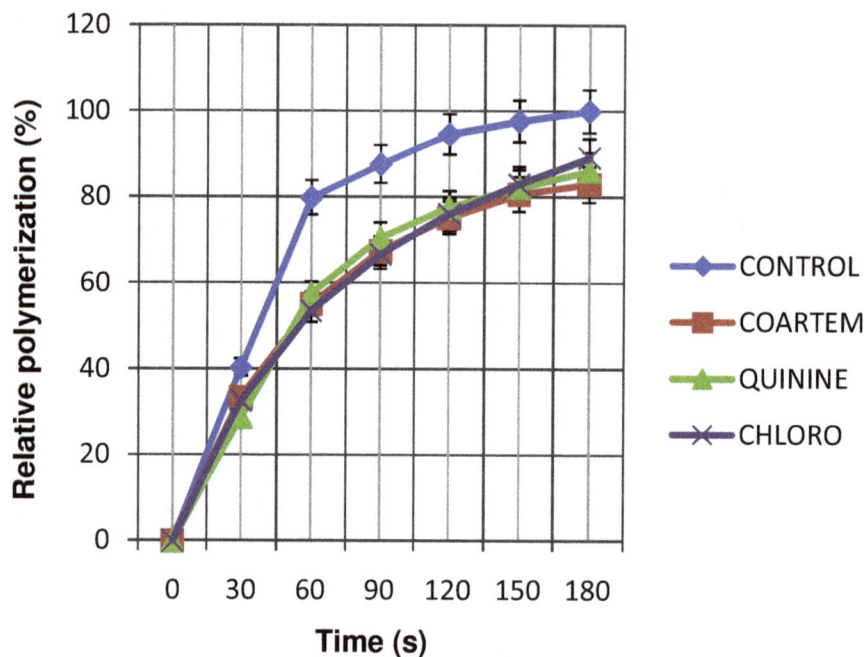

Figure 1. Percentage polymerization of HbS in the absence (Control Sample) and presence of three anti malarial drugs.

and test samples as the experimental time progressed. However, the absorbance of the polymerization mixture in the presence of the three anti malarial drugs was not significantly different (p<0.05) from the control sample at the 30th second. These values were indications that polymerization of HbS molecules occurred in the control sample and in the presence of the three anti malarial drugs (Figure 1). For instance, within the experimental time of 30-180 s, the relative polymerizations range between 33.33-82.95, 28.68-86.05% and 32.56-89.15% upon the introduction of Coartem, Quinine and Chloroquine phosphate, respectively.

The results presented in Table 2 showed the three anti malarial drugs caused significant (*p*<0.05) reduction in HbS polymerization in the following range: Coartem[TM] (17.05 - 31.07%), Quinine (13.95 - 28.85%) and

Chloroquine phosphate (10.85-33.01%).

Furthermore, whereas Chloroquine phosphate caused maximum inhibition of HbS polymerization at the 60th s (percentage inhibition= 33.01 ± 0.06%), the capacity diminished at the 180th s (percentage inhibition= 10.85 ± 0.06%). Generally, there was a dwindling capacity of the three anti malarials to inhibit HbS polymerization as the experimental time approached the 180th s.

The present study showed that the level of polymerization of HbS molecules was attenuated upon the introduction of the three anti malarial drugs in the polymerizing mixture. The pattern by which these drugs effected this inhibitory action was similar to phenylalanine (Ekeke and Shode, 1990; Anosike et al., 1991), methanol and water soluble extracts of dried fish (tilapia) and dried prawn (*Astacus red*) (Nwaoguikpe and Uwakwe, 2005),

Table 2. Percentage inhibition of HbS polymerization in the presence of three anti malarial drugs with time.

Drugs/ time (s)	Percentage inhibition of polymerization					
	30	60	90	120	150	180
CoartemTM	17.31 ± 0.09	31.07 ± 0.08	23.01 ± 0.08	20.49 ± 0.06	17.46 ± 0.08	17.05 ± 0.08
Quinine	28.85 ± 0.07	28.54 ± 0.10	19.47 ± 0.07	18.03 ± 0.08	15.87 ± 0.06	13.95 ± 0.08
Chloroquine P	19.23 ± 0.08	33.01 ± 0.06	23.89 ± 0.05	19.67 ± 0.10	15.08 ± 0.11	10.85 ± 0.06

methanol and water soluble extracts of *Cyperus esculentus* (tiger nut sedge) (Monago and Uwakwe, 2009).

These observations were obvious reflection of the capability of these anti-sickling agents to bind and shield the contact points of HbS monomers required for polymerization. Although these three anti malarials have been implicated in alteration of certain red blood cell elements in such a level that compromised physiochemical integrity and functionality of these cells (Chikezie et al., 2009a,b; Chikezie, 2009; Chikezie et al., 2008), our present findings showed they exhibited anti-sickling properties.

REFERENCES

Abdulmalik O, Safo MK, Chen Q, Yang J, Brugnara C (2005). 5-hydroxymethyl-2-furfural modifies intracellular sickle haemoglobin and inhibits sickling red blood cells. Br. J. Haematol., 128(4): 552-61.
Anosike EO, Uwakwe AA, Monanu MO, Ekeke GI (1991).Studies on human erythrocyte glutathione-S transferase from HbAA, HbAS and HbSS subjects. Biochem. Biomed. Acta, 50: 1051-1055.
Bindon J (2003). Natural Selection and Adaptation: Sickle Cell. http://www.as.ua.edu/ant/bindon/ant475/Sicklecell/Sicklecell.pdf. February 11, 2007.
Bownas J (2002). Genetic Profile: Sickle Cell Anemia. Nat. Instit. Health Pub. No. 96-4057.
Chang H, Ewert SM, Nagel RL (1983). Identification of 2-imidazolines as anti-sickling agents. Am. Soc. Pharmacol. Exp. Therapeut., 23(3): 731-734.
Chikezie PC (2009). Comparative methaemoglobin concentrations of three erythrocyte genotypes (HbAA, HbAS and HbSS) of male participants administered with five antimalarial drugs. Afr. J. Biochem. Res., 3(6): 266-271.
Chikezie PC (2008). Comparative *In vitro* osmotic stability of three human erythrocyte genotype (HbAA, HbAS and HbSS) in the presence of quinine and Chloroquine Phosphate. Int. Sci. Res. J., 1(2): 135-140.
Chikezie PC, Uwakwe AA, Monago CC (2009). Studies of human HbAA erythrocyte osmotic fragility index of non malarious blood in the presence of five antimalarial drugs. J. Cell Anim. Biol., 3(3): 039-043.

Chikezie PC, Uwakwe AA, Monago CC (2009). Glutathione S-transferase activity of three erythrocyte genotypes (HbAA, HbAS and HbSS) of male subjects/volunteers administered with Fansidar and Quinine. Afr. J. Biochem. Res., 3(5): 210-214.
Ekeke GI, Shode FC (1990). Phenylalanine is the predominant anti-sickling agent in Cajanus cajan seed extract. Plant Med., 56: 41-43.
Galbraith DA, Watts DC (1980). Changes in some cytoplasmic enzymes from red cells fractionated into age groups by centrifugation in Ficoll™/Triosil™ gradients. Comparison of normal human and patients with Duchenne muscular dystrophy. Biochem. J., 191:63-70.
Iwu MN, Igboko AO, Onwubiko H, Ndu UE (1988). Effect of cajanus cajan on gelation and oxygen affinity of sickle cell hemoglobin. J. Ethnopharm. 20: 99-104.
Kamber K, Poyiagi A, Delikonstantinos G (1984). Modifications in the activities of membrane-bound enzymes during *in vivo* ageing of human and rabbit erythrocytes. Comp. Biochem. Physiol. B., 77B: 95-99.
Koch AA, Olney RS, Yang Q (2002). Sickle Hemoglobin Allele and Sickle Cell Disease. Am. J. Epidemiol., 9: 839-845.
Martin DW (1983). Structure and function of a protein – haemoglobin. In: Martin, D.W; Mayes, P.A, Rodwell, V.W (eds). Harper's Review of Biochemistry. 9th Edition. Lange Medical Publications. California.
Mehanna AS (2001). Sickle cell anemia and antisickling agents then and now. Curr. Med. Chem. 8(2): 79-88.
Monago CC, Uwakwe AA (2009). Proximate composition and *in-vitro* anti sickling property of Nigerian *Cyperus esculentus* (tiger nut sedge). Trees Life J. 4(2). 1-6. www.TFLJournal.org.
Nwaoguikpe RN, Uwakwe AA (2005). The antisickling effects of dried fish (tilapia), dried prawn (*Astacus red)*. J. Appl. Sci. Environ. Mgt., 9(3): 115 – 119.
Statistical Analysis System (SAS) (2006). Package of 9.1 version.
Tsakiris S, Giannoulia-Karantana A, Simintzi I, Schulpis KH (2005).The effect of aspartame metabolites on human erythrocyte membrane acetylcholinesterase activity. Pharmacol. Res., 53: 1-5.
Uzoegwu PN, Onwurah AE (2003). Correlation of lipid peroxidation index with sickle haemoglobin concentrations in malarial-positive, -negative statuses of AA,AS and SS individuals from the University of Nigeria, Nsukka community. J. Biol. Res. Biotech., 1(1): 97-114.
Wainscoat JS (1987). The origin of mutant globin genes in human population. Acta. Haematol., 78: 154-158.
Weatheral DJ, Abdalla SH, Pippard K (1983). The anaemia of *Plasmodium falciparum* malaria. In malaria and the red cell. Ciba Symposium. London. Pitman., 94: 74-88.

29

Influence of the bioinsecticides, NeemAzal, on main body metabolites of the 3rd larval instar of the house fly *Musca domestica* (Diptera: Muscidae)

Mohammad A. Kassem[1], Tarek A. Mohammad[2] and Ahmad S. Bream[3]

[1]Department of Biochemistry, Faculty of Pharmacy, Omar El Mokhtar University, Al-Bieda, Libya.
[2]Department of Chemistry, Faculty of Science, Omar El Mokhtar University, Al-Bieda,Libya.
[3]Department of Zoology, Faculty of Sciences, Omar El Mokhtar University, Al-Bieda, Libya.

The present study was designed and conducted to determine the effects of Neemazal on the general body metabolism of the house fly *Musca domestica*. The second instar larvae of the house fly *M. domestica* (L.) were reared on artificial diet treated with the botanical extract, NeemAzal. Five concentration levels of this compound (1000, 500, 250, 125 and 62.5 ppm) were used. The main body metabolites, carbohydrates, proteins, lipids and cholesterol were determined during the early and late third larval instar. Various degrees of almost significant ($p < 0.001$) reducing effects were recorded for proteins, lipids and carbohydrates during the early and late third larval instar. These decrements paralleled to the increments of the concentration levels. On the other hand, results indicated a slight, but non significant, reducing action of NeemAzal on cholesterol of early larval instar. This action followed by significant increase of the same metabolite at the end of this instar comparing with control congeners, except the highest concentration (1000 ppm) which exhibit a non significant ($p > 0.01$) reducing action on cholesterol. It is quite clear from these results that disturbances of nutrient contents might affect larval growth and development.

Key words: *Musca domestica*, *Azadirachta indica*, NeemAzal, neem, main metabolite.

INTRODUCTION

Phytochemicals from the neem tree, *Azadirachta indica* Juss, have been the most extensively studied in the late decades among the efficient alternatives of synthetic pesticides (Ntalli et al., 2009). Recent advances dealing with the activity of neem products were reported in the comprehensive reviews by Mordue and Blackwell (1993). Extracts from neem seeds have been reported to have wide ranging biological activities against insects (Isman et al., 1990; Schmutterer, 1990; Ghoneim et al., 2001). These include feeding deterrence, oviposition retarding, growth regulation, development impairing, against several insect species (Rice et al., 1985; Barnby and Klocke, 1990; Schmutterer, 1990; AliNiazee et al., 1997; Prakash and Roa, 1997; Ghoneim et al., 2000). The major active ingredient in the neem seeds is azadiractin,

actually a mixture of seven tetranortriterpenoid isomers (Rembold et al., 1984). Pure azadirachtin is highly unstable, as well as the costly and tedious process of isolation, and the remote chances of synthesis of azadirachtin precludes the chance of utilization of the pure compound in insect pest management (Larson, 1987). The neem preparations are preferred over pure azadirachtin for field applications (Isman et al., 1990). NeemAzal is a neem seed preparation with the azadirachtin (the most active ingredient) content of 20%, and inexpensive production in addition to its safety toxicologically and environmentally (Kleeberg, 1992). The housefly *Musca domestica* is an endophilic eusynanthrope cosmopolitan insect, trophically and micro climatically related to man and his domestic animals. Houseflies are known as transmitters of human and animal diseases. In the present study, using the house fly *M. domestica* as an experimental insect, was undertaken to determine the effects of NeemAzal on the general

*Corresponding author: E-mail: moalikassem@yahoo.com.

Table 1. Effects of NeemAzal on total carbohydrate of the whole body tissue homogenate of early- and late-aged third larval instar of the house fly *M. domestica,* using feeding technique.

Concentration (ppm)	Total carbohydrate (mg/g fresh body weight)			
	Early third	Change (%)	Late third	Change (%)
1000	31.3 ± 1.2^{ns}	11.4	$24.2 \pm 2.5^{*}$	- 17.7
500	$22.7 \pm 1.8^{**}$	- 19.2	26.4 ± 2.6^{ns}	- 10.2
250	$23.1 \pm 1.6^{*}$	-17.8	26.1 ± 2.5^{ns}	-11.2
125	$23.1 \pm 1.2^{*}$	-17.8	25.6 ± 1.4^{ns}	-12.9
62.5	25.1 ± 0.5^{ns}	-10.7	$23.7 \pm 2.3^{*}$	-19.4
Control	28.1 ± 1.3	---	29.4 ± 2.3	---

ns: not significantly different (P >0.05), * significantly different (P <0.05), ** highly significantly different (P <0.01), *** very highly significantly different (P <0.001).

body metabolism to evaluate the potency of this new bioinsecticides in insect control.

MATERIALS AND METHODS

Tested insect

A culture of the house fly *M. domestica* was maintained for several generations under the conditions of 27 ± 2°C and 70 to 75% RH at entomology department, Faculty of Agriculture, Omar Al-Mukhtar University. The adult flies kept in 30 × 30 × 40 cm cages and provided with cotton pads soaked in 10% sucrose as food. Also, cotton pads were saturated with a milky solution in Petri dishes to serve as ovipositing sites. The larval diet used to rear is described in Bream et al. (1999).

NeemAzal treatments

The neem seed extract used herein was in the form of an emulsifiable concentrate containing 20% azadirachtin as the active ingredient. A concentration range of five levels was prepared using tap water as a solvent, 1000, 500, 250, 125, and 62.5 ppm. Four replicates (50 larvae or rep.) of early second instar larvae were continuously fed on an artificial diet treated with each of these levels of concentrations. The controls fed a diet free from NeemAzal.

Preparations for biochemical studies

At early and late third instar larvae age, approximately 20 individuals were pooled and weighed for estimating the total protein, lipid, carbohydrates and cholesterol content. These criteria were estimated in the whole body Homogenate. Three pools were used as replicates. The collected samples were hand-homogenized in a centrifuge tube containing 0.5 ml of saline solution. A further volume of 1.5 ml of the saline was added for washing to make a total volume of 2 ml of the saline in each concentration. After a 20 min centrifugation at 8000 rpm, (using cooling centrifuge) the clear supernatant was removed for use in the body metabolites using. A PyeUnicam SP6-450 Uv/Vis 50 spectrophotometer.

Metabolites assay

In the present study, the total protein content of the samples was assessed by the procedure of Stanbio LiquiColor kit No. 0250

(Proven Biuret methodology). The optical density was measured at 550 nm. Cholesterol content was determined at 500 nm according to the procedure recommended with Stanbio Cholesterol Kit No. 1010 (Cholesterol oxidase methodology). The total carbohydrate content was determined according to the method of Singh and Sinha (1977). Lipids were extracted and estimated according to the methods of Knight et al. (1972). The results were expressed in mg per g fresh body weight.

Statistical analysis

Data obtained were analyzed by the Student's t-distribution and refined by Bessel correction (Moroney, 1956) for the test significance of difference between means.

RESULTS

Results of Table 1 showed that the total body carbohydrate content was increased with the age of control larvae. A similar result was obtained for larvae treated with NeemAzal, except those treated with low and high concentration levels. The treatment with NeemAzal resulted in significant reduction in the carbohydrate contents.

As indicated in Table 2, the total body protein content was increased with the age of control and treated larvae. It appears from Table 2 that treated early instar larvae showed decrements in protein content paralleled to the increments of the concentration level. The change percentage reached -49.8 at the highest concentration level. A similar trend was shown during the end of this instar. Lipid content of control larvae increased during the instar examined (Table 3). The same tendency was detected in treated larvae, except in the case of larvar treated with the highest concentration level (1000 ppm). Generally, various degrees of reduction were recorded after treatment of second instar larvae with NeemAzal. The same tendency was detected in the larvae after treatment with NeemAzal. It is quiet clear from the results of Table 4 that NeemAzal had an inhibitory effect on cholesterol level in treated larvae, in early larval instar tested and vise versa for late-aged larvae examined.

Table 2. Effects of NeemAzal on total protein of the whole body tissue homogenate of early- and late-aged third larval instar of the house fly *M. domestica,* using feeding technique.

Concentration (ppm)	Total protein (mg/g fresh body weight)			
	Early third	Change (%)	Late third	Change (%)
1000	13.3 ± 2.7***	- 49.8	15.0 ± 1.4***	- 45.5
500	13.5 ± 3.1***	-49.1	17.1 ± 1.7***	-37.8
250	16.5 ± 3.3**	-37.7	19.5 ± 1.6**	-29.1
125	18.4 ± 2.2**	-30.6	21.0 ± 1.1*	-23.6
62.5	20.0 ± 2.1*	-24.5	22.3 ±1.3*	-18.9
Control	26.5 ± 1.4	---	27.5 ± 1.9	---

ns: not significantly different (P >0.05), * significantly different (P <0.05), ** highly significantly different (P <0.01), *** very highly significantly different (P <0.001).

Table 3. Effects of NeemAzal on total lipids of the whole body tissue homogenate of early- and late-aged third larval instar of the house fly *M. domestica,* using feeding technique.

Concentration (ppm)	Total lipid (mg/g fresh body weight)			
	Early third	Change (%)	Late third	Change (%)
1000	23.3 ± 2.4***	- 47.0	23.2 ± 2.4***	- 58.6
500	25.9 ± 2.0***	- 41.1	20.4 ± 3.6***	-63.6
250	28.6 ± 2.4***	- 35.0	27.7 ± 1.2***	-50.5
125	28.6 ± 2.1***	- 35.0	36.9 ± 2.5***	-34.1
62.5	32.4±3.3***	- 26.4	42.0± 2.6***	-25.0
Control	44.0 ± 1.2	---	56.0 ± 5.2	---

ns: not significantly different (P >0.05), * significantly different (P <0.05), ** highly significantly different (P <0.01), *** very highly significantly different (P <0.001).

Table 4. Effects of NeemAzal on total cholesterol of the whole body tissue homogenate of early- and late-aged third larval instar of the house fly *M. domestica,* using feeding technique.

Concentration (ppm)	Cholesterol (mg/g fresh body weight)			
	Early third	Change (%)	Late third	Change (%)
1000	0.82 ± 0.09 ns	-50.0	1.9 ± 0.23ns	-9.5
500	1.05 ±0.03 ns	-35.9	4.2 ± 0.29**	100
250	0.94 ± 0.03 ns	-42.7	4.9 ± 0.31***	133
125	0.94 ± 0.03 ns	-42.7	4.7 ± 0.42**	123.8
62.5	0.98 ± 0.05ns	-60.9	3.8 ± 0.35*	80.9
Control	1.64 ± 0.15	---	2.1 ± 0.27	---

ns: not significantly different (P >0.05), * significantly different (P <0.05), ** highly significantly different (P <0.01), *** very highly significantly different (P <0.001).

DISCUSSION

Effect on the total body carbohydrate content of third larval instar

Results of Table 1 showed that the total body carbohydrate content was increased with the age of control larvae. A similar result was obtained for larvae treated with neemazal, except those treated with low and high concentration levels. The treatment with NeemAzal resulted in significant reduction in the carbohydrate contents. At the early age, this reduction was significant except at higher and lower concentration levels. This reducing action of NeemAzal was extended to the late-age of these larvae. These results agree with that of Baker et al., (2002) after treatment of *Spodoptera littoralis* with different plant extracts; Abou El-Ela et al. (1993) who determined great reduction in carbohydrates during the pupal stage of the fly *Synthesomyia nudiseta* after larval treatment with some IGRs. Depending on the results of

Abo El-Ghar et al. (1995) on *Agrotis ipsilon* feeding of larvae on the petroleum ether extracts from *Ammi majus* and *Apium graveness* and acetone or and ethanol extracts from *Melia azedarach* and *Vinca rosea* caused a considerable reduction in the total carbohydrates of larval haemolymph. In addition, Khalaf (1998) estimated a significant reduction in the carbohydrate content in the whole pupal period of *Muscina stabulans* after larval treatments with volatile oils of *Cymbopogon citratus* and *Rosmarinus efficinalis*. On the other hand, Abou El-Ela et al. (1990) found that treatment of *M. domestica* larvae with Altosid (ZR-515) resulted in decreased carbohydrates in 1 day old pupae and some increments in 3 day and 5 day old pupae. On contrast significant increases of carbohydrate content were observed in larvae of *S. littoralis* by the JHA (Isopropyl 3, 7, 11-triethyl-2, 4-dodacadiote) (Ismail, 1980).

Effect on the total body protein content of third larval instar

Proteins are very complex and at the same time, they comprise most of the characteristics of living matter. They are present in all viable cells, in that they are the compounds which, as nucleoproteins, are essential to the process of cell division and, as enzymes and hormones, control many chemical reactions in the metabolism of cells. Protein synthesis is necessary for the maintenance of body growth and reproduction. Many factors have been implicated in the control of protein synthesis (Carlisle et al., 1987).

It appears from Table 2 that treated early instar larvae showed decrements in protein content paralleled to the increments of the concentration level. The change percentage reached -49.8 at the highest concentration level. A similar trend was shown during the end of this instar. These results agree with that of Bakr et al. (2002) using different plant extracts against *S. littoralis*. On the contrast, Amer (1990) determined increasing protein content in pupae of *S. littoralis* after larval treatment with mevalonic acid. Also, Basiouny (2000) estimated considerable increments of proteins throughout different developmental stages of *M. stabulans* by the chitin inhibitors IKI-7899 and XRD-473.

Effect on the total body lipid content of third larval instar

Lipids are important source of energy for insects. These are obtained from the diet or are synthesized by the insect itself (Gilby and Gilly, 1965). Lipid turnover in insects is regulated by neuroendocrine-controlled feedback loops (Downer, 1985). Lipids are essential structural components of the cell membrane and cuticle. They facilitate water conservation both by the formation of an impermeable cuticular barrier and by yielding metabolic water upon oxidation, and they include important hormones and pheromones (Gilbert, 1967).

Lipid content of control larvae increased during the instar examined (44.0 ± 1.2 and 56.0 ± 5.2 mg/g lipids of early- and late-aged third larval instar) (Table 3). The same tendency was detected in treated larvae, except in the case of larvar treated with the highest concentration level (1000 ppm). Generally, various degrees of reduction were recorded after treatment of second instar larvae with neemazal. This reduction was detected on early and late- aged larvae and it is found to be statistically significant with various degrees of change percentage. These results agreed with those of Baker et al. (2002) after treatment of *S. littoralis* larvae with different plant extracts and that of Bream (2002) after treatment of *Rhynchophorus ferrugineous* prepupae with Azadirachtine and Jojoba extracts. Synthesis of lipids by the fat body of *Choristoneura fumiferana* was diminished at all times examined during the 6 day experimental period after treatment with fenoxycarb (Mulye and Gordon, 1993). On the other hand, Ghoneim (1994) determined significant increments of lipid content throughout the pupal stage of *S. littoralis* by larval treatments with mevalonic acid and IKI-7899.

Effect on the total body cholesterol content of third larval instar

The synthesis of sterol from acetate has not yet been demonstrated in arthropods and, indeed, arthropods require a dietary sterol for several physiological functions and in some cases for life itself (Gilbert, 1967). The major functional sterol in arthropods is cholesterol. The roles of sterols in arthropods are likely manifold and include being precursors of the arthropods molting hormone (s) and components of subcellular membrane structures (Gilbert, 1967).

Cholesterol content of control larvae increased by the age (1.64 ± 0.15 and 2.1 ± 0.27 mg/g lipids of early-and late-aged larvae examined). The same tendency was detected in the larvae after treatment with NeemAzal. It is quite clear from the results of Table 4 that NeemAzal had an inhibitory effect on cholesterol level in treated larvae, in early larval instar tested and vise versa for late-aged larvae examined. This reducing effect was found to be not statistically significant. Lipid turnover in insects is regulated by neuroendocrine-controlled feedback loops (Downer, 1985). This may explain the decrements in cholesterol, which is one of the precursors of molting hormone and other hormones, required at that instar to form the complete pupal stage.

Conclusively, various degrees of reducing actions were recorded for proteins, lipids and carbohydrates during the early and late third larval instar. Also, this reducing action of NeemAzal was detected on cholesterol during the early instar larvae but followed by significant increase at the end of the same instar.

REFERENCES

Abo El-Ghar MR, Radwan HSA, Ammar IMA (1995). Some biochemical effects of plant extracts in the black cutworm *Agrotis ipsilon* (Hufn.). Bull. Ent. Soc. Egypt. Econ. Ser., 22: 85-97.

Abou El-Ela R, Guneidy AM, El-Shafei AM, Ghali OI (1990). Effects of Altosid (ZR515) on oxygen consumption, carbon dioxide out put and carbohydrate content of organophosphorous resistant strain of *Musca domestica*. J. Egypt. Soc. Parasit., 20(1): 307-318.

Abou El-Ela RG, Taha MA, Rashad SS, Khalaf AA (1993). Biochemical activity of three insect growth regulators on the fly, *Synthesomyia nudiseta* Wiilp (Diptera : Muscidae). J. Egypt. Ger. Soc. Zool., 12(D): 15-25.

AliNiazee MT, Al Humeyri A, Saeed M (1997). Laboratory and field evaluation of a neem insecticide against *Archips rosanus* L. (Lepi.: Tortricidae). Can. Ent., 129: 27-33.

Amer MS (1990). Effects of the anti-JH (Mevalonic acid) on the main metabolites of *Spodoptera Littoralis*. Egypt. J. Appl. Sci., 5(1): 82-91.

Bakr FR, El bermawy S, Emara S, Abulyazid I, Abdlwahab H (2002). Biochemical Studies On *Spodoptera littoralis* Developmental Stages After Larval Treatment With Different Botanical Extracts. Proceeding of 2nd Int. Conf. Plant Prot. Res. Inst., Cairo, Egypt, 2002, 1: 888-893.

Barnby MA, Klocke JA (1990). Effect of azadirachtin on levels of ecdysteroids and prothoracicotropic hormone-like activity in *Heliothis virescens* (Febr) larvae. J. Insect Physiol., 36(2): 125-131.

Basiouny AL (2000). Some physiological effects of certain insect growth regulators (IGRs) on the flase stable fly, *Muscina stabulans* (Fallen.) (Diptera : Muscidae) Unpublished Ph.D. Thesis, Fac. Sci., Al-Azhar Univ., Egypt.

Bream AS (2002). Metabolic responsiveness of the red palm weevil, *Rhynchophorus ferrugineus* (Curculionidae: Coleoptera) to certain plant extracts. Proceeding of the 2nd Int. Symp. On Ornamental Agriculture in Arid Zones, Al-Ain, UAE. pp. 1-12.

Bream AS, Ghoneim KS, Mohammed HA (1999). The effectiveness of nonsteroidal Ecdysone mimic, Tebufenozide, on development, reproduction and progeny of the house fly *Musca domestica* (Diptera: Muscidae). Med. Fac. Landbouww. Univ. Gent., 64(3a): 219-227.

Carlisle JL, Oughton B, Ampleford E (1987). feeding causes the appearance of a factor in the haemolymph that stimulates protein synthesis. J. Insect Physiol., 33(7): 493-499.

Downer RGH (1985). Lipid metabolism. In "Comprehensive Insect Physiology, Biochemistry and Pharmacology" Ceds:Kerkut, G.A and Gilbert, L.I), Pbl. Pergamon Press, Oxford, 10: 75-114.

Ghoneim KS (1994). Changes in food metabolism induced by the action of JHA, pyriproxyfen, during the last nymphal instar of Schistocerca gregaria (Forsk) (Orthoptera: Acrididae). J. Fac. Educ., Ain Shams Univ., 19: 937-954.

Ghoneim KS, Mohammed HA, Bream AS (2000). "Efficacy of the neem seed extract NeemAzal, on growth and development of the Egyptian cotton leafworm Spodoptera littoralis (Boisd): (Lepidoptera: Noctuidae)". J. Egypt Ger. Soc. Zool., 33(E): 161-179.

Ghoneim KS, Bream AS, Tanani MA, Nassar MM (2001). Efficacy of CGA-184699 and CGA-259205 on survival, growth and development the red palm weevil *Rhynchophorus ferrugineus* (Olivier) (Coleoptera: Curulionidae). Proceed. Sec. Int. Conf. Date Palms, Al-Ain, UAE, pp. 246-279.

Gilbert LI (1967), Lipid metabolism and function in insects. Adv. Insect Physiol., 4: 69-211.

Gilby Y, Gilby R (1965). Lipids and their metabolism in insects. Anu. Rev. Ent., 10: 141-160.

Ismail IE (1980). Physiological studies of analogues upon the cotton leafworm *Spodoptera littoralis* (Lep., Noctuidae). Unpublished Ph.D. Thesis, Fac. Sci., Cairo Univ., Egypt.

Isman MB, Koul O, Luczynsk A, Keminski J (1990). Insecticidal and antifeedant bioactivities of neem oil and their relationship to azadirachtin content. J. Agric. Food Chem., 38: 1406-1411.

Khalaf AF (1998). Biochemical and Physiological impacts of two volatile plant oil on Muscina stabulans (Dipt.: Muscidae). J. Egypt Ger. Soc. Zool., 27(E): 315-329.

Kleeberg H (1992). Properties of NeemAzal-F, a new botanical insecticide. In "Insecticides: Mechanism of action and resistance" (eds. Otto, D. and Weber, D.). Intercept Publ., Andover, England, pp. 87-94.

Knight JA, Anderson S, Jams MR (1972). Chemical basis of the sulphovanillin reaction of estimating total lipid. J. Clin. Chem., 18(3): 199-200.

Larson RO (1987). Development of Margosan-0, a pesticide from neem seed. Proc. 3rd Int. Neem Conf. Rauischolzhausen, 1983, pp: 461-470.

Mordue AJ, Blackwell A (1993). Azadirachtin. an update. J. Insect Physiol., 39: 903-924.

Moroney MJ (1956). Facts from Figures. Pbl. Pinguin Book Ltd. Harmondsworth, Middlesex, p. 228.

Mulye H, Gordan R (1993). Effects of fenoxycarb, a juvenile hormone analogue, on lipid metabolism of the Eastern spruce budworm, *Choristoneura fumiferana*. J. Insect Physiol., 39: 721-727.

Ntalli NG, Menkissoglu-Spiroudi U, Giannakou IO, Prophetou-Athanasiadou DA (2009). Efficacy evaluation of a neem (*Azadirachta indica* A. Juss) formulation against root-knot nematodes *Meloidogyne incognita*. Crop Protection, 28(6): 489-494.

Prakash A, Rao J (1997). Botanical Pesticides in Agriculture. Lewis Publishers (1st ed).) India, p. 450.

Rembold H, Forster H, Czoppelt CH, Rao PJ, Sieber KP (1984). The azadirachtin, a group of insect growth regulator from the neem tree. In "Natural Pesticides from the neem tree) Azadirachta indica. A. Juss) and other Tropical Plants " Schmutterer, H., and Ascher, K.R.S. (Eds.). Proc. 2nd int. Neem Conf. Ravischholzhausen, Germany 1983. p. 153.

Rice MJ; Sexton S, Esmail AM (1985). Antifeedant phytochemical blocks oviposition by sheep blowfly. J. Aust. Entomol. Soc., 24: 16.

Schmutterer H (1990). Properties and potentials of natural pesticides from the neem tree, *Azadirachta indica*. Ann. Rev. Entomol., 35: 271-297.

Singh NB, Sinha RN (1977). Carbohydrate, lipid and protein in the development stages of *Sitopholes orzae* and *Sitophelus granarius* Ann. Entomol. Soc. Am., 70: 107-111.

Effect of essential oil of the leaves of *Eucalyptus globulus* on heamatological parameters of wistar rats

OYESOMI, Tajudeen Oyesina[1]*, AJAO, Moyosore Salihu[2], OLAYAKI, Iuquman Aribidesi[4] and ADEKOMI, Damilare Adedayo[3]

[1]Department of Anatomy, Kampala International University, Dar es Salaam, Tanzania.
[2]School of Anatomical Sciences, Faculty of Health Sciences, University of the Witwatersrand, Johannesburg, South Africa.
[3]Department of Anatomy, Faculty of Basic Medical Sciences, College of Health Sciences, University of Ilorin, Ilorin, Kwara State.
[4]Department of Physiology, Faculty of Basic Medical Sciences, College of Health Sciences, University of Ilorin, Ilorin, Kwara State.

The study was designed to evaluate the effect of essential oil extract of *Eucalyptus globulus* on haematological parameters of wistar rats. Twenty-five adult wistar rats weighing between 80 to 130 g were used. The rats were divided into five groups; with group one as the control group. Increasing doses (12.5, 25.0, 50.0, 72.5 mg/ml) of the extract were administered orally daily to the other four groups for a period of four weeks. The animals were sacrificed and the blood collected for haematological parameters using automated haematological analyzer K-X-21 machine. The results indicate significant increases in level of Haemoglobin (Hb), White Blood Cell (WBC), and Red Blood Cell (RBC) but a decrease in the levels of the Mean corpuscular volume (MCV), mean corpuscular hemoglobin concentration (MCHC). The study confirmed that extract oil of eucalyptus globulus have some significant effects on haematological parameters of the wistar rats and these effects are dose dependent.

Key words: *Eucalyptus globulus*, oil extract, haematological parameters, wistar rats.

INTRODUCTION

Man in solving its numerous medical challenges have for ages depends on his immediate environment taking advantages of nature provisions of it beauty for live and survival. They have learnt to depend on plants and in some cases animals in providing solutions to the myriad of their health problems (Oliver, 1960). However, the increasing use of plants for the therapeutic and medicinal use warrants an adequate scientific investigation to confirm the suitability of plants or otherwise for the purpose for which they are used. Most of these medicinal plants are taking as vegetables, smoked leafs as tobacco, while the stems and roots are sometimes cooked for drinking.

Eucalyptus globulus is an ever green tree 40 to 70 m tall (Little, 1983) widely planted in the sub tropics. Its roots, stem, leaves and seed have been widely used in traditional folk medicine in many parts of West Africa countries. The plant fresh leaves are sometimes eating as vegetables, while the dry leaves were often smoked as cigarettes, in this case, for asthma treatment while the oil is used in the form of an aperitif as a digestive (Brooker et al., 1999). The stems and roots are cooked as medicinal agents across different ethnics groups within the country. The medicinal uses of *Eucalyptus* are in the treatment of abscess, arthritis, boil, bronchitis, burns, catarrh, diabetes, and dysentery (Watt et al., 1962; Duke and Wain, 1981; List and Horhammer, 1969). It is also to be useful in the various treatments of lung ailment, malaria, bladder and liver infection (Boukef et al., 1976). However, the mechanisms by which extract of *E. globulus* exert it all these activities are not well

*Corresponding author. E-mail: drstoyesomi@yahoo.com or oyesina1@gmail.com.

understood. The chemical composition of *E. globulus* are eucalyptol (cineol), terpineol, sequiterpene, alcohol, aliphatics aldehyde, isoamyl alcohol, ethanol and terpenes (Morton, 1981).

Erythrocytes (red blood cell) which are anucleate matured cells are loaded with the oxygen carrying proteins known as haemoglobin. The normal concentration of erythrocytes in adult blood is approximately 3.9-5.5 million per micro litre in women and 4.1 to 6 million per micro litre in men (Junqueira et al., 2005). Human erythrocytes life span in circulation ranges from 90 to 120 days, while the worn out red blood cells are removed from the circulation by macrophages of the spleen and bone marrows. The effects of *Eucalyptus* globules on haematological parameters are not documented.

Leukocytes (white blood cells) are involved in the cellular and humoral defense mechanisms of the body are responsible for fight against foreign agents. The estimated total number of leukocytes in the blood varies according to age, sex and physiological conditions of the body. In normal healthy adults, they range from 6,000 to 10,000 leukocytes per micro litre blood (Junqueira et al., 2005). Majorities of these white blood cells migrate to the tissue, where they perform multiple functions as tissue macrophage and mostly died by apoptosis.

Blood platelets (thrombocytes) are nonnucleated disk like cell fragments 2 to 4 μm in diameter. Platelets originate from the fragmentation of giant polypoid megakaryocytes that reside in the bone marrow. Platelets count range from 200,000 to 400,000 per micro litre of blood. Platelets have a life span of about 10 days. Platelets function; the role of platelets in controlling haemorrhage can be summarized as primary aggregation, secondary aggregation, blood coagulation, clot retraction and clot removal (Junqueira et al., 2005). The study was designed to see the effect of *Eucalyptus* globules on the haematological parameters using wistar rats.

MATERIALS AND METHODS

Animals

Twenty five adult Wistar rats weighing between 80 to 130 g were used for the study. The rats were purchased from the animal house of Faculty of Pharmacy, Obafemi Awolowo University Ile-Ife, Nigeria. They were bred for weeks in the animal house of Department of Anatomy, University of Ilorin. The rats were fed with pellets grower mash obtained from Bendel feed mill, Yoruba road, Ilorin and water *ad libitum* during the breeding period designed to acclimatize the rats. They were exposed to 12 h of light and darkness per day. The animals were care for in accordance with the recommendations provided in the "Guide for the Care and Use of Laboratory Animals" prepared by the National Academy of Sciences (NIH, 1985).

The rats were randomly grouped into five groups. A: control group, and groups 1 to 4 were the experimental groups. Each of the rats were marked at the tail with different colours of pen marker and put into different segments of the cage, according to their group. The rats were sacrificed after four weeks of extract administration

using cervical dislocation. Blood from each rat was collected into labelled heparinised bottle to prevent coagulation of the blood and analyzed for the haematological parameters.

Preparation of extract

Fresh leaves of the plant *E. globulus* were collected around the Department of Anatomy, University of Ilorin, Nigeria. Botanical identification of the plants was done at the herbarium of the Department of Botany, Faculty of Sciences, University of Ilorin. The fresh leaves were dried under laboratory condition as this helps prevent destruction of active constituents of the plant, which may occur on exposure to radiation, and drying lasted for a week.

The plant extract was prepared by the process of hydro distillation using the Clevenger apparatus in which the grinded plant material was heated. The evaporated oil were condensed and decanted into sample bottles and refrigerated.

Administration of extract

Administration of the aqueous extract was done orally by means of calibrated syringe with attached rubber cannula. The control group received normal saline and $DMSO_4$. The experimental groups of 1, 2, 3, and 4 received extract of *E. globulus* at doses of 12.5, 25.0, 50.0, and 72.5 mg/ml respectively. The assigned doses per groups were administered daily and lasted for duration of four weeks.

Haematological parameters analysis

Evaluation of the haematological parameters was carried out using automated haematological analyzed K-X-21 made by Symex, Kobe, Japan. Sample of blood from the Wistar rats in heparinized bottle were analyzed using this machine for accuracy. Each sample was run twice and the average value calculated and recorded. The co-efficient of error of the analyzer machine is less than 5%. Data obtained were presented as mean ± standard deviation and in some cases; the use t-test was employed for comparism and the level of significance was predetermined as $p \leq 0.05$.

RESULTS

There was steady increase in the haemoglobin concentration and the estimated total red blood cell counts across the concentration gradient with increasing concentrations of the oil extract. The control group have a mean haemoglobin concentration of 9.63 ± 0.3 mg/dl and 12.1 ± 0.2 mg/dl for the group 4 with high dose of the oil extract. The estimated total red blood cell counts for the control group was 4.45 ± 0.04 ($\times 10^6$) and 6.16 ± 0.15 ($\times 10^6$) for the group 4. There was statistical significant difference between them (Table 1).

The mean haemoglobin concentration for the control group was 21.5 ± 0.5 pg and for the group 1 was 14.5 ± 0.5 pg which statistically significant. These values increases with the increase in the concentration of the oil extract from group 2 to group 4. Similar pattern were observed for the mean corpuscular haemoglobin concentration (Table 2).

The estimated total white blood count for the control group was 5.55 ± 0.5 × 10^9/L and that for the group 4 was

Table 1. Haemoglobin concentration (mg/dl) and total estimated Red Blood Cells counts (x10^6) in adult wistar rats.

Experimental group	Haemoglobin concentration (mg/dl)	Red blood cell counts (x10^6)
Control	9.63 ± 0.3	4.45 ± 0.04
Group 1	9.70 ± 0.2	4.52 ± 0.01
Group 2	9.90 ± 0.6	4.79 ± 0.01
Group 3	10.30 ± 0.25	4.87 ± 0.02
Group 4	12.10 ± 0.2*	6.16 ± 0.15*

(*) indicates statistical significant at p ≤ 0.05. Data were expressed as mean ± standard deviation.

Table 2. Mean Corpuscular Volume, Mean Corpuscular Haemoglobin , and Mean Corpuscular Haemoglobin Concentration in adult wistar rats.

Experimental group	Mean corpuscular volume (MCV)	Mean corpuscular haemoglobin (MCH)	Mean corpuscular haemoglobin concentration (MCHC)
Control	53.5 ± 0.5	21.5 ± 0.5	39.01 ± 1.00
Group 1	57.5 ± 0.5	14.5 ± 0.5*	24.00 ± 0.01*
Group 2	60.5 ± 0.5	16.0 ± 0.1	28.50 ± 0.01
Group 3	62.5 ± 0.5	16.5 ± 0.5	34.00 ± 0.02
Group 4	63.0 ±1.0*	17.5 ± 0.5	36.50 ± 0.50

(*) indicates statistical significant at p ≤ 0.05. Data were expressed as mean ± standard deviation.

Table 3. White blood cells count (x10^9/l), estimated Neutrophil counts (%), and estimated total Lymphocyte counts (%) in adult wistar rats.

Experimental group	White blood cells count (x10^9/l)	Neutrophil counts (%)	Lymphocyte counts (%)
Control	5.55 ± 0.50	10.0 ± 0.10	81.5 ± 0.5
Group 1	5.65 ± 0.05	10.0 ± 0.10	84.5 ± 0.1
Group 2	6.70 ± 0.10	10.0 ± 0.10	88.0 ± 1.0
Group 3	7.20 ± 0.10*	12.0 ± 0.10	90.0 ± 0.1
Group 4	8.30 ± 0.10*	15.0 ± 0.10*	91.5 ± 05*

(*) indicates statistical significant at p ≤ 0.05. Data were expressed as mean ± standard deviation.

8.3 ± 0.3 × 10^9/L. The neutrophil counts was stable in the control, group 1 and 2 but steadily increases from group 3 to group 4 and this was significant. Similar observations were not recorded in the lymphocytes counts which show a linear increase along the concentration gradients of the administered oil extract (Table 3).

DISCUSSION

The study demonstrated the effect of varied concentration of oil extract of E. globulus on the haematological parameters in adult Wistar rats. The apparent increase in the haemoglobin concentrations across the experimental groups was dose dependent and this may be due to increased iron concentration present in the extract. This finding was corroborated by Osawa and Namiki (2005) that observed increased iron concentrations of various

extract of E. globulus. Though, the haemoglobin value was 9.6 mg/dl on the average for the control group, the steady rise in the values with increase in concentrations of the extract across the experimental groups may also be due to the presence of some of the phytochemical contents of eucalyptus globulus that may have increased the size of the red blood cells. This observation was corroborated Yakubu et al., 2008; Oyesomi and Ajao, 2011 where the roles of phytochemical agents in reproductive hormones and testis were demonstrated respectively.

The reduction of the mean corpuscular volume (MCV) and that of the mean corpuscular haemoglobin concentration (MCHC) by the oil extract of Eucalyptus globulus across the increased concentration gradient administered to wistar rats may be due to the decrease in size of the red blood cells produced which may partly explained the findings of the increase in the estimated total red blood cell counts that was recorded from the study. Though, some of the constituents of the extract may stimulate production of blood cells; these may be immature and may be of irregular shapes and sizes. The mechanisms by which this carried out is not fully understood and beyond the scope of the present study. However, this observation was corroborated by Osawa and Namiki (2005) and Medubi et al. (2010) in their various related studies.

The present study demonstrated a gradual increase in the estimated total white blood cell count (WBC) from that of those of the control groups and these appears to dose dependent. This may be result from the immune busting activities of some this medicinal plants as demonstrated by Adefolaju et al. (2009) where aqueous extract of plant have Hepatoprotective activity in rats. However, the selective busting of the lymphocytes component of the differential counts was not clearly understood and that post a future challenge for research. In conclusion, the study shows that the essential oil extract of *E. globulus* administered at increasing dosage used as outlined in the present study for the duration of one month enhanced the haemopoietic activities in wistar rats. The mechanisms for the observed increased may be due to the presence of some constituents of iron which are of great importance in the production of blood.

REFERENCES

Adefolaju GA, Ajao MS, Olatunji L A, Enaibe BU, Musa MG (2009). Hepatoprotective effect of aqueous extract of water leaf (*Talinum Triangulare*) on carbon tetrachloride (CCL4) induced liver damage in wistar rats. Int. J. Pathol., 8(1): 1-9.

Boukef K, Balanshad G, Lallemand M, Brenard P (1976). Study of flavonic heterosides and aglycones isolated from the leaves of *Eucalyptus globulus* (Hot Abstract 47:1899).

Brooker SG, Cammbie RC, Cooper RC (1999). New Zealand medicinal plants. Heinemann.

Duke JA, Wain KK (1981). Medicinal plants of the world. Computer index with more than 85,000 entries vols.

Junqueira LC, Carnerio L, Robert OK (2005). Textbook of basic histology, Lange publication, 9th Edition, Pg. 218-230.

List PH, Horhammer L (1969). Harger's Handbuck der pharmaceutischen praxis. 2(6). Springer-Veilag, Berlin.

Little ELJ (1983). Common fuel wood crops: A handheld book for their identification.

Medubi LJ, Ukwenya VO, Aderinto OT, Makanjuola VO, Ojo OA, Bamidele O, Ajao MS (2010). Effects of administration of ethanolic root extract of *Jatropha Gossypifolia* and predenisolone on the kidneys of wistar rats. Electron. J. Biomed., 2: 41- 48.

Morton JF (1981). Atlas of medicinal plants of Middle America, Bahamas to Yuscan.cc, Thomas, Springfeild, L.L.

National Institutes of Health (1985). Guide for the Care and Use of Laboratory Animals: DHEW Publication (NIH), revised. Office of Science and Health Reports, DRR/NIH, Bethesda, USA.

Oliver B (1960). Medicinal plants in Nigeria. Ibadan College of Arts and Sciences and Technology, Ibadan. p 358.

Osawa T, Namiki (2005). A novel type of anti-oxidant isolated from leaf wax of *Eucalyptus* leaves; and its suitabilities as an anticoagulant for biochemical and haematological analysis. Afr. J. Bio., 4(7): 679-689.

Oyesomi TO, Ajao MS (2011). Histological effect of aqueous extract of *Anacardium occidentale* (cashew) stems bark on adult wistar rat testis. Med. Prac. Rev., 2(7): 73-77.

Watt JM, Breyer-Brandwork MJ (1962). "Medicinal and poisonous plants of southern and eastern Africa", E&S Livingstone Edinburgh.

Yakubu MT, Akanji MA, Oladiji AT, Olatinwo AWO, Adesokan AA, Yakubu MO, Owoyele BVO, Sunmonu TO, Ajao MS (2008). Effect of *Cnidoscolous aconitifolius* (Miller) I. M. Johnston Leaf extract on Reproductive Hormones of Female Rats. Iranian J. Reprod. Med., 6(3): 149-155.

Screening of antioxidant activity, total phenolics and gas chromatograph and mass spectrometer (GC-MS) study of delonix regia

P. Maria jancy Rani[1]*, P. S. M. Kannan[2] and S. Kumaravel[3]

[1]Department of Chemistry, Dravidian University, Srinivasavanam, Chittoor District, Kuppam-517425, Andhra Pradesh, India.
[2]PRIST University, Vallam, Thanjavur-613 403, Tamil Nadu, India.
[3]Food Testing Laboratory, Indian Institute of Crop Processing Technology, Pudukkottai Road, Thanjavur-613 005. Tamil Nadu, India.

The present study was carried out for identification of the phytochemicals present in the *Delonix regia* leaves and also evaluates the total phenols, total flavonoids and antioxidant activity of the leaf extract. Total phenols were carried out by Folin Ciocalteu method and the phenolic content was 16.00 mg/100 g of gallic acid equivalent (GE). Antioxidant activity was evaluated by 2,2-Diphenyl-1-Picrylhydrazyl (DPPH) method and the leaves of *D. regia* showed 10.73 mg/100 g of ascorbic acid equivalent antioxidant capacity (AEAC). The gas chromatograph and mass spectrometer (GC-MS) study was also carried out and it showed the presence of phytochemicals like phytol (RT: 15.49), Cumarin 7, 8-dihyddro-7-hydroxy-6-methoxy-8-oxo (RT: 15.92), Squalene (RT: 25.41) and Vitamine (RT: 29.97).

Key words: Total phenols, total flavonoids, antioxidant activity, 2,2-Diphenyl-1-Picrylhydrazyl (DPPH), gas chromatograph and mass spectrometer (GC-MS), delonix regia.

INTRODUCTION

The *Delonix regia* belongs to the family of Caesalpinioideae. The members of the genus are flowering trees, native to the East Africa, has been used in traditional Indian medicine for the treatment of rheumatism, stomach disorders. (Thirugnanam et al., 2003) and its leaves are used in the treatment of bronchitis and pheumonia in infants. Leaf extracts of *D. regia* are reported for strong anti-inflammatory activity. (Sethuraman et al., 1986). The medicinal properties of plants have been investigated in the recent scientific developments throughout the world, due to their potential antioxidant activities, no side effects and economic viability (Auudy et al., 2003). Recently there has been an upsurge of interest in the therapeutic potentials of medicinal plants as antioxidants in reducing such free radical induced tissue injury. *D. regia* contains many polyphenolic compounds, terpenoids, tannins, cardiac glycosides, anthroquinones. Since these compounds have an antioxidant potential. The antioxidant potency of *D. regia* was investigated by employing various established *in vitro* systems. (Om Prakash and Yamini, 2007). The present study was carried out to study the flavonoids, antioxidant activity of *D. regia* and the chemical constituents were studied by GC-MS.

*Corresponding author. E-mail: mariajancyrani@yahoo.in.

Abbreviations: GE, Gallic acid equivalent; DPPH, 2,2-diphenyl-1-picrylhydrazyl; AEAC, acid equivalent antioxidant capacity; GC-MS, gas chromatograph and mass spectrometer; UV, ultraviolet; NIST, National Institute Standard and Technology; O.D, optical density.

MATERIALS AND METHODS

Collection and processing of plant material

The leaves of the plant *D. regia* collected from Thanjavur District in the month of July, 2010 and authenticated by Dr. John Britto, Rapinet Herbarium, ST. Joseph's College, Tiruchirappalli. The leaves were cleansed and shade dried for a week and grounded into uniform powder. 1 g of plant material was added to 20 ml of aqueous ethanol (20%, v/v) for 18 h at room temperature. The extracts were filtered and used for the estimation of total phenols and antioxidant activity.

Total phenols

0.5 ml of freshly prepared sample was taken and diluted with 8 ml of distilled water. 0.5 ml of Folin Ciocalteu Reagent (1 N) was added and kept at 40°C for 10 min. 1 ml of Sodium Carbonate (20%) was added and kept in dark for one hour. The color was read at 650 nm using Shimadzu ultraviolet (UV)-1650 Spectrophotmeter (Malick and Singh, 1980). The same procedure was repeated for all standard gallic acid solutions and standard curve obtained. The sample concentration was calculated as GE.

Total flavonoids

0.5 ml of aqueous extract of sample is diluted with 3.5 ml of distilled water at zero time and 0.3 ml of 5% sodium nitrate was added to the tubes. After 5 min, 0.3 ml of aluminium chloride (10%) was added to all the tubes. At the 6 min, 2 ml of sodium hydroxide (1 M) was added to the mixture. Immediately, the contents of the reaction mixture were diluted with 2.4 ml of distilled water and mixed thoroughly. Absorbance of the mixture was determined at 510 nm versus a prepared blank immediately. Gallic acid was used as the standard compound for quantification of total flavonoids as mg/100g (Zhisen et al., 1999).

Antioxidant activity

DPPH method

0.1 ml of the freshly prepared sample was taken in test tubes. 6 ml of DPPH solution (0.1 mM) was added and the tubes kept in dark for one hour. The color was read at 517 nm. The difference in the optical density (O.D) of DPPH solution and DPPH solution+ sample was calculated. The decrease in OD with sample addition is used for calculation of the antioxidant activity. Ascorbic acid standards were prepared in different concentrations and antioxidant was determined as ascorbic AEAC mg/100 g of sample (Koleva et al., 2002).

GC-MS analysis

Preparation of extract

Leaves of *D. regia* were shade dried. 20 g of the powdered leaves were soaked in 95% ethanol for 12 h. The extracts were then filtered through Whatmann filter paper No.41 along with 2 gm sodium sulfate to remove the sediments and traces of water in the filtrate. Before filtering, the filter paper along with sodium sulphate was wetted with 95% ethanol. The filtrate was then concentrated by bubbling nitrogen gas into the solution. The extract contained both polar and non-polar phytocomponents of the plant material used. 2 µl of these solutions was employed for GC-MS analysis (Merlin et al., 2009).

GC analysis

GC-MS analysis was carried out on a GC clarus 500 Perkin Elmer system comprising a AOC-20i autosampler and GC gas interfaced to a MS instrument employing the following conditions: column Elite-1 fused silica capillary column (30 × 0.25 mm ID × 1 µM df,

composed of 100% Dimethyl poly diloxane), operating in electron impact mode at 70 eV; helium (99.999%) was used as carrier gas at a constant flow of 1 ml /min and an injection volume of 0.5 µl was employed (split ratio of 10:1) injector temperature 250°C; ion-source temperature 280°C. The oven temperature was programmed from 110°C (isothermal for 2 min), with an increase of 10°C/min, to 200°C, then 5°C/min to 280°C, ending with a 9 min isothermal at 280°C. Mass spectra were taken at 70 eV; a scan interval of 0.5 s and fragments from 40 to 450 Da. Total GC running time is 36 min.

Identification of components

Interpretation on mass spectrum GC-MS was conducted using the database of National Institute Standard and Technology (NIST) having more than 62,000 patterns. The spectrum of the unknown component was compared with the spectrum of the known components stored in the NIST library. The name, molecular weight and structure of the components of the test materials were ascertained.

RESULTS AND DISCUSSION

Total phenolics and flavonoid content in the leaves of Delonix regia

It has been recognized that flavonoids show antioxidant activity and their effects on human nutrition and health are considerable. The mechanisms of action of flavonoids are through scavenging or chelating process (Kessler et al., 2003). Phenolic compounds are a class of antioxidant agents which act as free radical terminators (Om Prakash and Yamini, 2007). The flavonoid contents of the extracts in terms of GE (Table 1). Total phenolic content of the ethanolic extract of *D. regia* leaves is 16.00 mg/ 100 g of GE. The value of phenolic content indicates that the plant has antioxidant activity.

GC-MS study

The GC-MS study of *D. regia* leaves has shown many phytochemicals which contributes to the medicinal activity of the plant (Tables 2 and 3). The major components which present Benzenetriol (RT: 6.74), Butyl 8-methylnonyl ester (RT: 13.51), Lupeol (RT: 35.90) and Vitamin E (RT: 29.97). The other compounds like Hexadecanoic acid, 2-hydroxy-1- (RT: 21.20), 1, 6-Anhydro-a-D-glucopyranose (RT: 8.21) and 1,3,5-Benzenetriol also present in the leaves of *D. regia* (Figure 1). Figures 2, 3 and 4 shows mass spectrum and structure of phytol, lupeol and coumarin, 7,8-dihydro-7-hydroxy-6-methoxy-8-oxo- compound which is suggested to be a diterpenoid, triterpenoid and coumarin compound and is used as an anticancer, anti-inflammatory, antioxidant, antimicrobial and diuretic.

Conclusion

The   study   clearly   indicates   that the leaf extract was

Table 1. Total phenolics, flavonoids and antioxidant activity in the leaves of *Delonix regia*.

S/N	Parameter analyzed	Values obtained
1	Total phenols (mg/100 g) GE*	16.00
2	Total flavonoids (mg/100 g) GE*	0.20
3	Antioxidant activity (mg/100 g) AEAC**	10.73

The values are mean value of three replicates. Gallic acid equivalent, ascorbic acid equivalent antioxidant capacity.

Table 2. Phytocomponents identified in the ethanolic extract of the leaves of *Delonix regia* by GC-MS.

RT	Name of the compound	Peak area (%)
2.17	Butane, 1,1-diethoxy-2-methyl-	0.49
3.03	Propane, 1,1,3-triethoxy-	0.47
6.74	1,2,3-Benzenetriol[Synonyms: Pyrogallol]	50.51
8.21	1,6-Anhydro-á-D-glucopyranose(levoglucosan)	2.05
9.43	1,3,5-Benzenetriol	1.73
11.16	3-O-Methyl-d-glucose	40.17
12.03	3,7,11,15-Tetramethyl-2-hexadecen-1-ol	0.81
13.51	1,2-Benzenedicarboxylic acid, butyl 8-methylnonyl ester	0.14
15.49	Phytol	1.59
15.92	Coumarin, 7,8-dihydro-7-hydroxy-6-methoxy-8-oxo-	0.18
21.20	Hexadecanoic acid, 2-hydroxy-1 (hydroxymethyl)ethyl ester	0.10
25.41	Squalene	0.10
29.97	Vitamin E	1.19
35.90	Lupeol	0.47

Table 3. Activity of phytocomponents identified in *Delonix regia* extract- by GC-MS.

RT	Name of the compound	Compound nature	**Activity
2.17	Butane, 1,1-diethoxy-2-methyl-	Ether compound	No activity reported
3.03	Propane, 1,1,3-triethoxy-	Ether compound	No activity reported
6.74	1,2,3-Benzenetriol [Synonyms: Pyrogallol]	Pyrogallol	Antioxidant, antiseptic antibacterial, antidermatitic fungicide, pesticide, antimutaginic dye candidicide
8.21	1,6-Anhydro-á-D-glucopyranose (levoglucosan)	Sugar moiety	Preservative
9.43	1,3,5-Benzenetriol	Poly Hydroxy compound	Antioxidant, antiseptic antibacterial, antidermatitic fungicide, pesticide, antimutaginic dye candidicide

Table 3. Contd.

11.16	3-O-Methyl-d-glucose	Sugar moiety	Preservative
12.03	3,7,11,15-Tetramethyl-2-hexadecen-1-ol	Terpene alcohol	Antimicrobial, antiinflammatory
13.51	1,2-Benzenedicarboxylic acid, butyl 8-methylnonyl ester	Plasticizer compound	Antimicrobial, antifouling
15.49	Phytol	Diterpene	Antimicrobial, antiinflammatory, anticancer diuretic
15.92	Coumarin, 7,8-dihydro-7-hydroxy-6-methoxy-8-oxo-	Coumarin compound	Antimicrobial
21.20	Hexadecanoic acid, 2-hydroxy-1-(hydroxymethyl)ethyl ester	Fatty acid ester	No activity reported
25.41	Squalene	Triterpene	Antibacterial, antioxidant, antitumour, cancer preventive, immunostimulant, chemo preventive and lipoxygenase-inhibitor pesticide
29.97	Vitamin E	Vitamin compound	Antiageing, analgesic, antidiabatic antiinflammatory, antioxidant, antidermatitic, antileukemic, antitumour, anticancer, hepatoprotective, hypocholesterolemic antiulcerogenic, vasodilator, antispasmodic, antibronchiti and anticoronary
35.90	Lupeol	Triterpenoid compound	Antimicrobial, antiinflammatory and anticancer

**Source: Dr. Duke's phytochemical and ethnobotanical databases (Online database).

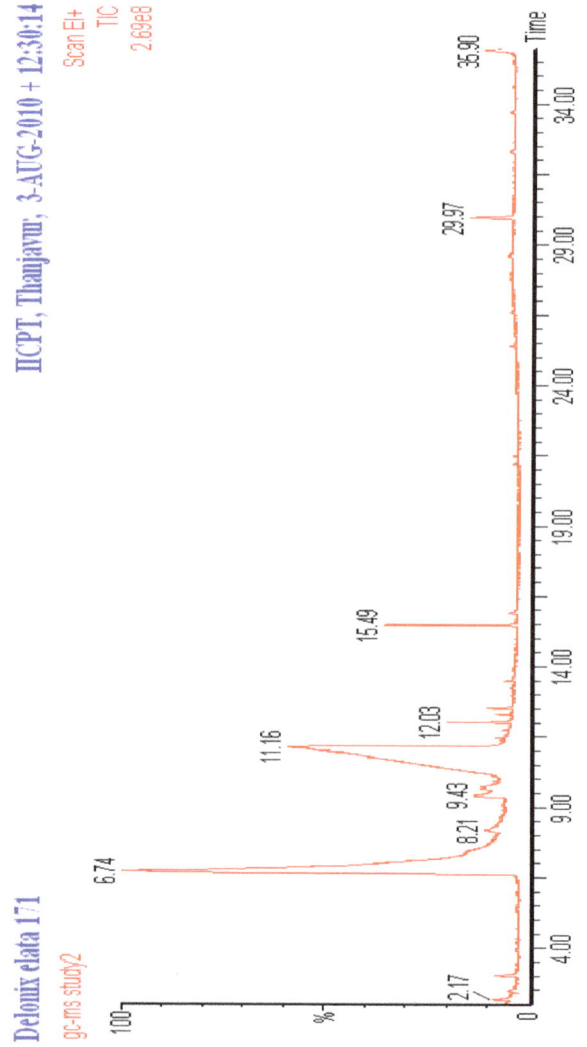

Figure 1. Chromatogram of *Delonix regia* leaves by GC-MS.

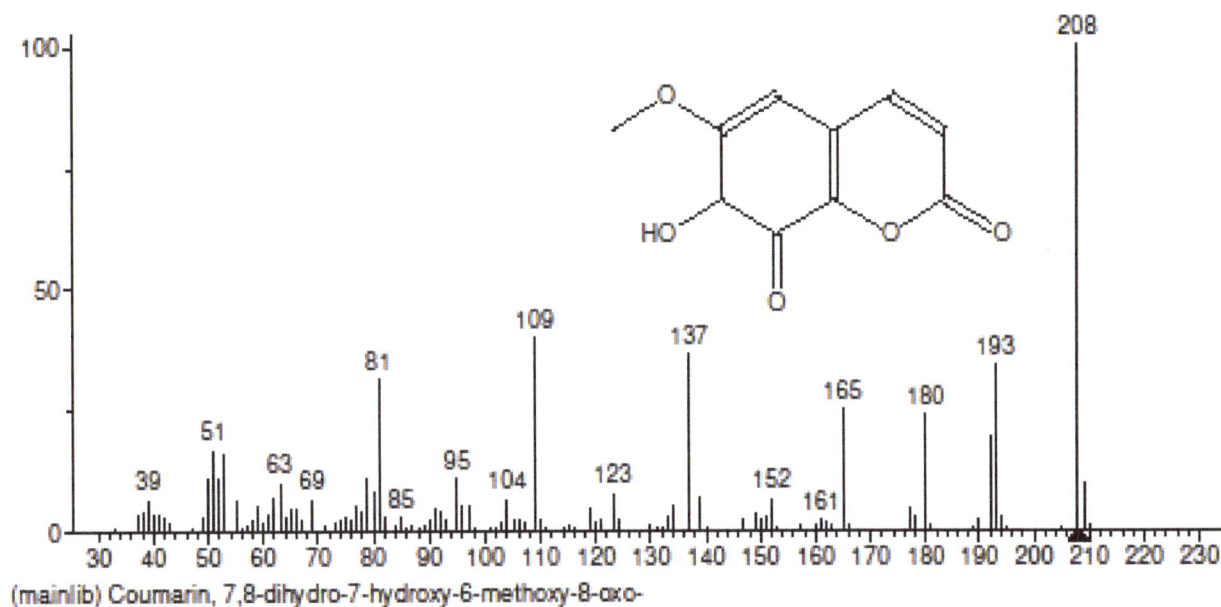

(mainlib) Coumarin, 7,8-dihydro-7-hydroxy-6-methoxy-8-oxo-

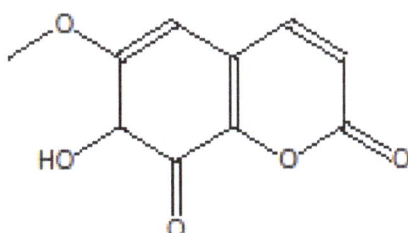

Name: Coumarin, 7,8-dihydro-7-hydroxy-6-methoxy-8-oxo-
Formula: $C_{10}H_8O_5$
MW: 208 CAS#: N/A NIST#: 263136 ID#: 114276 DB: mainlib
Other DBs: None
Contributor: A.A.Kutin, Moscow, Russia
10 largest peaks:
  208 999 | 109 398 | 137 363 | 193 340 | 81 311 | 165 251 | 180 239 | 192 194 | 51 167 | 53 161 |
Synonyms:
no synonyms.

Figure 2. Mass spectrum and structure of coumarin, 7, 8-dihydro-7-hydroxy-6-methoxy-8-oxo compound.

normal in antioxidants, phenolics and flavonoids. The GC-MS study also showed many Phytochemicals Benzenetriol, Butyl 8-methylnonyl ester, Lupeol, VitaminE, Hexadecanoic acid, 2-hydroxy-1- ,1,6-Anhydro-a-D-glucopyranose, 1,3,5-Benzenetriol, Phytol, Coumarin, 7,8-dihydro-7-hydroxy-6-methoxy-8-oxo-, phytol, lupeol and Squalene which contributes the activities like antimicrobial, antioxidant, anticancer, hypercholesterolemic, antiulcerogenic and other activities. This investigation has helped to identify the compounds present in the leaves of *D. regia*, a hitherto uninvestigated species. This investigation has helped to identify the compounds present in the leaves of *D. regia*, evaluation of pharmacological activity in the ethanol extract is in progress.

ACKNOWLEDGEMENT

The authors thank very much Dr. K. Alagusundaram, Director, Indian Institute of Crop Processing Technology, Thanjavur for providing all the facilities and support used to carry out the work.

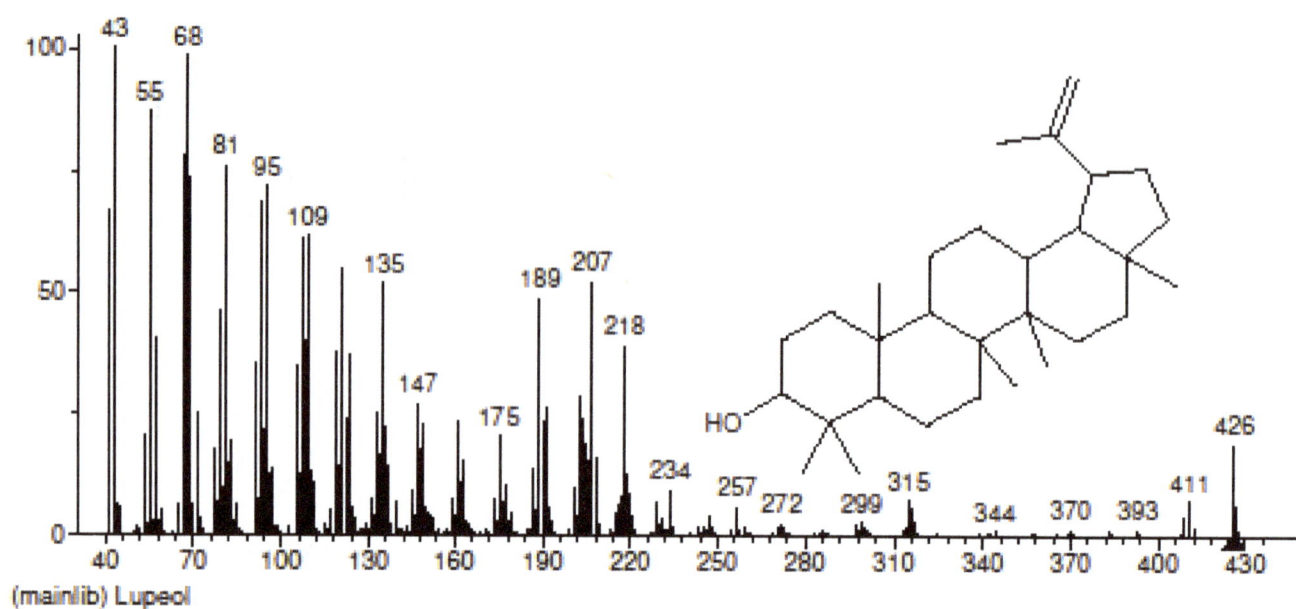

(mainlib) Lupeol

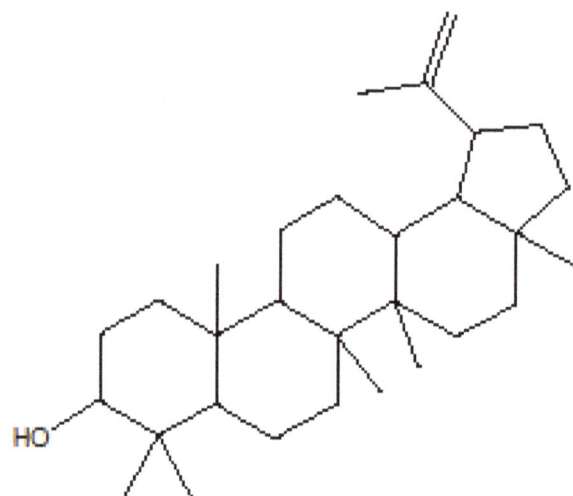

Name: Lupeol
Formula: $C_{30}H_{50}O$
MW: 426 CAS#: 545-47-1 NIST#: 124852 ID#: 7258 DB: mainlib
Other DBs: None
Contributor: H. Fales, LC, NHLBI, NIH, Bethesda, MD 20892
10 largest peaks:
  43 999 | 68 983 | 55 869 | 67 782 | 81 759 | 69 733 | 95 716 | 93 684 | 41 667 | 109 613 |
Synonyms:
1.Lup-20(29)-en-3-ol, (3.beta.)-
2.Lup-20(29)-en-3.beta.-ol
3..beta.-Viscol
4.Clerodol
5.Fagarasterol
6.Fagarsterol
7.Lupenol
8.Monogynol B
9.Triterpene lupeol

Figure 3. Mass spectrum and structure of pentacyclic lupeol.

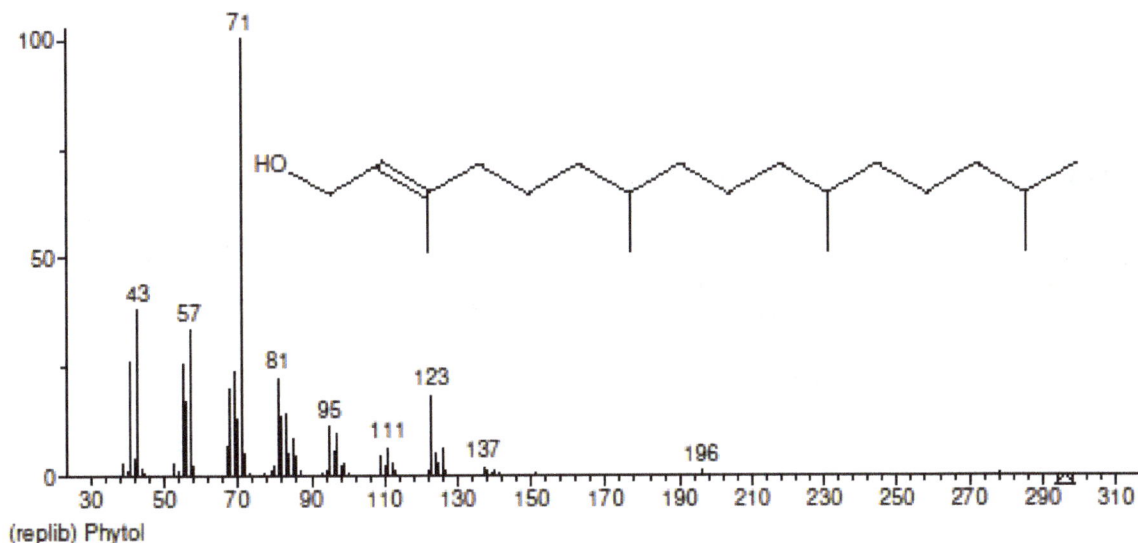

Name: Phytol
Formula: $C_{20}H_{40}O$
MW: 296 CAS#: 150-86-7 NIST#: 108727 ID#: 8157 DB: replib
Other DBs: None
Contributor: Philip Morris R&D
10 largest peaks:
    71 999 | 43 381 | 57 334 | 41 260 | 55 259 | 69 239 | 81 223 | 68 199 | 123 184 | 56 169 |
Synonyms:
1.2-Hexadecen-1-ol, 3,7,11,15-tetramethyl-, [R-[R*,R*-(E)]]-
2.trans-Phytol
3.3,7,11,15-Tetramethyl-2-hexadecen-1-ol
4.(2E)-3,7,11,15-Tetramethyl-2-hexadecen-1-ol #

Figure 4. Mass spectrum and structure of phytol.

REFERENCES

Auudy B, Ferreira F, Blasina L, Lafon F, Arredondo F, Dajas R, Tripathi PC (2003). Screening of antioxidant activity of three Indian medicinal plants, traditionally used for the management of neurodegenerative diseases. J. Ethnopharmacol., 84: 131-138(s).

Kessler M, Ubeaud G, Jung L (2003). Anti-and pro-oxidant activity of rutin and quercetin derivatives. J. Pharm. Pharmacol., 55: 131-142.

Koleva II, Van Beek TA, Linssen JPH, de Groot A, Evstatieva LN (2002). Screening of plant extracts for antioxidant activity: a comparative study on three testing methods. Phytochem. Anal. 13: 8-17.

Malick CP, Singh MB (1980). In: Plant Enzymology and Histo Enzymology, Kalyani Publishers, New Delhi, p. 286.

Merlin NJ, Parthasarathy V, Manavalan R, Kumaravel S (2009). Chemical Investigation of Aerial Parts of Gmelina asiatica Linn by GC-MS. Pharmacognosy Res., 1(3): 152-156.

Om Prakash T, Yamini BT (2007). Antioxidant properties of different fractions of Vitex negundo Linn. Food Chem., 100(3): 1170-1176.

Sethuraman MG, Sulochana N (1986). The anto-inflammatory activity of Delonix elata. Curr. Sci., 55: 343.

Thirugnanam S, Mooligai M (2003). (Tamil) Trichy: Selvi Publishers, p. 33, 117, 131, 139, 147.

Zhisen J, Meng CT, Jianming W (1999). Food Chem., 64: 555-559.

Changes in haemorrheologic and fibrinolytic activities upon hypertension and diabetic chemotherapy in Calabar diabetic residents, Nigeria

M. S. Edem, A. O. Emeribe and J. O. Akpotuzor*

Department of Heamatology, University of Calabar Teaching Hospital, Calabar, Cross River State, Nigeria.

This research was carried out to assess the effect of hypertension on haemorrheologic and fibrinolytic activities in fifty (50) diabetics resident in Calabar municipality and the values obtained were compared with those of fifty (50) age and sex-matched non diabetics in the same locality. Relative plasma viscosity plasma fibrinogen concentration, euglobulin lysis time and fasting blood sugar were determined using standard methods. The relative plasma viscosity, plasma fibrinogen concentration, euglobulin lysis time and the fasting blood sugar values were significantly higher in diabetics ($P < 0.05$) when compared with the controls. Correlation between RPV and duration of diabetes was positive and significant ($r = 0.323$, $p < 0.05$). Also, correlation between fasting blood sugar and plasma fibrinogen concentration was positive and significant ($r = 0.635$, $p < 0.05$).There was no significant increase in RPV, PFC and ELT of type I diabetes when compared with type II diabetes ($P > 0.05$). The RPV and ELT of diabetics with hypertension was increased, but showed no significant difference ($P > 0.05$) with that of those who had no hypertension. However, the PFC of diabetics with hypertension was significantly increased ($P < 0.05$) when compared with that of the control group. The diabetics who were on oral hypoglycaemic agents and insulin showed no significant difference ($P > 0.05$) in RPV, PFC and ELT when compared with those who were on combination therapy of oral hypoglycaemic agents (glanil, glucophage), but there was significant increase when RPV, PFC and ELT values were compared with the controls ($P < 0.05$). This work shows that defective rheology and poor fibrin clearing may be the contributory factor to vascular and thrombotic complications seen in diabetics.

Key words: Hypertension, haemorrheologic, fibrinolytic, diabetes mellitus.

INTRODUCTION

Diabetes mellitus, a syndrome characterized by chronic hyperglycemia due to absolute or relative deficiency of insulin is estimated to afflict over 170 million people world wide (Wokoma, 2002) and this represents about 2% of the world's population. In Nigeria, about 1 - 7% of the population is affected, with over 90% of these being non-insulin dependent (Fabiyi et al., 2002). Diabetes has been classified based on the clinical staging and etiology into type I and type II. Type I or insulin dependent diabetes mellitus (IDDM) or juvenile on-set diabetes is characterized by beta cell destruction leading to absolute

insulin deficiency, which may be due to auto-immune mechanism and patients are prone to ketoacidosis (WHO, 1999). Type II or non-insulin dependent diabetes mellitus (NDDM) or adult onset diabetes is predominated by insulin resistance with relative insulin deficiency or an insulin secretary defect. Here, ketoacidosis is infrequent and obesity is a predisposing factor (WHO, 1999).

Most adverse diabetes outcomes are a result of vascular complications both at macro vascular and micro vascular levels (Hogan et al., 2003). Macro vascular complications are more common and up to 80% of patients with type II diabetes will develop or die of macrovascular disease (Evans et al., 2002). Atheroscle-rosis an example of macro vascular disorder has been recognized as a major cause of mortality in diabetic population (Benett, 1999) and also implicated in the

*Corresponding author. E mail: akpotuzor@yahoo.com.

Table 1. Comparison of fasting blood sugar (FBS), relative plasma viscosity (RPV), plasma fibrinogen concentration (PFC) and euglobulin lysis (ELT) in diabetic and non-diabetic subjects.

Subjects	FBS (mg/100 ml)	RPV	PFC (g/l)	ELT (min)
Diabetics (n = 50)	184.14 ± 87.30	1.65 ± 0.08	5.22 ± 1.89	268.4 + 56.9
Non-diabetics (n = 50)	69.48 ± 6.30	1.49 + 0.06	2.50 + 0.65	147.6 + 46.96
P value	< 0.05	< 0.05	< 0.05	< 0.05

circulatory disturbances seen in diabetics (Colwell et al., 1992).

Plasma viscosity in nephrotic patients and haemorrheologic and fibrinolytic activity in hypertensive Nigerians have been studied (Oviasu et al., 1998, Aigbe and Famodu, 1999). Again reports on alteration in platelet count and activity, coagulopathy and fibrinolytic aberration in diabetic subjects have also been made (Colwell et al., 1992; Adediran et al., 2004) with scanty information on haemostatic changes in diabetic Nigerians particularly those who are on chemotherapy as well as those with hypertension. The increasing prevalence of diabetes and the associated financial burden on our nation's economy is a major challenge for the health sector. Therefore preventive measures may be as important as measures towards discovering a permanent cure for the disease. This work is aimed at determining the plasma viscosity, euglobulin lysis time, and fibrinogen levels in diabetics on chemotherapy as well as those with hypertension and comparing them with non-diabetics who are nomotensive residents in Calabar Municipality. It is believed that this will aid in a better understanding of the haemorrheology of diabetic patients.

MATERIALS AND METHODS

A total of fifty (50) diabetic subjects were included in this study. They were 20 males and 30 females aged between 35 - 75 years attending the diabetic clinic of University of Calabar Teaching Hospital. Of these, 11 (6 females and 5 males) diabetics were with hypertension, while 39 (24 females and 15 males) were without hypertension. Three (3) subjects were on monotherapy (insulin) while the remaining 47 subjects were either on insulin and oral hypoglycaemic agents or a combination therapy of oral hypoglycaemic agents (glanil, glucophage). Diabetes in this study had a fasting plasma glucose levels greater than 126 mg/100 ml in two or more occasions (WHO, 1999). As at the time of study, there were no signs of coexisting disease capable of explaining the abnormally high concentration of glucose as observed from their hospital folders (medical history) and their personal data obtained via a comprehensive questionnaire after due approval from the ethical committee of the hospital. The subjects had diabetes mellitus for an average of 1 – 7 years. Hypertension in this study was defined as a diastolic pressure of 90 mmHg or above, measured while the subject was sitting. The mean of two blood pressure readings was recorded after the subject sat for about 10 min. Fifty age-matched non-diabetic apparently healthy volunteers (30 males and 20 females) living in Calabar municipality were used as controls in this study. They were selected from blood donors, staff of University of Calabar Teaching Hospital and workers of Calabar municipal council. Informed consent was obtained from all

the participants.

Seven milliliters (7 ml) of venous blood was collected from each subject and 4.5 ml was added to 0.5 ml of sodium citrate anticoagulant (31.3 g/l) for coagulation studies while 2 ml was dispensed into fluoride oxalate bottle for the determination of fasting blood sugar. For the coagulation studies, the whole blood was spun at 3000 revolution per minute for 10 min to obtain platelet poor plasma required for the analysis. Tests were performed within 3 h of sample collection and in duplicates. Standard methods of Haugie, (1986), Reid and Ugwu, (1987), Ingram's and Hills (1976) and Nelson (1944) were employed for the determination of euglobulin lysis time, relative plasma viscosity, plasma fibrinogen concentration and fasting blood sugar levels respectively. The results were expressed as mean ± standard deviation and students' t-test for paired means was used for statistical comparison

RESULT

Table 1 shows the means and standard deviations of the various parameters that were analyzed. The mean fasting blood sugar of diabetic patients 184.14 ± 87.30 mg/100 ml, (normal range 63 - 99 mg/100) was significantly higher than that of the control subjects which was 69.48 ± 6.3 mg/100 ml. The mean relative plasma viscosity of diabetics was 1.65 ± 0.08 (normal range, 1.47 - 1.86) and this showed a significant increase (p < 0.05) when compared with the control subjects (1.49 ± 0.06). The mean fibrinogen level in diabetics was 5.22 ± 89 g/l, (normal range, 1.5 - 4.0 g/l) and this was found to be significantly higher (p < 0.05) than that of the control subjects (2.5 ± 0.65 g/l). The euglobulin lyses time measured in minutes was 268.4 ± 56.9 (90 - 240 min) for the diabetics and was significantly higher (p < 0.05) than 147.6 ± 46.96 obtained for the controls.

Table 2 shows mean values of the patients with type I diabetes in comparison with those who had type II diabetes. The values obtained showed no statistical difference (P > 0.05). Figure 1 shows a positive and significant correlation between RPV and duration of diabetes among the diabetic subjects (r = 0.323, p < 0.05). Figure 2 shows a positive and significant correlation between fasting blood sugar and plasma fibrinogen concentration among the diabetic subjects (r = 0.635, p < 0.05).

Table 3 shows haemorrheologic and fibrinolytic activities among diabetics with respect to presence of hypertension. The diabetics with hypertension had mean relative plasma viscosity of 1.69 ± 0.08 and this showed no statistical difference when compared with 1.64 ± 0.07

Table 2. Haemorrheologic and fibrinolytic activities in (50) diabetics based on type of diabetes mellitus.

Type of DM	RPV	PFC (g/l)	ELT (min)
Type I (N = 16)	1.635 ± 0.07	4.8 ± 1.97	265.93 ± 55.65
Type II (N = 34)	1.66 ± 0.08	5.41 ± 1.84	269.56 ± 58.41
P value	> 0.05	> 0.05	> 0.05

Table 3. Haemorrheologic and fibrinolytic activities among diabetic patients with respect to presence of hypertension.

Subjects	RPV	PFC (g/l)	ELT (min)
Diabetics with hypertension (n = 11)	1.69 ± 0.08	6.27 ± 1.85	278.64 ± 55.05
Diabetics without hypertension (n = 39)	1.64 ± 0.07	4.92 ± 1.81	263.97 ± 58.39
P value	> 0.05	< 0.05	> 0.05

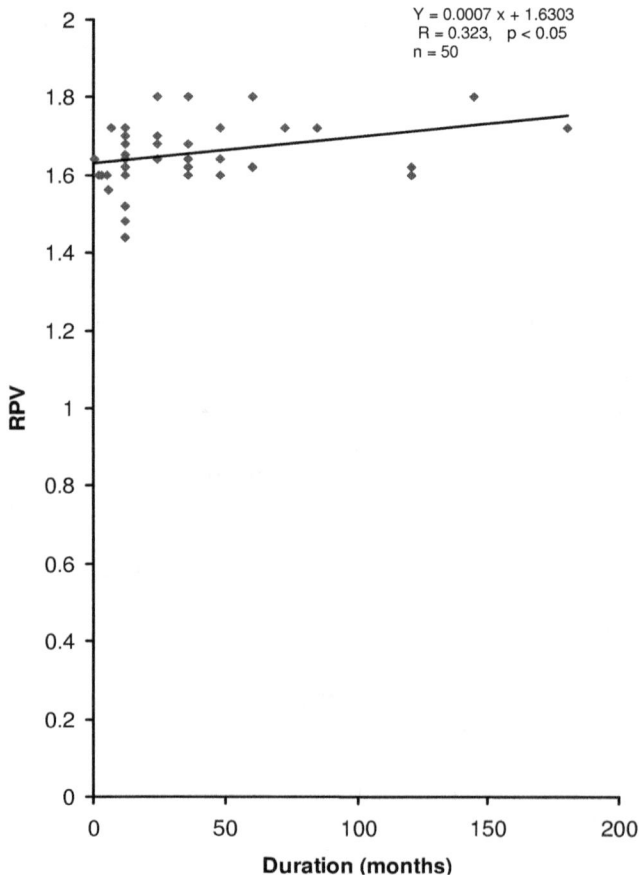

Figure 1. Correlation graph between relative plasma viscosity and duration of diabetes.

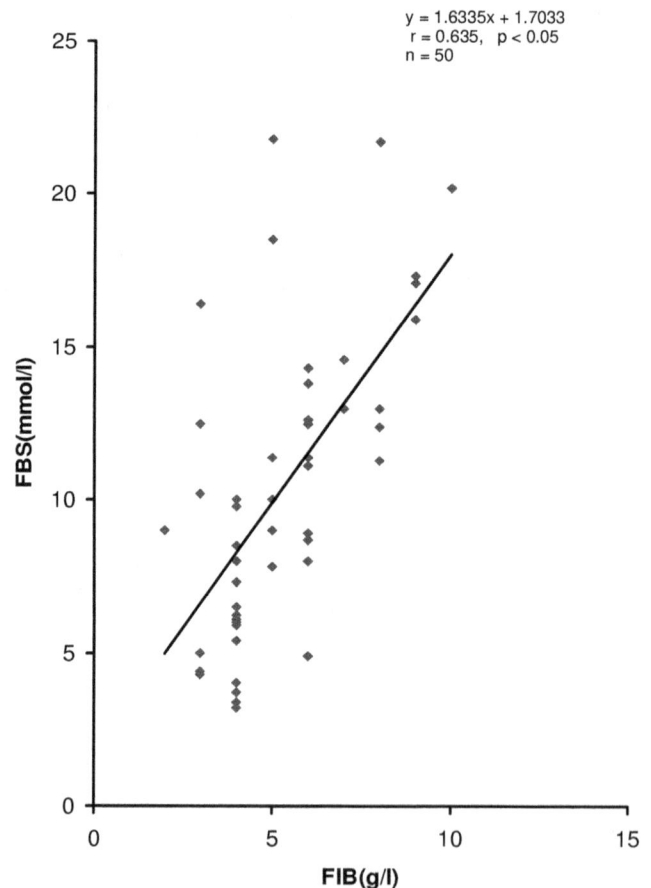

Figure 2. Correlation graph between fasting blood sugar and plasma fibrinogen concentration in diabetic subjects.

obtained for the diabetics without hypertension (p > 0.05). The mean fibrinogen level of diabetics with hypertension was 6.27 ± 1.85 g/l and that of the diabetics without hypertension was 4.92 ± 1.81g/l. There was a significant difference between them (p < 0.05). The mean ELT of diabetics with hypertension was 278.64 ± 55.05 min and this showed no statistical difference (p > 0.05) with 263.97 ± 58.39 min obtained for diabetics without hypertension. Table 4 shows haemorrheologic and fibrinolytic activities among diabetics subjects based on

Table 4. Haemorrheologic and fibrinolytic activities among diabetic patients based on chemotherapy.

Chemotherapy	RPV	PFC (g/l)	ELT (min)
A: Monotherapy (insulin) (n = 3)	1.61± 0.01	4.0 ± 1.41	253.33 ± 71.12
B: Oral hypoglycaemic agents and insulin (n = 6)	1.63 ± 0.09	5.5 ± 2.15	271.67 ± 57.07
C: Combination therapy of Oral hypoglycaemic agents (glanil, glucophage) (n = 41)	1.66 ± 0.08	5.39 ± 1.86	272.68 ± 58.16
D: Non diabetics (n = 50)	1.49 ± 0.06	2.50 ± 0.65	147.60 ± 46.96
P value (A vs. B)	> 0.05	> 0.05	> 0.05
(B vs. C)	> 0.05	> 0.05	> 0.05
(C vs. D)	< 0.05	< 0.05	> 0.05
(D vs. A)	< 0.05	< 0.05	< 0.05

chemotherapy. The diabetics on insulin injection alone had mean RPV of 1.61 ± 0.01, PFC of 4.0 ± 1.41 g/l, and ELT of 253.33 ± 71.12 min while those on oral hypoglycaemic agents and insulin injection (n = 6) had mean RPV of 1.63 ± 0.09, PFC of 5.5 ± 2.15 g/l, and ELT of 271.67 ± 57. 07 min. Those on oral combination therapy of hypoglycaemic agents alone like glanil and glucophage (n = 41) had mean RPV of 1.66 ± 0.08, PFC of 5.39 ± 1.86 g/l and ELT of 272.68 ± 58.16 min. No significant difference was observed when the parameters were compared in pairs ($p > 0.05$). However when either of the means obtained from the three groups of diabetics were compared with mean RPV (1.49 ± 0.06), PFC (2.50 ± 0.65 g/l) and ELT (147.6 ± 46.96 min) obtained from the non-diabetic subjects, there was significant difference between them ($p < 0.05$).

DISCUSSION

In this study, relative plasma viscosity (RPV), plasma fibrinogen concentration (PFC) and euglobulin lysis times (ELT) were determined in diabetic and control subjects. Significantly increased RPV (1.65 ± 0.08), PFC (5.22 ± 1.89g/l), and prolonged ELT (268.4 ± 56.9 min) were observed when compared with apparently healthy controls who had 1.49 ± 0.06, 2.50 ± 0.65 g/l, and 147.60 + 46.96 min respectively. This implies defective fibrinolysis and hyperviscous plasma states in diabetics. This raised plasma viscosity causes sluggish flow in microcirculation and results in insufficient tissue perfusion, which could predispose diabetics to high risk for peripheral arterial and heart disease (Baker, 1991). The findings in this study also agree with previous work done by Aigbe and Famodu (1999), who observed that increased plasma viscosity may increase the peripheral resistance, thus promoting elevation of blood pressure. This research observed a significant and positive correlation ($r = 0.323$, $p < 0.05$) between RPV and duration of diabetes. The reason for this type of finding may be as a result of the varying plasma protein concentrations at different durations of diabetes. This agrees with the findings of Aigbe and Famodu (1999) who reported that RPV showed inconsistent changes

within the first five years but increased there after.

Fibrinogen levels (6.27 ± 1.85 g/l) showed significant increase ($P < 0.05$) in diabetics with hypertension when compared with those who had no hypertension (4.92 ± 1.81 g/l). The hyperfibrinogenaemia and concomitant hyperviscous plasma observed in this study could either play a role in the pathogenesis of the hypertension or be the consequence of the hypertension itself (Reid and Ojogwu, 1992). Infiltration of the vessel wall by fibrinogen due to increased blood viscosity, increased platelet aggregation and thrombus formation and increased fibrin formation are several pathways by which acute or chronic increase in fibrinogen levels can lead to a cardiovascular event (Vinik et al., 2001) however, persistence of fibrin deposited in the vessel wall is believed to play an important role in the development of hypertension. A defective rheology and fibrin clearing mechanisms may contribute to the aetiology of vascular complication in hypertensive patients especially in the long term (Aigbe and Famodu, 1999). Fibrinogen level in this study had a significant positive correlation with fasting blood sugar among the diabetic subjects ($r = 0.635$, $p < 0.05$). Previous studies in the general as well as in smaller populations have also reported positive correlation between plasma fibrinogen and glucose concentrations. It is therefore possible that spontaneous hyperglycaemia contributed to fibrinogen synthesis under basal condition (Festa et al., 2002).

The ELT of diabetics with hypertension was apparently higher (278.64 ± 55.05 min) than that of the subjects with only diabetes (263.97 ± 58.39 min) though no statistical significance was observed. Adediran et al. (2004) had reported prolonged ELT in diabetic population as an additional risk factor for thromboembolic diseases. A defective rheology and fibrin-clearing mechanism may contribute to the aetiology of vascular complications. This study also observed that there was no significant difference in RPV, PFC, and ELT of type I when compared with type II diabetes. The haemorrheologic and fibrinolytic activities in diabetics showed significant difference ($P < 0.05$) when compared with that of control subjects despite chemotherapy. However, when the comparison was made between those on monotherapy (insulin alone) and combination therapy (insulin with oral hypoglycaemic agents) and those taking combination of oral hypoglycaemic agents only (glanil, glucophage and metformin) no significant difference was observed in all the parameters ($P > 0.05$). The diabetic patients still had increased values in RPV, ELT and PFC when compared with the control subjects despite being on one form of hypoglycaemic agent or the other. This work has given base line values of RPV, PFC and ELT in diabetics and control subjects in this locality. It has also established that there is a significant increase in haemorrheology and reduced fibrinolytic activities in diabetic subjects when compared with the non-diabetic (control) subjects. In conclusion, the study observed higher defective rheology and fibrin clearing in diabetes with hypertension in Calabar. Furthermore mode of chemotherapy plays little or no role on the rheology and fibrin clearing in diabetes. This defective rheology and fibrin clearing seen in diabetes may be contributory factors to the vascular and thrombotic complications usually observed in diabetic subjects.

REFERENCES

Adediran IA, Ikem RT, Borisade MF (2004). Fibrinolytic activity in Nigerian diabetics. Post Graduate Med. J. 80: 610-612.
Aigbe A, Famodu AA (1999). Haemorrheologic and fibrinolytic activities in Nigerians. Clin. Hemorrheol. Microcirculation 21: 415 -410.
Baker IA (1991). Fibrinogen, viscosity and white blood cell counts are major risk factors for ischeamic heart disease. The caerphilly and speed well collaborative heart disease studies. Circulation 83: 836-844.
Bennett PH (1999). Impact of the new WHO classification and diagnostic criteria. Diabetes Obesity and Metabolism 1: 51–56.
Colwell JA, Lyons TJ, Klein RL (1992). New concepts about the pathogenesis of atherosclerosis and thrombosis in diabetes mellitus. In: Levin ME, O'Neal LW, Bowker JH, eds. The diabetic foot St Louis: Mosby Yearbook 79–114.

Evans JM, Wang J, Morris AD (2002). Comparison of cardiovascular risk between patients with type 2 diabetes and those who had had a myocardial infraction: Cross sectional and cohort studies. Br. Med. J. 324: 939-942.
Fabiyi AK, Kolawole BA, Adeshinto O, Ikem RT (2002). The impact of knowledge attitude practice and belief of type 2 Nigerian diabetes patients on drug compliance. Diab. Int. 291: 15-17.
Festa AD, Agostino R Jr, Tracy RP, Haffner SM (2002). The insulin resistance atherosclerosis study, elevated levels of acute protein and plasminogen activator inhibitor predict the development of type 2 diabetes. Diabetes 4: 1131-1137.
Hogan P, Dall T, Nikolov P (2003). Economic cost of diabetes in the US in 2002. Diabetic Care 26: 917-932.
Haugie C (1986). Euglobulin lysis time method. In William J W, Benller E, Esley AJ, Lichman N (Eds.), Haematology (3rd edition) (pp. 1670-1678). Singapore: McGraw Hill.
Ingram GIC, Hills M (1976). International committee communications. Reference methods for one stage prothrombin time test on human blood. Thrombo Haemostat. 36: 337-338.
Nelson N (1944). Glucose in blood and cerebrospinal fluid (copper reduction). J. Biochem. 153: 375-376.
Oviasu E, Famodu AA, Ojeh EA (1998). Plasma Viscosity in nephrotic Nigerians. Clin. Haemorrheol. 15: 897-899.
Reid HL, Ojogwu LT (1992). Hyperfibrinogenemia and hyperviscous plasma in hypertensive. Afr. Angiol. 5: 249-256.
Reid HL, Ugwu AC (1987). A sample technique for rapid determination of plasma viscosity. Nig. J. Physiol. Sci. 3: 45-48.
Vinik AI, Erbas T, Park TS (2001). Platelet dysfunction in type 2 diabetes. Diabetes Care 24: 1476–1485.
Wokoma FS (2002). Diabetes and hypertension in Africa, an overview. Diab. Int. 12(12): 36-40.
World Health Organization (1999). Diabetes mellitus and its complications. Report of WHO consultation (part 1): Diagnosis and classification of diabetes mellitus. Bulletin of World Health Organization 99(2): 1-58.

Impact of *Anthocleista vogelii* root bark ethanolic extract on weight reduction in high carbohydrate diet induced obesity in male wistar rats

Anyanwu, G. O.[1], Onyeneke, E. C.[2], Usunobun, U.[3] and Adegbegi, A. J.[4]

[1]Department of Biochemistry, Bingham University, Karu, Nasarawa State, Nigeria.
[2]Department of Biochemistry, University of Benin, Benin City, Edo State, Nigeria.
[3]Department of Biochemistry, Benson Idahosa University, Benin City, Edo State, Nigeria.
[4]Department of Science Laboratory Technology, Rufus Giwa Polytechnic, Owo, Ondo State, Nigeria.

Obesity increases the risk of developing a number of diseases such as insulin resistance, type 2 diabetes, hypertension, hypercholesterolemia, stroke and heart attack. The aim of this study was to investigate the impact of ethanolic root bark extract of *Anthocleista vogelii* on weight reduction in high carbohydrate diet (HCD) induced obesity in male Wistar rats. Thirty male Wistar rats were housed in three steel cages containing 10 rats each. For the period of obesity induction, Group 1 was fed with normal pellet diet (NPD), while Groups 2 and 3 were fed with HCD for 14 weeks. During the 4 weeks treatment period, Group 1 was fed with NPD, Groups 2 and 3 were fed with HCD, and only Group 3 received 500 mg/kg b.w *A. vogelii* extract. The ethanolic root bark extract of *A. vogelii* significantly decreased (P<0.05) food intake, body weight, total fat mass, adiposity index and low density lipoprotein cholesterol, but showed no significant difference (P<0.05) in body mass index, total cholesterol, triglycerides, high density lipoprotein cholesterol, very low density lipoprotein cholesterol when compared with the HCD obese control. The results indicated that the ethanolic root bark extract of *Anthocleista vogelii* has potential to reduce weight in animals.

Key words: *Anthocleista vogelii,* obesity, body mass index, total fat mass, adiposity index, lipid profile, high carbohydrate diet.

INTRODUCTION

Obesity is a medical condition in which excess body fat has accumulated to the extent that it may have an adverse effect on health, leading to reduced life expectancy and/or increased health problems (WHO, 2000). Obesity also increases the risk of developing a number of chronic diseases including: insulin resistance, Type-2 diabetes, high blood pressure, hypercholestero-lemia, stroke, heart attack, congestive heart failure, gallstones, gout and gouty arthritis, osteoarthritis, sleep apnea and mortality.

The change in the average weight of the population is

occurring quickly, and within a few generations the bell-curve of human-weight distribution has shifted toward greater weight (Power and Schulkin, 2008). This suggests that aside genetic susceptibility, other factors like over-eating, high carbohydrate or fat diet, frequency of eating, slow metabolism, sedentary lifestyle and medications are playing important role in the increased prevalence of obesity.

Contemporary Western solid diets are frequently high in fats and sugars, and may be energy dense (Archer et al., 2007). High carbohydrate and fat diets are fast becoming the trend in developing countries like Nigeria. These solid dietary components are increasingly consumed in association with sweetened drinks, snacks and other liquid formulations. For example, evidence suggests that the consumption of sugar-sweetened soft drinks by children more than doubled between 1965 and 1996 (Cavadini et al., 2000). However, the role of carbohydrates in weight gain is not clear. For several decades obesity has been on the rise, although dietary fat has fallen and caloric intake for most pediatric groups, at least in the US, has been steady (Troiano et al., 2000). Simple carbo-hydrate intake may be to blame, particularly in children (Slyper, 2004), who regularly consume fruit juices and snacks like potato chips.

There are two main types of effective weight-loss medications: appetite suppressants (sibutramine) and lipase disruptors (Orlistat). As with almost all medications, weight-loss drugs have side effects, they are expensive, and are not safe for everyone. Traditional medicinal plants are often cheaper, locally available, and easily consumable (raw or as simple medicinal preparations) and have less adverse effect.

Some traditional healers have claimed that some medicinal plants in Nigeria like Anthocleista vogelii could be used to treat obesity. In Nigeria, A. vogelii is known by different names in different languages: 'Mpoto' (Igbo), apa oro, sapo (Yoruba) and kwari (Hausa). A. vogelii is a tree 6 to 20 m high, trunk 15–55 cm diameter, twigs with spines, leaves usually to 40 cm long, even to 150 cm, by 24 to 45 cm wide, inflorescence terminal with white sweet-scented flowers; of the closed-forest or mature regenerated jungle, usually in south west (Burkill, 1985).

Although, A. vogelii have not been scientifically proven to have anti-obesity potential, they have been found to be medicinally useful in other ailments. The seed and bark are used as purgative and antidote for snake bite. The bark and root are used in the healing of dropsy, swellings, oedema, gout and venereal diseases. The leaf-bud serves as antidotes for venomous stings, bites (Burkill, 1985). Alaribe et al. (2012) demonstrated the antiplasmodial effects of petroleum ether extract of leaf of A. vogelii. The in vivo antimalarial activity of ethanolic leaf and stem bark extracts of Anthocleista djalonensis (another species) was reported by Anita et al. (2009). In a different study, Ateufack et al. (2010) reported that the aqueous and methanol extracts of the stem bark of A. vogelii possesses

spasmogenic activity on both ileal and stomach smooth muscle fragments. The aim of this study was to investigate the impact of ethanolic root bark extract of A. vogelii on weight reduction in high carbohydrate diet (HCD) induced obesity in male Wistar rats.

MATERIALS AND METHODS

Plant material

The fresh root bark of A. vogelii Planch was collected from a farm land located in Umuekwune community, Imo State, Nigeria. The authentication of the plants was done by a taxonomist at the Forestry Research Institute of Nigeria (FRIN), Ibadan, Oyo State. A voucher specimen of the plant was deposited in the Herbarium, FRIN and the Herbarium, University of Benin, Nigeria.

Preparation of plant extract

The fresh root bark of the plant was washed, chopped into pieces and air-dried at room temperature. The dried plant part was milled into powder and weighed. 250 g of the plant powder was soaked in 500 ml of absolute ethanol separately in a container for 72 h with intermittent shaking. Then, it was filtered through Whatman No. 1 filter paper. The resulting filtrate was evaporated under reduced pressure using a rotary evaporator and there after freeze dried to get powder. The yield was stored in a refrigerator (4°C) till when needed.

Phytochemical screening

The phytochemical analysis of the root bark of A. vogelii plant was carried out using standard methods of analysis according to methods of Treatise and Evans (1989) and Sofowora (1993).

Acute toxicity (LD$_{50}$) of plants

The acute toxicity of the ethanolic extract of A. vogelii was carried out as described by Shah et al. (1997) and Burger et al. (2005). Thirty five albino rats were divided into seven groups of five (5) rats each weighing between (180-200 g). The rats were subjected to 24 h fasting (with only water) before the extract was administered. The extract was suspended in distilled water and administered in doses of 200, 400, 800, 1600, 3200 and 6400 mg/kg body weight orally. The seventh group served as control and received only distilled water. The rats were observed for signs of toxicity and mortality for the first critical 4 h and then for each hour for the next 12 h, followed by 6 hourly intervals for the next 56 h giving a total of 72 h observations, thereafter daily for 7 days.

Experimental animals

In this study, thirty male Wistar rats weighing 85 ± 5 g was used. The rats were bought from Anatomy Department, University of Benin and moved to the animal house of the Department of Biochemistry, University of Benin, Nigeria. All animals were housed in steel cages and each cage contained 10 rats. Rats were maintained under controlled temperature (±23°C) and a 12:12 h light/dark cycle. During the two weeks acclimatization period, the rats had free access to tap water and normal pellet diet (NPD) until they were assigned to individual groups. This work was carried out in accordance with the guidelines of Faculty of Life Sciences at University of Benin for animal use.

Composition of experimental diet

The normal pellet diet had 5% fat, 60 % carbohydrates and 30% protein, while the HCD contained 5% fat, 80% carbohydrates and 10% protein, the fibre, mineral and vitamin content were of the same quantity for both diets. The food/supplement used to compose the diet based on the different classes of food include: carbohydrate (processed cassava locally called *garri*), protein (dried *bonga* fish), fat (butter), fibre, vitamins and mineral mixture were obtained from a retail goods manufactured by SuperMax Nig. Ltd.

Induction of obesity in the rats

The thirty rats were randomly assigned into three groups of ten. The first group fed on NPD, while obesity was induced in the second and third group by feeding on HCD for 14 weeks.

Experimental design and animal grouping

After the period of obesity induction, Group 1 was fed the NPD, Groups 2 and 3 were fed with HCD, and only Group 3 received 500 mg/kg b.w *A. vogelii* as treatment for 4 weeks. The 500 mg/kg b.w dose of *A. vogelii* extract was reached from careful study of few works done by other researchers (Mbiantcha et al., 2013; Ogbonnia et al., 2011) and a pilot dose dependent study done in our laboratory. The *A. vogelii* extract was suspended in normal saline and then it was administered orally to the rats in Group 3, whereas the control groups (Groups 1 and 2) received normal saline for 4 weeks using a gavage tube. During the 4 weeks of treatment, all the rats had free access to their diets and water.

Food intake and body weight measurement

The daily food intake of the rats was measured in the morning using a weighing balance. Food intake was calculated by subtracting the amount of food left over in each cage (that is, the refusal and spillage for the individual solid diets) from the measured amount of food provided at the previous day (gm/day/cage). The mean of food intake was represented in gm/day/group. The body weight of the rat was measured weekly in grams (g).

Anthropometrical determinations

The body weight of the rat was measured weekly in grams (g). The body length (nose-to-anus length) was determined weekly in centimeter (cm) in all rats. The body weight and body length was used to determine the body mass index (BMI) as described by Noveli et al. (2007):

Body mass index (BMI) = body weight (g)/length^2 (cm^2)

Blood sample preparation

At the end of the experiment, rats were fasted for 12 to 14 h. Blood was collected by cardiac puncture from the rats at fasting state after being anesthetized with chloroform. The blood samples were collected in plain tubes, allowed to coagulate at room temperature and centrifuged at 3500 rpm for 15 min at room temperature for separation of serum. The clear, non-haemolysed supernatant was

separated using clean dry Pasteur pipette and stored at -20°C. Serum was used to assay for the lipid profile levels of the rats.

Adipose tissue dissection and fat mass determination

After abdominal incision, five different white adipose depots (two subcutaneous and three intra-abdominal) and interscapular BAT were harvested from each rat. The WAT and BAT were dried on separate filter papers and weighed in grams (g). The total fat mass was the combined weight of the WAT and BAT.

Adiposity index

Adipose tissue was isolated from the epididymal, visceral and retroperitoneal pad. These were dried on filter paper and weighed (g.) Adiposity index was determined by the sum of epididymal, visceral and retroperitoneal fat weights divided by body weight × 100, and expressed as adiposity percentage (Taylor and Phillips, 1996).

Lipid profile assay

Serum was used to assay for the following parameters: total cholesterol (Trinder, 1969), triglycerides (Tietz, 1990), high density lipoprotein cholesterol (Tietz, 1976), very low density lipoprotein cholesterol (triglycerides/5) and low density lipoprotein cholesterol (Friedewald et al., 1972).

Serum glucose

Serum was used to assay for glucose as described by Trinder (1969).

Insulin ELISA assay

The quantitative measurement of insulin in serum was performed using a solid phase enzyme-linked immunosorbent assay (ELISA) based on the sandwich principle according to the manufacturer's instructions (DRG Insulin ELISA kit, DRG International, Inc. USA). The microtitre wells were coated with a monoclonal antibody directed towards a unique antigenic site on the insulin molecule. An aliquot of serum sample containing endogenous insulin was incubated in the coated well with enzyme conjugate, which is an anti-insulin antibody conjugated with biotin. After incubation, the unbound conjugate was washed off. During the second incubation step Streptavidin Peroxidase Enzyme Complex binds to the biotin-anti-insulin antibody. The amount of bound HRP complex was proportional to the concentration of insulin in the sample.

Leptin (sandwich) ELISA assay

The quantitative measurement of leptin in serum was performed using a solid phase enzyme-linked immunosorbent assay (ELISA) based on the sandwich principle according to the manufacturer's instructions (DRG Leptin ELISA kit, DRG International, Inc. USA). The microtitre wells were coated with a monoclonal antibody directed towards a unique antigenic site on a leptin molecule. An aliquot of sample containing endogenous leptin was incubated in the coated well with a specific biotinylated monoclonal anti leptin antibody. A sandwich complex was formed. After incubation the unbound material was washed off and a Streptavidin Peroxidase Enzyme Complex was added for detection of the bound leptin.

Table 1. Food intake of rats during obesity treatment with *A. vogelii.*

Group	Food intake (g)				
	Week 14	Week 15	Week 16	Week 17	Week 18
Normal control	64.84 ± 2.55^a	60.54 ± 1.76^a	63.60 ± 1.71^b	68.19 ± 2.28^a	63.13 ± 2.11^b
HCD obese control	54.16 ± 2.99^b	50.67 ± 2.43^c	69.76 ± 3.07^a	70.54 ± 2.35^a	76.72 ± 1.64^a
HCD + *A. vogelii*	61.92 ± 3.24^a	55.33 ± 1.24^{ab}	63.37 ± 0.72^b	46.02 ± 1.31^b	40.21 ± 1.22^c

Values are expressed as means ± SEM. Means in the same column not sharing common letter(s) are significantly different ($p < 0.05$).

Table 2. Body weight of rats during obesity treatment with *A. vogelii.*

Group	Body weight (g)				
	Week 14	Week 15	Week 16	Week 17	Week 18
Normal control	270.33 ± 6.39^b	274.00 ± 4.93^c	275.67 ± 8.09^b	290.33 ± 8.41^b	293.67 ± 9.17^e
HCD obese control	352.67 ± 9.21^a	368.33 ± 9.94^a	365.00 ± 11.02^a	389.00 ± 6.35^a	399.33 ± 7.06^a
HCD + *A. vogelii*	350.00 ± 3.51^a	360.33 ± 5.36^{ab}	359.33 ± 0.88^a	368.33 ± 4.37^a	368.33 ± 6.12^b

Values are expressed as means ± SEM. Means in the same column not sharing common letter(s) are significantly different ($p < 0.05$).

Table 3. BMI of rats during obesity treatment with *A. vogelii.*

Group	BMI (g/cm^2)				
	Week 14	Week 15	Week 16	Week 17	Week 18
Normal control	0.74 ± 0.05^b	0.71 ± 0.02^b	0.66 ± 0.01^b	0.66 ± 0.01^b	0.64 ± 0.00^c
HCD obese control	0.99 ± 0.03^a	0.96 ± 0.04^{ab}	0.87 ± 0.01^a	0.89 ± 0.01^a	0.86 ± 0.01^{ab}
HCD + *A. vogelii*	1.04 ± 0.03^a	0.99 ± 0.04^a	0.83 ± 0.01^a	0.84 ± 0.01^{ab}	0.84 ± 0.02^b

Values are expressed as means ± SEM. Means in the same column not sharing common letter(s) are significantly different ($p < 0.05$).

Having added the substrate solution, the intensity of colour developed was proportional to the concentration of leptin in the sample.

Statistical analysis

The experimental results were expressed as the mean ± S.E.M. Statistical significance of difference in parameters amongst groups was determined by One way ANOVA followed by Duncan's multiple range test. $P<0.05$ was considered to be significant.

RESULTS

Phytochemical screening and acute toxicity (LD$_{50}$) of plant

The ethanolic root bark extract of *A. vogelii* contained alkaloid, saponin, tannin, steroid and cardiac glycosides. Acute toxicity study revealed general weakness and sluggishness as the major behavioral changes observed in the rats at 6400 mg/kg b. wt. oral dose. These behavioral changes disappeared after 1 h of observation.

No death was recorded at any of the doses administered. Oral LD$_{50}$ was therefore not determined because mortality was not observed.

Obesity treatment period of the rats

The food intake of the HCD obese control was increased significantly as compared to the normal control at the 18th week. The food intake of the HCD treated group was significantly decreased as compared to the HCD obese control throughout the treatment period (Table 1).

The body weight of HCD obese control group maintained significantly increased body weight as compared to the normal control through out the treatment period. Again, the body weight of group treated with *A. vogelii* was significantly decreased as compared to the HCD obese control at the 18th week (Table 2).

The BMI of the HCD obese control group remained significantly increased as compared to the normal control at the end of the treatment. But there was no significant difference within the HCD obese and that treated with

Table 4. Fat mass of rats during obesity treatment with *A. vogelii*.

Group	BAT (g)	WAT (g)	Total fat mass (g)	Adiposity index
Normal control	0.93 ± 0.02^{ab}	19.65 ± 0.88^{b}	20.58 ± 0.50^{c}	3.29 ± 0.18^{c}
HCD obese control	0.93 ± 0.00^{ab}	28.50 ± 0.68^{a}	29.43 ± 0.68^{a}	4.63 ± 0.11^{a}
HCD + *A. vogelii*	1.03 ± 0.08^{a}	20.27 ± 0.74^{b}	21.30 ± 0.68^{bc}	2.78 ± 0.16^{b}

Values are expressed as means ± SEM. Means in the same column not sharing common letter(s) are significantly different (p < 0.05). BAT= brown adipose tissue; WAT= white adipose tissue.

Table 5. Lipid profile of rats during obesity treatment with *A. vogelii*.

Group	TG (mg/dl)	TC (mg/dl)	HDL-C (mg/dl)	VLDL-C (mg/dl)	LDL-C (mg/dl)
Normal control	45.58 ± 0.91^{c}	43.78 ± 1.45^{b}	29.97 ± 1.70^{ab}	9.12 ± 0.11^{c}	4.68 ± 0.50^{b}
HCD obese control	83.57 ± 5.00^{ab}	54.50 ± 2.09^{a}	23.98 ± 1.51^{e}	16.71 ± 1.00^{ab}	13.82 ± 1.39^{a}
HCD + *A. vogelii*	91.47 ± 5.27^{ab}	51.60 ± 1.61^{ab}	25.78 ± 0.69^{bc}	18.29 ± 1.05^{a}	7.53 ± 1.0^{b}

Values are expressed as means ± SEM. Means in the same column not sharing common letter(s) are significantly different (p < 0.05). TG = triglycerides, TC = total cholesterol, HDL-C = high density lipoprotein cholesterol, VLDL-C = very low density lipoprotein cholesterol, LDL-C = low density lipoprotein cholesterol.

A.vogelii (Table 3).

The HCD obese control group had significantly greater total fat mass and adiposity index as compared to the normal control. The HCD treated group had significantly decreased total fat mass and adiposity index as compared to the HCD obese control. The HCD group treated with *A. vogelii* had the least total fat mass and adiposity index throughout the treatment period (Table 4).

As shown in Table 5, the HCD obese rats had significantly increased serum triglycerides, total cholesterol, VLDL cholesterol and LDL cholesterol, with a decrease in HDL cholesterol which was significant as compared to the normal control. Only the LDL cholesterol level of the HCD treated with *A. vogelii* was significantly decreased when compared with the HCD obese control.

The glucose, insulin and leptin level of the HCD obese control was significantly increased as compared to the normal control. With exception of insulin, the glucose and leptin levels of the HCD obese rats treated with *A. vogelii* showed significant decrease when compared with the HCD obese control.

DISCUSSION

The ethanolic root bark extract of *A. vogelii* contained alkaloid, saponin, tannin, steroid and cardiac glycosides. The phytochemistry of all the parts of *A. vogelii* has been shown to have alkaloids; glycosides, saponins and steroids (Burkill, 1985). A synergistic relationship amongst phytochemicals has been adduced to be responsible for the overall beneficial effect derivable from plants (Liu, 2004). Saponins are capable of neutralizing some enzymes in the intestine that can become harmful,

building the immune system and promoting wound healing (Akinmoladun et al., 2007). Also, saponins also promote wound healing (Okwu and Okwu, 2004).

Acute toxicity study of the ethanolic root bark extract of *A. vogelii* revealed that the extract was not toxic to the rats at 6400 mg/kg b. wt. oral dose. Although, the rats exhibited some behavioral changes as stated above, these changes disappeared after 1 h of observation. In a different toxicity study on *A. vogelii,* no lethality was observed at 2000 mg/kg body weight i.p. in mice (Alaribe et al., 2012).

Obesity is a chronic metabolic disorder that occurs from the imbalance between energy intake and energy expenditure which is followed by increased body weight, enlarged fat mass and elevated lipid concentration in blood. Food intake and body weight are direct measures of obesity (Haslam and James, 2005). The food intake and body weight of the HCD obese rats increased significantly as compared to the normal control. Thus, it showed that the rats in the HCD obese group were obese. However, *A. vogelii* caused significant decrease in food intake and body weight in HCD induced obese rats during the treatment period. The reduction in the food intake in the HCD obese group treated with *A. vogelii* may be due to its ability to suppress the animals' appetite indicating action of bioactive components like saponin, and this was similar to the results reported by Chidrawar (2011).

The BMI of the HCD obese control remained significantly increased as compared to the normal control at the end of the treatment; this corroborates the work done by Novelli et al. (2007), which showed that high caloric diet increased significantly the BMI of rats as compared to the control (standard diet). Although there

Table 6. Blood glucose, insulin and leptin of rats treated with *A. vogelii* extracts.

Group	Glucose (mg/dl)	Insulin (µIU/ml)	Leptin (ng/ml)
Normal control	65.00 ± 2.89^c	2.33 ± 0.03^b	2.63 ± 0.15^b
HCD obese control	89.67 ± 3.18^a	2.50 ± 0.06^a	2.86 ± 0.09^a
HCD + *A. vogelii*	79.33 ± 0.33^b	2.50 ± 0.06^a	2.30 ± 0.12^b

Values are expressed as means ± SEM. Means in the same column not sharing common letter(s) are significantly different ($p < 0.05$).

was a slight decrease in the BMI of the rats treated with *A. vogelii*, there was no significant difference between the HCD obese control and HCD obese rats treated with *A. vogelii* (Table 3).

Obesity is a condition of abnormal or excessive fat accumulation in adipose tissue. So, by measuring fat mass in the rats, we can directly correlate fat mass with obesity. In this study, total fat mass included: WAT and interscapular BAT of each rat. It was discovered that the HCD obese control group had significantly greater total fat mass as compared to the normal control. The *A. vogelii* extract significantly decreased the WAT and total fat mass of HCD obese rats.

Adiposity index was determined by the sum of epididymal, visceral and retroperitoneal fat weights divided by body weight × 100, and expressed as adiposity percentage (Taylor and Phillips, 1996). The HCD obese control group had significantly higher adiposity index as compared to the normal control rats, which was similar to the results found in other studies (Jeyakumar et al., 2009). The *A. vogelii* extract produced a significant decrease in the total fat mass and adiposity index as compared to the obese controls.

Some of the chemical constituents, such as saponins, flavonoids, and some triterpenoids, have been reported for their antiobesity effect in various plants (Yun, 2010). A wide variety of plants possess pancreatic lipase inhibitory effects, including *Panax japonicus* (Han et al., 2005), *Platy-codi radix* (Han et al., 2000), *Salacia reticulata* (Kishino et al., 2006), *Nelumbo nucifera* (Ono et al., 2006). These pancreatic lipase inhibitory phytochemicals include mainly saponins, polyphenols, flavonoids and caffeine (Kim and Kang, 2005; Moreno et al., 2006; Shimoda et al., 2006). Saponins are known bioactive substances that can reduce the uptake of cholesterol and glucose at the gut through intra-lumenal physiochemical interaction (Price et al., 1987). Saponins as a class of natural products are also involved in complexation with cholesterol to form pores in cell membrane bilayers (Francis et al., 2002) as such may be used as anticholesterol agents or cholesterol lowering agent. Thus, it is suggestive that the *A. vogelii* root bark extract used in this present study, which contains saponins reduced the fat accumulation in the treated obese rats by inhibiting the activity of pancreatic lipase.

In the present study, there was a significant increase in

serum triglycerides, total cholesterol, VLDL cholesterol and LDL cholesterol, with a decrease in HDL cholesterol (which was not significant) of the HCD obese rats. Also, it is well known that excess sugar in the human diet can be converted both into glycerol and fatty acids and, thus, into lipids such as triglycerides (Murray et al., 2006). This explain why the triglyceride level and VLDL cholesterol in the HCD obese rats is significantly higher than the normal control. The *A. vogelii* extract had no significant difference in the TG, TC and VLDL as compared to the HCD obese rats.

The increased level of LDL cholesterol is a common feature of Obesity. LDL cholesterol is found to cause endothelial damage, oxidative stress and inflammation, which further aggravates obesity (Mistry et al., 2011). This necessitates the elevated LDL level in the HCD obese control. It has been observed that increase in lipid level particularly LDL-cholesterol is predictive for coronary events such as atherosclerosis and coronary heart disease (Blake et al., 2002). So, it is necessary to reduce LDL cholesterol level in obesity to treat as well as protect the disease from expansion (Shibano et al., 1992). The *A. vogelii* root bark extract had significantly decreased LDL cholesterol in the HCD obese treated group.

Serum glucose, insulin and leptin concentrations were significantly higher in HCD obese group as compared to the normal control. Leptin is a hormone produced mainly by adipocytes, and is involved in controlling body weight by increasing both satiety and energy expenditure (Tentolouris et al., 2008; Vigueras-Villaseñor et al., 2011). Leptin levels are excessively high in obese people as a result of leptin resistance, which is associated with weight gain. In this present study, the HCD obese rats showed significant increase in leptin concentration. This result corroborates other studies that show high leptin levels in models of rodent DIO obesity (Ghanayem et al., 2010; Tentolouris et al., 2008; Fam et al., 2007).

The extract of *A. vogelii* significantly decreased glucose and leptin concentration in the treated rats, but there was no significant difference insulin concentration between the treated and HCD obese rats. Leptin concentration is related to the amount and distribution of body fat such that the higher the body weights, the higher the leptin concentration in human and rodents (Aizawa-Abe et al., 2000). This implies that decrease in body weight or body fat will bring about a corresponding decrease in leptin concentration in the rats as seen in Group 3 of this study (Table 6). Also, this suggests that *A. vogelii* extract might have bioactive component(s) that decreased the accumulation of fats and as such reduce the body weight of HCD obese rats. *A. vogelii* extract caused decreased glucose level in the treated rats, this could be attributed to its ability to facilitate glucose entrance into cells and increase insulin sensitivity. *A. vogelii* might as well contain bioactive components that possess hypoglycemic effects.

Conclusion

A number of medicinal plants, including crude extracts of these plants can induce body weight reduction and prevent diet-induced obesity. The ethanolic root bark extract of *A. vogelii* significantly decreased food intake, body weight, total fat mass, adiposity index and low density lipoprotein cholesterol. Thus the ethanolic root bark extract of *A. vogelii* possess some potential to reduce weight and therefore, might be explored in the treatment of obesity. However, there is need for further research on extensive identification of the active ingredients of the plant and the role they play in weight reduction.

REFERENCES

Aizawa-Abe M, Ogawa Y, Masuzaki H, Ebinara K, Satoh N, Iwai H, Matsuoka N, Hayashi T, Hosoda K, Inoue G, Yoshimara Y, Nakao K (2000). Pathophysiological role of leptin in obesity-related hypertension. J. Clin. Invest. 105(9):1243-1252.

Akinmoladun AC, Ibukun EO, Afor E, Akinrinlola BL, Onibon TR., Akinboboye AO, Obuotor EM, Farombi EO (2007). Chemical constituents and antioxidant activity of *Alstonia boonei*. Afr. J. Biotechnol. 6 (10): 1197-1201.

Alaribe CSA, Coker HAB, Shode FO, Ayoola G, Adesegun SA, Barimo J, Anyim EI, Anyakora C (2012). Antiplasmodial and Phytochemical Investigations of Leaf Extract of *Anthocleista vogelii* (Planch). J. Nat. Prod. 5: 60-67.

Anita SB, Jude EO, Emmanuel IE, Francis UU, Emmanuel B (2009). Evaluation of the *in vivo* antimalarial activity of ethanolic leaf and stem bark extracts of *Anthocleista djalonensis*. Indian J. Pharmacol. 41(6): 258–261.

Archer ZA, Corneloup J, Rayner DV, Barrett P, Moar KM, Mercer JG (2007). Solid and liquid obesogenic diets induce obesity and counter-regulatory changes in hypothalamic gene expression in juvenile Sprague-dawley rats. J. Nutr. 137:1483-1490.

Ateufack G, Nguelefack TB, Tane P, Kamany A (2010). Spasmogenic activity of the aqueous and methanol extracts of the stem bark of *Anthocleista vogelii* planch (Loganiaceae) in rats. Pharmacologyonline 1:86-101.

Burger C, Fischer DR., Cordenunzzi DA, Batschauer de Borba AP, Filho VC, Soares dos Santos AR (2005). Acute and subacute toxicity of the hydroalcoholic extract from *Wedelia paludosa* (*Acmela brasilinsis*) (Asteraceae) in mice. J. Pharma. Sci. 8: 370-373.

Burkill HM (1985). The useful plants of West Tropical Africa. 2nd edn. Vol. 2. The White friars Press limited, Great Britain. pp. 352-364.

Cavadini C, Siega-Riz AM, Popkin BM (2000). US adolescent food intake trends from 1965 to 1996. Arch. Dis. Child. 83:18–24.

Chidrawar VR, Patel KN, Sheth NR, Shiromwar SS, Trivedi P (2011). Antiobesity effect of *Stellaria media* against drug induced obesity in Swiss albino mice. Ayu. 32(4): 576–584.

Fam BC, Morris MJ, Hansen MJ, Kebede M, Andrikopoulos S, Proietto J, Thorburn AW (2007). Modulation of central leptin sensitivity and energy balance in rat model of diet-induced obesity. Diabetes Obes. Metab. 9(6):840-852.

Francis C, George G, Zohar K, Harinder PS, Makhar LM, Klaus B (2002). The biological action of saponins in animal system: a review. Br. J. Nutr. 88(6):587-605l.

Friedewald WT, Levy RI, Fredrickson DS (1972). Estimation of the concentration of low-density lipoprotein cholesterol in plasma, without the use of preparative ultracentrifuge. Clin. Chem. 18:499-502.

Ghanayem BI, Bai R, Kissling GE, Travlos G, Hoffler U (2010). Diet-induced obesity in male mice is associated with reduced fertility and potentiation of acrylamide-induced reproductive toxicity. Biol. Reprod. 82(1):94-104.

Han LK, Xu BJ, Kimura Y, Zheng Y, Okuda H (2000). *Platycodi radix* affects lipid metabolism in mice with high fat diet-induced obesity. J. Nutr. 130 (11): 2760-2764.

Han LK, Zheng YN, Yoshikawa M, Okuda H, Kimura Y (2005). Anti-obesity effects of chikusetsusaponins isolated from *Panax japonicus* rhizome. BMC Complement. Altern. Med. 5(9):1-10.

Haslam DW, James WP (2005). Obesity. Lancet 366 (9492): 1197–209.

Jeyakumar SM, Lopamudra P, Padmini S, Balakrishna N, Giridharan NV, Vajreswari A (2009). Fatty acid desaturation index correlates with body mass and adiposity indices of obesity in Wistar NIN obese mutant rats strains WNIN/Ob and WNIN/GR-Ob. Nutr. Metab. 6:27.

Kim HY, Kang MH (2005). Screening of Korean medicinal plants for lipase inhibitory activity. Phytother. Res. 19 (4): 359-361.

Kishino E, Ito T, Fujita K, Kiuchi Y (2006). A Mixture of the *Salacia reticulata* (Kotala Himbutu) aqueous extract and Cyclodextrin reduces the accumulation of visceral fat mass in mice and rats with high-fat diet-induced obesity. J. Nutr. 136 (2): 433-439.

Liu RH (2004). Potential synergy of phytochemicals in cancer prevention: mechanism of action. J. Nutr. 134: 34795-34855.

Mbiantcha M, Nguessom KO, Ateufack G, Oumar M, Kamanyi A (2013). Analgesic properties and toxicological profile of aqueous extract of the stem bark of *Anthocleista vogelii* planch (Loganiaceae). IJPCBS 3(1): 1-12.

Mistry KG, Deshpande SS, Shah GB, Gohil PV (2011). Effect of Sarpogrelate in high fat diet induced obesity in rats. Asian J. Pharm. Biol. Res. 1(4): 441-446.

Moreno DA, Ilic N, Poulev A, Raskin I (2006). Effects of *Arachis hypogaea* nutshell extract on lipid metabolic enzymes and obesity parameters. Life Sci. 78 (24): 2797-2803.

Murray RK, Granner DK, Mayes PA, Rodwell WV (2003). Harper's Illustrated Biochemistry. 26th edn., McGraw-Hill Companies, USA.

Novelli ELB, Diniz YS, Galhardi CM, Ebaid GMX., Rodrigues HG, Mani F, Fernandes AAH, Cicogna, AC, Novelli Filho JLVB (2007). Anthropometrical parameters and markers of obesity in rats. Lab. Anim. 41: 111–119.

Ogbonnia SO, Mbaka GO, Anyika EN, Emordi JE, Nwakakwa N (2011). An Evaluation of Acute and Subchronic Toxicities of a Nigerian Polyherbal Tea Remedy. Pak. J. Nutr. 10 (11): 1022-1028.

Okwu DE, Okwu ME (2004). Chemical composition of *Spondias mombin* plant parts. J. Sustain. Agric. Environ. 6: 140-147.

Ono Y, Hattori E, Fukaya Y, Imai S, Ohizumi Y (2006). Anti-obesity effect of *Nelumbo nucifera* leaf extract in mice and rats. J. Ethnopharmacol. 106 (2): 238-244.

Power ML, Schulkin J (2008). Sex differences in fat storage, fat metabolism, and the health risks from obesity: possible evolutionary origins. Br. J. Nutr. 99(5):931-940.

Price KR, Johnson LI, Feriwick H (1987). Chemical and biological significance of saponin in food science. Nutrition 26:127-135.

Shah MA, Garg SK, Garg KM, (1997). Subacute toxicity studies on Pendimethalin in rats. Int. J. Pharmacol. 29: 322-324.

Shibano T, Tanaka T, Morishima Y, Yasuoka M, Fujii F (1992). Pharmacological profile of a new 5-hydroxytryptamine 2 receptor antagonist, DV-7028. Arch. Int. Pharmacodyn. Ther. 319:114–128.

Shimoda H, Seki E, Aitani M. (2006). Inhibitory effect of green coffee bean extract on fat accumulation and body weight gain in mice. BMC Complement. Altern. Med. 6 (9): 1-9.

Slyper AH (2004). The pediatric obesity epidemic: causes and controversies. J. Clin. Endocrinol. Metab. 89 (6):2540-7.

Sofowora (1993). A medicinal plants and traditional medicine in Africa. Spectrum Books Ltd, Ibadan. p. 289.

Taylor BA, Phillips SJ (1996). Detection of obesity QTLs on mouse chromosomes 1 and 7 by selective DNA pooling. Genomics 34(3):389-398.

Tentolouris N, Pavlatos S, Kokkinos A, Perrea D, Pagoni S, Katsilambros N (2008). Diet-induced thermogenesis and substrate oxidation are not different between lean and obese women after two different isocaloric meals, one rich in protein and one rich in fat. Metabolism 57(3):313-320.

Tietz NW (1976). Fundamentals of Clinical Chemistry. 2nd edn., W.B. Saunders Company, Philadelphia, USA pp. 878-878.

Tietz NW (1990). Clinical Guide to Laboratory Tests, 2nd edn., W. B. Saunders Company, Philadephia, USA. pp. 554-556.

Treatise GE, Evans WC (1989). Pharmacognosy.11th edn. Brailliar Tiridel Can. Macmillan publishers. Ltd, Ibadan. p. 289.

Trinder P (1969). Determination of glucose in blood using glucose oxidase with an alternative oxygen acceptor. Ann. Clin. Biochem. 6: 24-27.

Troiano RP, Briefel RR, Carroll MD, Bialostosky K (2000). Energy and fat intakes of children and adolescents in the United States: data from the national health and nutrition examination surveys. Am. J. Clin. Nutr. 72:1343-1353.

Vigueras-Villaseñor RM, Rojas-Castañeda JC, Chávez-Saldaña M, Gutiérrez-Pérez O, García-Cruz ME, Cuevas-Alpuche O, Reyes-Romero MM, Zambrano E (2011). Alterations in the spermatic function generated by obesity in rats. Acta Histochem. 113(2):214-20.

WHO (2000). World Health Organization: Part I: The problem of overweight and obesity. In World Health Organization, Obesity: preventing and managing the global epidemic. Geneva: WHO Technical Report Series; 2000.

Yun JW (2010). Possible anti-obesity therapeutics from nature – A review. Phytochemistry 71:1625-1641.

Permissions

List of Contributors

E. U Etuk
Department of Pharmacology, College of Health Sciences, Usmanu Danfodiyo University, Sokoto, Nigeria

U. U Francis
Department of Pharmacognosy and natural medicine, University of Uyo, Nigeria

I Garba
Department of Pharmacology, College of Health Sciences, Usmanu Danfodiyo University, Sokoto, Nigeria

Kh. Sh. Hamadah
Department of Zoology, Faculty of Science, Al-Azhar University, Madenit Nasr, Cairo, Egypt

K. S. Ghoneim
Department of Zoology, Faculty of Science, Al-Azhar University, Madenit Nasr, Cairo, Egypt

M. A. Tanani
Department of Zoology, Faculty of Science, Al-Azhar University, Madenit Nasr, Cairo, Egypt

Shailah Abdullah
Department of Biochemistry, Faculty of Medicine, Universiti Kebangsaan Malaysia, Jalan Raja Muda Abdul Aziz, 50300 Kuala Lumpur, Malaysia

Siti Amalina Zainal Abidin
Department of Biochemistry, Faculty of Medicine, Universiti Kebangsaan Malaysia, Jalan Raja Muda Abdul Aziz, 50300 Kuala Lumpur, Malaysia

Noor Azian Murad
Centre of Lipids and Engineering and Applied Research, Universiti Teknologi Malaysia, Jalan Semarak, 50300 Kuala Lumpur, Malaysia

Suzana Makpol
Department of Biochemistry, Faculty of Medicine, Universiti Kebangsaan Malaysia, Jalan Raja Muda Abdul Aziz, 50300 Kuala Lumpur, Malaysia

WanZurinah Wan Ngah
Department of Biochemistry, Faculty of Medicine, Universiti Kebangsaan Malaysia, Jalan Raja Muda Abdul Aziz, 50300 Kuala Lumpur, Malaysia

Yasmin Anum Mohd Yusof
Department of Biochemistry, Faculty of Medicine, Universiti Kebangsaan Malaysia, Jalan Raja Muda Abdul Aziz, 50300 Kuala Lumpur, Malaysia

B. O. George
Department of Biochemistry, Faculty of Science, Delta State University, Abraka, Delta State, Nigeria

E. Osioma
Department of Biochemistry, Faculty of Science, University of Ilorin, Kwara State, Nigeria

J. Okpoghono
Department of Biochemistry, Faculty of Science, Delta State University, Abraka, Delta State, Nigeria

O. O. Aina
Department of Biochemistry, Nigeria Institute of Medical Research, Yaba, Lagos State, Nigeria

Chunli Mei
Department of Neurology, China - Japan Friendship Hospital, Jilin University, Changchun, 130012, China
Beihua University, Jilin 132013, China

Jinting He
Department of Neurology, China - Japan Friendship Hospital, Jilin University, Changchun, 130012, China

Jing Mang
Department of Neurology, China - Japan Friendship Hospital, Jilin University, Changchun, 130012, China

GuihuaXu
Department of Neurology, China - Japan Friendship Hospital, Jilin University, Changchun, 130012, China

Zhongshu Li
Department of Neurology, China - Japan Friendship Hospital, Jilin University, Changchun, 130012, China

Wenzhao Liang
Department of Neurology, China - Japan Friendship Hospital, Jilin University, Changchun, 130012, China

Zhongxin Xu
Department of Neurology, China - Japan Friendship Hospital, Jilin University, Changchun, 130012, China

K. S. Prashanth
Research Scholar, Dr M.G.R University, Chennai Tamil Nadu, India

T. R. S. Chouhan
Ethica Matrix CRO Pvt Ltd, Hyderabad, India

Snehalatha Nadiger
Department of Biotechnology, New Horizon College of Engineering, Outer Ring Road, Panathur post Bangalore-560087, Karnataka(s) India

Ghorbani Masoud
Department of Research and Development, Pasteur Institute of Iran (Research and Production complex), 25th km Tehran Karaj Highway, Tehran, Iran

Mehdi Shafiee Ardestani
Department of Hepatitis and Aids, Pasteur Institute of Iran, Tehran, Iran

Mohammad A. Al-Fararjeh
Department of the Medical Laboratory Technology, Hashemite University. Zarqa, Jordan

Nader Jaradat
Department of Medical Technology, Faculty of Allied Medical Sciences, Zarqa University, P. O. Box 132222-Zarqa 13110- Jordan

Abdulrahim Aljamal
Department of Medical Technology, Faculty of Allied Medical Sciences, Zarqa University, P. O. Box 132222-Zarqa 13110- Jordan

Jean Sakandé
Laboratory of Biochemistry, University of Ouagadougou, Burkina Faso

Josiane B Kaboré
Laboratory of Biochemistry, University of Ouagadougou, Burkina Faso

Elie Kabré
Laboratory of Biochemistry, University of Ouagadougou, Burkina Faso

Boblwendé Sakandé
Laboratory of Philadelphie Clinic of Ouagadougou, 09 BP 863 Ouagadougou, Burkina Faso

Mamadou Sawadogo
Laboratory of Biochemistry, University of Ouagadougou, Burkina Faso

N. P. Minh
Department of Food Technology, Ho Chi Minh City University of Technology, 268 Ly Thuong Kiet Street, District 10, Ho Chi Minh City, Vietnam

T. B. Lam
Department of Food Technology, Ho Chi Minh City University of Technology, 268 Ly Thuong Kiet Street, District 10, Ho Chi Minh City, Vietnam

T. T. D. Trang
Department of Food Technology, Ho Chi Minh City University of Technology, 268 Ly Thuong Kiet Street, District 10, Ho Chi Minh City, Vietnam

O. O Babalola
Department of Biochemistry, Faculty of Science, Obafemi Awolowo University Ile Ife Nigeria

S. O Babajide
Department of Science Laboratory Technology, Moshood Abiola Polytechnic, P. M. B 2210, Abeokuta, Nigeria

Ruiqiang Ma
College of Biological Sciences, China Agricultural University, Beijing 100193, China
Biotechnology Research Institute, Chinese Academy of Agricultural Science, Beijing 100081, China

Ying Zhang
Biotechnology Research Institute, Chinese Academy of Agricultural Science, Beijing 100081, China
Zhengzhou Fruit Research Institute, Chinese Academy of Agriculture Sciences, Henan 450009, China

Haozhou Hong
Biotechnology Research Institute, Chinese Academy of Agricultural Science, Beijing 100081, China

Wei Lu
Biotechnology Research Institute, Chinese Academy of Agricultural Science, Beijing 100081, China

Wei Zhang
Biotechnology Research Institute, Chinese Academy of Agricultural Science, Beijing 100081, China

Min Lin
Biotechnology Research Institute, Chinese Academy of Agricultural Science, Beijing 100081, China

Ming Chen
Biotechnology Research Institute, Chinese Academy of Agricultural Science, Beijing 100081, China

Olufemi Bamidele
Department of Biochemistry, The Federal University of Technology, P.M.B. 704, Akure, Nigeria

Joshua Ajele
Department of Biochemistry, The Federal University of Technology, P.M.B. 704, Akure, Nigeria

Ayodele Kolawole
Department of Biochemistry, The Federal University of Technology, P.M.B. 704, Akure, Nigeria

Akinkuolere Oluwafemi
Department of Biology, The Federal University of Technology, P.M.B. 704, Akure, Nigeria

C. Nwangwu Spencer
Department of Biochemistry, Faculty of Basic Medical Sciences, Igbinedion University, P. M. B. 0006, Okada, Nigeria

Ike Francisca
Department of Biochemistry, Faculty of Basic Medical Sciences, Igbinedion University, P. M. B. 0006, Okada, Nigeria

Olley Misan
Pathology Department, Igbinedion University Teaching Hospital, P. M. B. 0006, Okada, Nigeria

M. Oke James
College of Pharmacy, Department of Pharmaceutical Chemistry, Igbinedion University, P. M. B. 0006, Okada, Nigeria

Uhunmwangho Esosa
Department of Biochemistry, Faculty of Basic Medical Sciences, Igbinedion University, P. M. B. 0006, Okada, Nigeria

O. F. Amegor
Department of Medical Laboratory Sciences, Igbinedion University, P. M. B. 0006, Okada, Nigeria

Ubaoji Kingsley
Department of Biochemistry, Faculty of Natural Sciences, Nnamdi Azikiwe University Awka, Nigeria

Nwangwu Udoka
Department of Biochemistry, Faculty of Natural Sciences, Nnamdi Azikiwe University Awka, Nigeria

Hala A. Awney
Department of Environmental Studies, Institute of Graduate Studies and Research, Alexandria University, Alexandria, Egypt

Vaclav Vetvicka
Department of Pathology, University of Louisville, Louisville, KY 40202, USA

Zuzana Vancikova
1st Medical Faculty, Department of Pediatrics, Thomayer University Hospital, Prague, Czech Republic

A. Ologundudu
Department of Biochemistry, Adekunle Ajasin University, P.M.B. 001, Akungba Akoko, Ondo State, Nigeria

A. O. Ologundudu
Department of Biochemistry, Adekunle Ajasin University, P.M.B. 001, Akungba Akoko, Ondo State, Nigeria

I. A. Ololade
Department of Chemistry and Industrial Chemistry, Adekunle Ajasin University, P.M.B.001, Akungba Akoko, Ondo State, Nigeria

F. O. Obi
Department of Biochemistry, Faculty of Life Sciences, University of Benin, Benin City, Edo State, Nigeria

I. O. Omotuyi
Department of Biochemistry, Adekunle Ajasin University, Akungba-Akoko, Ondo State, Nigeria

A. Ologundudu
Department of Biochemistry, Adekunle Ajasin University, Akungba-Akoko, Ondo State, Nigeria

T. O. Okugbo
Department of Basic Sciences (Biochemistry option), Benson Idahosa University, G.R.A. Benin City, Edo State, Nigeria

K. A. Oluyemi
Department of Biology/Biotechnology, William Paterson University, Wayne, New Jersey, USA

O. E. Omitade
Department of Biochemistry, University of Port-Harcourt, Port-Harcourt, Rivers State, Nigeria

Saeed Khalaji
Department of Animal Science, College of Agriculture and Natural Resource, University of Tehran, Karaj31587-11167, Iran

Mojtaba Zaghari
Department of Animal Science, College of Agriculture and Natural Resource, University of Tehran, Karaj31587-11167, Iran

Somaye Nezafati
Department of Animal Science, College of Agriculture and Natural Resource, University of Tehran, Karaj31587-11167, Iran

S. Adekunle Adeniran
Department of Biochemistry, Ladoke Akintola University of Technology, Ogbomoso Oyo State, Nigeria

M. H. Awwad
Faculty of Science, Benha University Benha, Egypt

Samy A. Abd El-Azim
Faculty of Pharmacy, Cairo University Cairo, Egypt

F. A. M. Marzouk
Labelled Compounds Department, Hot Laboratory Center, Atomic Energy Authority. P. O. Box 13759, Cairo, Egypt

E. A. El-Ghany
Labelled Compounds Department, Hot Laboratory Center, Atomic Energy Authority. P. O. Box 13759, Cairo, Egypt

M. A. Barakat
Faculty of Pharmacy, Cairo University Cairo, Egypt

Cosmas O. Ujowundu
Biochemistry Department, Federal University of Technology, Owerri, Imo State, Nigeria

Agwu I. Ukoha
Biochemistry Department, Federal University of Technology, Owerri, Imo State, Nigeria

Comfort N. Agha
Biochemistry Department, Federal University of Technology, Owerri, Imo State, Nigeria

Ngwu Nwachukwu
Biochemistry Department, Federal University of Technology, Owerri, Imo State, Nigeria

Kalu O. Igwe
Biochemistry Department, Federal University of Technology, Owerri, Imo State, Nigeria

M. O. Akiibinu
Department of Chemical Pathology and Immunology, College of Health Sciences, Olabisi Onabanjo University, Ago- Iwoye, Ogun State, Nigeria

A. O. Ogundahunsi
Department of Chemical Pathology and Immunology, College of Health Sciences, Olabisi Onabanjo University, Ago- Iwoye, Ogun State, Nigeria

O. I. Kareem
Department of Nuclear Medicine, University College Hospital, Ibadan, Nigeria

A. A. Adesiyan
Department of Biomedical Sciences, Ladoke Akintola University of Technology, Ogbomosho, Oyo State, Nigeria

B. O. Idonije
Department of Chemical Pathology, College of Medicine, Ambrose Alli University, Ekpoma, Edo State, Nigeria

F. A. A. Adeniyi
Department of Chemical Pathology, College of Medicine, University of Ibadan, Ibadan, Nigeria

C. O. Ibegbulem
Department of Biochemistry, Federal University of Technology, Owerri, Nigeria

N. C. Agha
Department of Biochemistry, Federal University of Technology, Owerri, Nigeria

I. Emeka-Nwabunnia
Department of Biotechnology, Federal University of Technology, Owerri, Nigeria

Jinlian Hua
Key Laboratory for Reproductive Physiology and Embryo Biotechnology of Agriculture, Shaanxi Centre of Stem Cells Engineering and Technology, Ministry of China, Shaanxi, China
Key Lab for Agriculture Molecular Biotechnology Centre, Northwest, A and F University, Yangling, Shaanxi, 712100 China

Pubin Qiu
Key Laboratory for Reproductive Physiology and Embryo Biotechnology of Agriculture, Shaanxi Centre of Stem Cells Engineering and Technology, Ministry of China, Shaanxi, China
Key Lab for Agriculture Molecular Biotechnology Centre, Northwest, A and F University, Yangling, Shaanxi, 712100 China

Haijing Zhu
Key Laboratory for Reproductive Physiology and Embryo Biotechnology of Agriculture, Shaanxi Centre of Stem Cells Engineering and Technology, Ministry of China, Shaanxi, China
Key Lab for Agriculture Molecular Biotechnology Centre, Northwest, A and F University, Yangling, Shaanxi, 712100 China

Hui Cao
Key Laboratory for Reproductive Physiology and Embryo Biotechnology of Agriculture, Shaanxi Centre of Stem Cells Engineering and Technology, Ministry of China, Shaanxi, China
Key Lab for Agriculture Molecular Biotechnology Centre, Northwest, A and F University, Yangling, Shaanxi, 712100 China

Fang Wang
Key Laboratory for Reproductive Physiology and Embryo Biotechnology of Agriculture, Shaanxi Centre of Stem Cells Engineering and Technology, Ministry of China, Shaanxi, China
Key Lab for Agriculture Molecular Biotechnology Centre, Northwest, A and F University, Yangling, Shaanxi, 712100 China

Wei Li
Beike Bio-Technology Co., Ltd of Jiangsu Province, Taizhou, Jiangsu Province, 225300 China

Abdelmula M. Abdella
Faculty of Medical Laboratory Science, Al-Neelain University, Khartoum, Sudan

Al-Fadhil E. Omer
Faculty of Medical Laboratory Science, Al-Neelain University, Khartoum, Sudan

Badruldeen H. Al-Aabed
Faculty of Medicine, University of Science and Technology, Khartoum, Sudan

Balwant Rai
Prabhu Dayal Memorial (PDM) Dental College and Research Institute - Haryana - Bahadurgarh-India

Jasdeep Kaur
BJS dental collage, India

Simmi Kharb
BJS dental collage, India

Rajnish Jain
BJS dental collage, India

S. C. Anand
BJS dental collage, India

Jaipaul Singh
BJS dental collage, India

P. C. Chikezie
Department of Biochemistry, Imo State University, P. M. B. 2000, Owerri, Imo State, Nigeria

C. M. Chikezie
Department of Biochemistry, Imo State University, P. M. B. 2000, Owerri, Imo State, Nigeria

P. I. Amaragbulem
Department of Biochemistry, Imo State University, P. M. B. 2000, Owerri, Imo State, Nigeria

Mohammad A. Kassem
Department of Biochemistry, Faculty of Pharmacy, Omar El Mokhtar University, Al-Bieda, Libya

Tarek A. Mohammad
Department of Chemistry, Faculty of Science, Omar El Mokhtar University, Al-Bieda, Libya

Ahmad S. Bream
Department of Zoology, Faculty of Sciences, Omar El Mokhtar University, Al-Bieda, Libya

OYESOMI, Tajudeen Oyesina
Department of Anatomy, Kampala International University, Dar es Salaam, Tanzania

AJAO, Moyosore Salihu
School of Anatomical Sciences, Faculty of Health Sciences, University of the Witwatersrand, Johannesburg, South Africa

OLAYAKI, luquman Aribidesi
Department of Physiology, Faculty of Basic Medical Sciences, College of Health Sciences, University of Ilorin, Ilorin, Kwara State

ADEKOMI, Damilare Adedayo
Department of Anatomy, Faculty of Basic Medical Sciences, College of Health Sciences, University of Ilorin, Ilorin, Kwara State

P. Maria jancy Rani
Department of Chemistry, Dravidian University, Srinivasavanam, Chittoor District, Kuppam-517425, Andhra Pradesh, India

P. S. M. Kannan
PRIST University, Vallam, Thanjavur-613 403, Tamil Nadu, India

S. Kumaravel
Food Testing Laboratory, Indian Institute of Crop Processing Technology, Pudukkottai Road, Thanjavur-613 005 Tamil Nadu, India

M. S. Edem
Department of Heamatology, University of Calabar Teaching Hospital, Calabar, Cross River State, Nigeria

A. O. Emeribe
Department of Heamatology, University of Calabar Teaching Hospital, Calabar, Cross River State, Nigeria

J. O. Akpotuzor
Department of Heamatology, University of Calabar Teaching Hospital, Calabar, Cross River State, Nigeria

G. O. Anyanwu
Department of Biochemistry, Bingham University, Karu, Nasarawa State, Nigeria

E. C. Onyeneke
Department of Biochemistry, University of Benin, Benin City, Edo State, Nigeria

U. Usunobun
Department of Biochemistry, Benson Idahosa University, Benin City, Edo State, Nigeria

A. J. Adegbegi
Department of Science Laboratory Technology, Rufus Giwa Polytechnic, Owo, Ondo State, Nigeria